Polarographic Oxygen Sensors

Aquatic and Physiological Applications

Edited by
E. Gnaiger and H. Forstner

With 142 Figures

Springer-Verlag
Berlin Heidelberg New York 1983

Dr. ERICH GNAIGER
Dr. HELLMUTH FORSTNER

Institut für Zoologie
Abteilung Zoophysiologie
Universität Innsbruck
Peter-Mayr-Straße 1a
A-6020 Innsbruck, Austria

ISBN 3-540-11654-0 Springer-Verlag Berlin Heidelberg New York
ISBN 0-387-11654-0 Springer-Verlag New York Heidelberg Berlin

Library of Congress Cataloging in Publication Data. Main entry under title: Polarographic oxygen sensors. Bibliography: p. Includes index. 1. Oxygen – Analysis. 2. Polarograph and polarography. I. Gnaiger, E. (Erich), 1952–. II. Forstner, H. (Hellmuth), 1935–. QP535.01P58. 1983. 574.19′214. 82-19419.

Offsetprinting and bookbinding: Brühlsche Universitätsdruckerei, Giessen.
2131/3130-543210

Preface: A Biologist's View

> Es liegt ein tiefes und gründliches Glück darin,
> daß die Wissenschaft Dinge ermittelt, die stand-
> halten und die immer wieder den Grund zu neuen
> Ermittlungen abgeben: − es könnte ja anders sein!
>
> *Friedrich Nietzsche*
> Die fröhliche Wissenschaft

Molecular oxygen comprises about 20% of our atmosphere, but less than 5% of this amount is dissolved at equilibrium in water. As a consequence of this low solubility in seawater, in freshwater and in aqueous body fluids, living cells are subjected to a universal problem of low oxygen availability. Thus biologically relevant measurement of dissolved oxygen extends from a cellular scale where sensors with diameters of less than 10^{-6} m are employed, to the oceanic scale where in situ measurements are performed at depths beyond 10^4 m.

This book covers the wide spectrum of aquatic and physiological applications of polarographic oxygen sensors. It is intended as a basic introduction for the student, and as a readily available compilation of detailed information for the specialist. Assessments of various methods for monitoring aquatic environments and physiological processes aid in overcoming practical problems frequently encountered in the laboratory and in the field. Concomitant with the provision of experimental guidelines, topics of bioenergetics are addressed on the basis of respiratory oxygen exchange and related physiological mechanisms.

A respiratory physiologist would specify polarographic oxygen sensors as perfect oxygen conformers, their respiratory rate being linearly dependent on p_{O_2}. He would find that their oxygen consumption is proportional to the area of their oxygen-transducing membrane, with a Q_{10} of 1.4 and a constant of proportionality (about 2 nmol O_2 h^{-1} mm^{-2}) which might be typical of, for example, fish eggs. Although the increasing unpredictability of their oxygen consumption with increasing age would be understandable, he might be surprised that, despite their vulnerability to desiccation, oxygen consumption remained the same in air as in water. In these respects polarographic oxygen sensors are not different from organisms, except that microsensors consume oxygen proportional to their diameter instead of area. But do we know that microorganisms do not behave similarly?

These apparent phenomena are explained on the basis of physico-chemical principles in Part I. Why is this understanding so important? Since oxygen concentration per unit partial pressure (c_{O_2}/p_{O_2} = solubility) is low in water, small concentration changes become apparent as large changes of p_{O_2}, signalled by the sensor. This feature pre-destines polarographic oxygen sensors for aquatic application: The sensor is 20 to 30 times more sensitive to respiratory rates in water than in the same volume of air.

The operational principle of polarographic oxygen sensors is based on oxygen diffusion to the polarized cathode, where oxygen reduction (in micromoles O_2 per second) generates the electrical signal (in amperes). According to Faraday's law the ratio is calculated as 2.591 μmol O_2 s^{-1} A^{-1} (= 9.328 nmol O_2 h^{-1} μA^{-1}). Variations in the construction of polarographic oxygen sensors are mainly related to the problem of how oxygen diffuses to the cathode. In this respect the actual design of every sensor necessarily involves a compromise, since optimising particular functions, such as sensitivity and response time, detracts from others, such as stability and stirring requirements. The optimal design therefore depends upon the application.

The many new commercial oxygen sensors that have come onto the market during the past few years include, to some extent, original ideas, or are merely copies of existing sensors. For instance, the information that a polarographic oxygen sensor for marine applications (InterOcean Systems, San Diego) is virtually the same sensor at a price six times that of the original (Ingold, CH) probably comes too late for some purchasers.

Proficiency in solving respirometric problems facilitates research into the physiological mechanisms and functional interpretations of gas exchange. With this in mind, respirometric and in situ monitoring methods, of which the polarographic oxygen sensor forms an integral part, are outlined in the detail that is of practical importance but usually neglected (Parts II and III). Methods related to studies of oxygen exchange are also dealt with. Responses to environmental variables and to toxicological or pharmacological agents, metabolic patterns and biological rhythms can be resolved by automatic long-term monitoring of oxygen consumption with polarographic oxygen sensors. In combination with simultaneous measurements of, e.g., locomotory activity, ventilatory rates, chloroplast migration, biomass production and notably heat dissipation such investigations can be most fruitful.

It is noteworthy that bioenergetics and the recognition of oxygen have a common root in Lavoisier's ingeneous concept of combustion and his classical experiments on direct and indirect calorimetry. Keeping in line with the extension of the direct and indirect calorimetric approach to the bioenergetics of aquatic animals, the thermodynamic interpretation of oxygen consumption in aquatic organisms is revised in an Appendix and more closely aligned to the needs of ecological energetics. Several case studies illustrate the practical application of the methods and point out concurrent conceptual advances.

The respirometers employed in these studies are no longer simply closed or open systems. Further differentiation has given rise to specific types such as the "rubber mask", "twin-flow" or "slurp gun" respirometers. Various types have invaded lakes, coastal regions and even the deep sea, yet respirometers are still most common in the shallow, constant temperature water baths on the bench. Restocking with new variations and their association with other instumental groups seems rewarding, especially in some laboratories where respirometers have become fossilized or extinct.

Previous summaries of the applications of polarographic oxygen sensors in medicine and cellular physiology can be found in Kessler M. et al., eds. (Oxygen Supply. Urban Schwarzenberg, München−Berlin− Wien, 1973) and Fatt I. (Polarographic Oxygen Sensors. CRC Press, Cleveland, 1975). For an extended theoretical discussion the reader is referred to Hitchman L.M. (Measurement of dissolved oxygen. Wiley, New York and Orbisphere Laboratories, Geneva, 1978).

We hope that the joy and benefits of interdisciplinary communication as experienced by the editors of this book may be shared by our readers. Ideas and technological advances in one field may unexpectedly provide the key to solving problems in other apparently unrelated disciplines (compare e.g. Chaps. I.4 and III.2, or p. 117 and 247). Thus perhaps even those readers whose special knowledge or special needs are neglected or inadequately dealt with may find some inspiration.

The plan for this book originated during a workshop held at the Institut für Zoophysiologie der Universität Innsbruck, Austria, in October 1978. Several methodological developments and the editorial work were supported by the *Fonds zur Förderung der wissenschaftlichen Forschung in Österreich,* projects no. 2919, 2939, 3307 and 3917 (Univ. Prof. Dr. W. Wieser, Univ. Innsbruck, principal investigator), by the *Forschungsförderungsbeitrag der Vorarlberger Landesregierung* and by a British Council Scholarship at the Institute for Marine Environmental Research, Plymouth, England. I thank all contributors for their encouraging cooperation. I want to express my special gratitude to Mrs. J. Wieser for her experienced help in improving the English style of several manuscripts, and I thank my colleagues and friends for their support and fruitful comments.

Innsbruck − Plymouth   Erich Gnaiger

Contributors

You will find the addresses at the beginning of the respective contributions

Contents

Part II   Laboratory Applications of Polarographic Oxygen Sensors

Part III   Field Applications of Polarographic Oxygen Sensors

Part I   Principles of Sensor Design and Operation

Chapter I.1 Factors Influencing the Stability of Polarographic Oxygen Sensors

J.M. Hale[1]

1 Introduction

An ideal polarographic oxygen sensor (POS) exhibits a time-independent relationship between the current it delivers, throughout its specified operating temperature range, and the activity of oxygen contacting its membrane. All applications of POS require stability in some measure, and long-term monitoring applications demand stability over periods up to 1 year.

Available sensors approach more or less closely to this ideal, depending upon their makeup, their design, and the evolution of certain thermodynamic variables of the system. This article presents a systematic examination of those factors which influence the steady-state current output of POS, includes an attempt to weigh the relative importance of these various factors, lists the symptoms which enable the user to recognize individual effects, and gives cures, when these are possible, for the troubles.

At the outset a distinction must be drawn between a drift of steady-state sensitivity, and the short-term transient approach to stabilization following any major change in operating conditions, such as a change of membrane, of temperature, of applied voltage, or of oxygen activity. Such a transient change of sensitivity is easily recognizable, and operating procedure is easily adapted to it. It is usually sufficient to wait for a certain predetermined time, the stabilization time, before resuming measurements after such a perturbation. Typically, the stabilization time lasts for several minutes.

After this stabilization time has elapsed, the output from the POS settles down to a level which can appear to be constant over a period of several hours, but which nevertheless can drift noticeably, at constant oxygen activity, over a period of months. It is this drift which is discussed in detail in this section.

2 Theory of the Sensitivity of POS

The causes of instability of POS are best discussed with reference to a theory of the dependence of POS-sensitivity upon relevant dimensions and other parameters. Such

1 Orbisphere Laboratories, 3, Chemin de Mancy, CH–1222 Vésenaz, Switzerland

Polarographic Oxygen Sensors (ed. by Gnaiger/Forstner)
© Springer-Verlag Berlin Heidelberg 1983

theories have been elaborated by a number of workers specializing in different fields [1–5, 9, 11, 13, 18, 23]. In drawing these works together, therefore, the opportunity has been taken here to generalize the results and to extend their significance.

2.1 Oxygen Fugacity, Partial Pressure, Activity and Concentration

Theories of POS sensitivity are based upon Faraday's laws and the diffusion equation, but also upon an assumption that thermodynamic equilibrium exists at the two-phase boundaries: membrane/sample and membrane/electrolyte solution. This assumption leads to the boundary conditions, employed in the solution of the diffusion equation, that the chemical potential, or equivalently the fugacity of oxygen, is the same on each side of the two-phase boundaries. When this assumption is valid, it is the fugacity of oxygen which is really measured by a POS.

The fugacity of a component in a nonideal solution is a true measure of its escaping tendency [10, 20]. It is equal to the partial vapor pressure of the component if the vapor behaves as an ideal gas, as is the case for oxygen at the low pressures normally monitored with POS; hence it is correct to say that the signal from a POS is a measure of the partial pressure of oxygen contacting the membrane. Deviations of the fugacity from partial pressure are discussed in standard works on thermodynamics.

It is also correct to say that the signal from a POS is a measure of the relative activity a_s of oxygen in a solution, since the activity of a gas is proportional to its fugacity by definition:

$$a_s = f_{O_2}/f_{O_2}^{\circ}. \tag{1}$$

$f_{O_2}^{\circ}$ is the fugacity of a freely chosen standard state of the gas, and is therefore a constant. The activity is a dimensionless quantity (cf. Chap. I.3).

The user of a POS should always bear in wind, however, that the *concentration*, c_s, of oxygen in a solution can vary independently of the fugacity. These quantities are related through the activity coefficient y_s of oxygen in the particular sample medium

$$a_s = y_s \cdot c_s/c_s^{\circ} = f_{O_2}/f_{O_2}^{\circ} \tag{2}$$

and y_s can vary, at constant fugacity, due to variations in the concentrations of other components of the sample solution [24, 25]. This phenomenon has long been understood, and accounted for, in the field of oceanography, where oxygen measurements are made in saline water and substantial variations of y_s occur, but is less appreciated in other fields. These points are relevant when testing the stability of POS, by comparison with the results of a chemical titration method such as that due to Winkler ([26], App. D), because titrations determine concentrations rather than fugacities. In this case the composition of the sample solution must be maintained constant during the test, to keep y_s constant.

2.2 The Oxygen Diffusion Impedances of the Sample, the Membrane, and the Electrolyte Solution

The electrical signal I_1 [A] delivered by a POS is proportional, by Faraday's laws, to the flux of oxygen J_{O_2} [mol m^{-2} s^{-1}], arriving at the cathode and originating in the sample:

$$I_1 = nFA\, J_{O_2}.$$ (3)

Here $n = 4$ is the number of electrons added to each oxygen molecule arriving at the cathode, F [C mol^{-1}] is the Faraday, and A [m^2] is the area of the cathode. Further discussion will be restricted to the "one-dimensional" case, when the cathode is sufficiently large relative to the membrane thickness for edge effects to be ignored. Conditions at the cathode are so arranged that the fugacity of oxygen is reduced to zero at its surface; then the flux of oxygen is maximized, and is determined in the steady state by the resistance to oxygen diffusion offered by the sample, the membrane and the electrolyte layer (cf. Chap. I.3).

$$J_{O_2} = f_{O_2} / [(z_s/D_s S_s) + (z_m/D_m S_m) + (z_e/D_e S_e)].$$ (4)

Here f_{O_2} [kPa] is the fugacity of oxygen in the bulk of the sample, and the three terms in the denominator represent the diffusion impedances of the boundary layer in the sample, the membrane, and the electrolyte layer respectively. Z [m] is the thickness of the layer, D [m^2 s^{-1}] is the diffusion coefficient of oxygen and S [mol m^{-3} kPa^{-1}] the solubility of oxygen in the layer. The boundary layer is that region of the sample, adjacent to the sensing membrane, within which the concentration of oxygen is lower than in the bulk of the sample due to consumption by the sensor itself.

Clearly, the output from the POS can be stable only if the total diffusion impedance is constant, and the causes of instability must be sought among those factors which can influence the magnitudes of the individual contributions to this impedance. These factors are considered below.

The largest contribution to the total diffusion impedance is normally that due to the membrane. Table 1 lists magnitudes of the relevent parameters, and the derived values of the corresponding diffusion impedances, $z_m/D_m \cdot S_m$, for six commonly used membrane materials.

In aqueous electrolytes at 25°C, the oxygen solubility S_e and diffusion coefficient D_e are of the order of 8.5 · 10^{-3} mol m^{-3} kPa^{-1} [17] and 2 · 10^{-9} m^2 s^{-1} [27] respectively. Ideally, the electrolyte layer should have a thickness z_e of 1 μm, so that its diffusion impedance would be approximately 6.0 m^2 s kPa mol^{-1}.

In the sample, assumed for example to be pure water at 25°C, $S_s = 0.0125$ mol m^{-3} kPa^{-1} [17] and $D_s = 2.6 \cdot 10^{-9}$ m^2 s^{-1} ([14], Chap. I.4). The thickness z_s of the boundary layer, however, depends upon the type and vigor of agitation of the sample. One convenient example which can be treated in detail with exact mathematical methods is that in which the sensing face of the POS is disk-shaped and is rotated relative to the sample fluid about an axis through the center of the disk. In this case the boundary layer thickness is uniform and is given by [16]:

$$z_s = 17\, D_s^{1/3}\, \nu_s^{1/6}\, w^{-1/2},$$ (5)

Table 1. Solubility and diffusion parameters for 1 mil (0.001 inch or 25.4 μm) membranes of various materials, at 25°C

Material	D_m [m² s⁻¹]	S_m [mol m⁻³ kPa⁻¹]	z_m/D_mS_m [m² kPa s mol⁻¹]	$0.53\, z_m^2/D_m$ [s]
Polytetrafluoro-ethylene (PTFE)	2.54×10^{-11}	0.106	0.943×10^7	25.4
Perfluoroalkoxy (PFA)	2.69×10^{-11}	0.093	1.02×10^7	24.0
Fluorinated ethylene-propylene (FEP)	1.73×10^{-11}	0.088	1.67×10^7	37.3
Ethylene-tetra-fluoroethylene (Tefzel)[a]	1.34×10^{-11}	0.017	1.12×10^8	48.1
Polytrifluoro-monochloro ethylene	3.43×10^{-13}	0.040	18.5×10^8	1881

[a] Tefzel is a registered trademark of DuPont de Nemours

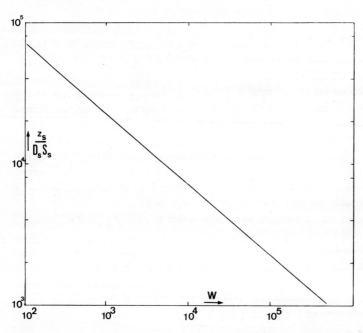

Fig. 1. Dependence of the diffusion impedance of the boundary layer in the sample upon rotation rate of the stirrer (see text)

where ν_s [m² s⁻¹] is the kinematic viscosity of the sample fluid and w is the rotation speed in rotations per minute. It is clear that the thickness and diffusion impedance of the boundary layer decrease with increase of rotation rate. Figure 1 illustrates this conclusion for $\nu_s = 10^{-6}$ m² s⁻¹, other parameters having previously quoted magnitudes typical of an aqueous sample at 25°C. Further studies on the subject of diffusion limitations due to the sample are reported in [23].

2.3 Mechanical Requirements for Stability

In order to achieve highly stable output from a POS, conditions should be chosen such that the membrane-diffusion impedance is precisely determined and outweighs the contributions of the electrolyte layer and of the sample by as great a margin as possible. This means that the electrolyte and boundary layer thicknesses must be less than certain critical values, so that membrane tension and sample agitation must be carefully controlled. These critical values become smaller as the membrane permeability increases or as the membrane thickness decreases.

In order to express these requirements more quantitatively, Eqs. (1) and (2) are rewritten in the form:

$$I_1/I_m = (1 + r)^{-1},$$

(6)

where

$$I_m = nFAD_m S_m f_{O_2}/z_m$$

(7)

and

$$r = [(z_s/D_s S_s) + (z_e/D_e S_e)] / (z_m/D_m S_m).$$

(8)

I_m would be the magnitude of the sensor current if the membrane alone were responsible for limiting the supply of oxygen to the cathode, and r is the ratio of the sum of the diffusion impedances of the electrolyte layer and the sample, to the diffusion impedance of the membrane.

Clearly, $I_1 = I_m$ when r = 0, and for any value of r greater than zero, the current I_1 is smaller than I_m. Figure 2 plots the dependence of the ratio of the sensor current to the "membrane" current, upon the ratio of diffusion impedances r. For small values of r, $I_1/I_m \cong 1 - r$.

If often happens that POS are operated under conditions for which the diffusion impedances of the electrolyte and/or boundary layers are poorly controlled and, therefore, variable. Variations of 100% are not unlikely. Figure 2 shows that such variations will result in poor stability of the output from the sensor if r is large, but will have little influence if r is small.

2.4 Electrochemical Requirements for Stability

In connection with Eq. (1), it was mentioned that the fugacity of oxygen at the cathode surface must be zero during operation of a POS. For this to be so, the rate

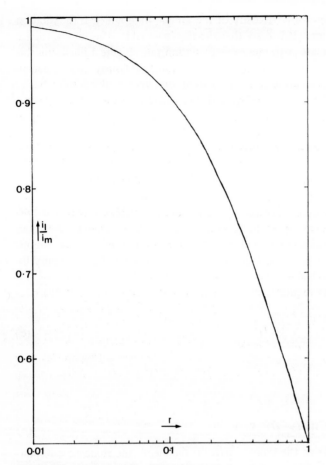

Fig. 2. The influence of the diffusion impedance of the electrolyte and boundary layers upon the current output from a POS

constant k for the electrochemical conversion of oxygen to hydroxide ions at the cathode surface must be large. If k is insufficiently large, the output current is smaller than the idealized current considered previously, being given by

$$I_1 = nFA f_{O_2} / [(1/kS_e) + (z_s/D_s S_s) + (z_m/D_m S_m) + (z_e/D_e S_e)]. \qquad (9)$$

Electrochemical rate constants depend upon the interfacial voltage difference U_i between the cathode and the electrolyte, in an exponential manner [22]:

$$k \sim k_o \exp (\alpha e U_i / RT), \qquad (10)$$

where α is a dimensionless constant ($o < \alpha < 4$), e is the charge on a proton, R is the gas constant and T the absolute temperature. By design, the applied voltage is made sufficiently large initially for the current to become independent of this voltage. Yet there are certain phenomena which may occur during operation of the sensor, which can influence the magnitude of the rate constant k, and hence of the sensor output.

POS are most commonly operated with a constant total applied voltage, which is shared between the anode/electrolyte interface, the cathode/electrolyte interface, and the electrolyte conductor. Ideally, the electrolyte solution (Chap. I.5) should be chosen such that it represents an almost perfect conductor, and its share of the applied voltage is virtually zero. Also ideally, the anode should be operated under practically equilibrium conditions so that its interfacial voltage is fixed by the thermodynamic parameters of the system. Ideally, therefore, the cathode interfacial voltage U_i should remain constant.

The electrolytic conductor may take an increasing share of the applied voltage with the passage of time, either because it becomes progressively depleted of salt due to the anode reaction, or because it becomes gradually diluted with sample water due to faulty sealing of the membrane to the sensor body (Chap. I.6).

The anode interface may take a variable fraction of the applied voltage, either because it becomes increasingly blocked by the deposition of nonporous insulating products, or because the penetration of electroactive vapors into the cell from the sample interfere with the intended anode reaction.

In the presence of such phenomena, the cathode voltage U_i is also variable, usually in the sense that it decreases with time, causing a downward drift in sensor output.

The oxygen electrode reaction is also subject to interference by deposits on the cathode surface, either of electroplated metals (the anode metal is frequently plated onto the cathode during operation), or of organic substances adsorbed from the electrolyte, or of products of side reactions at the cathode of interfering vapors from the sample. Again, a downward drift of sensor output at constant oxygen activity is the usual consequence.

3 The Factors Responsible for Instability of POS

Two types of instability are observed during the operation of POS [6, 19, 21]. Often, there is a gradual monotonic drift of sensitivity — usually in the sense that a decreasing current per unit oxygen activity is output, with the passage of time. Secondly, there are positive and negative fluctuations of sensitivity superimposed on the general drift, some of very short duration, of the order of seconds or minutes, and others of longer time constant of the order of days. Table 2 lists various causes for each of these observable symptoms.

The effects of most of these causes of instability can be eliminated, or at least minimized, by careful design.

3.1 Overstressed or Understessed Membranes

The most important single factor determining the sensitivity of a POS is the membrane tension. Because the membranes used for POS are normally thin, and partially elastic, their thicknesses, and hence the impedances they offer to oxygen diffusion, can be substantially altered if they become stretched when being mounted over the sensing

Table 2. Causes of instability of POS

General drift	Osmotic effects (downward)
	Release of membrane stress (downward)
	Leakage or evaporation of a gross excess of electrolyte (upward)
	Deposition of anode metal at edge of cathode causing increase of cathode area (upward)
	Membrane erosion (upward)
	Membrane clogging (downward)
	Anode blocking (downward)
	Dilution of electrolyte with sample water (downward)
	Cathode deactivation by anode metal or poisons
Shert-term fluctuations	Poorly stabilized anode supply voltage
	Poorly compensated temperature variations
	Flow rate variations of sample
	Gravitational (orientational) effects
	Applied pressure effects
	Vibrational effects
Intermediate-term fluctuations	Barometric pressure effects
	Residual current variations

face. One of two deleterious effects may then occur. Either the membrane may gradually slip under the restraining means, to return to its original thickness, or it may suffer "cold flow" as a result of the constant stress, with a consequent change of thickness. These thickness changes often continue during a period of months following mounting and are observed as a gradual drift of output signal. Experience suggests that true stability can only be achieved if membrane stretching is avoided during mounting.

At the same time it is necessary to create conditions, during the membrane mounting operation, which do not permit a thick electrolyte layer to remain between the membrane and the sensing cathode. This requirement follows from the rule that the diffusion impedance of the electrolyte layer should be as small as possible relative to that of the membrane.

Membrane mounting is therefore a rather delicate operation, which can be accomplished reproducibly only with a mechanical aid. Such a tool can be simple, but its design is not as trivial as it would seem at first sight.

3.2 Membrane Erosion and Membrane Clogging

Two problems encountered when operating POS in harsh environments are the erosion of the membrane by suspended solid particles, and the deposition on the membrane of films of algae, microorganisms, or grease [6]. Erosion of the membrane leads to erroneously high measurements, because of the connection between sensitivity and the reciprocal of membrane thickness. Clogging of the membrane produces erroneously low measurements.

Membrane erosion can be countered by using thicker, tougher membranes, whenever the consequent increase of response time of the sensor is of minor importance, and by filtering or settling out suspended particles in other circumstances.

The rate of membrane clogging has been found to vary greatly in different culture media [6]. It has been suggested that attachment of strongly acidic groups (e.g., sulfonate $-$ HSO_3^-) to the surface of the membrane contacting the sample medium might be an effective means of prevention of clogging [8], because these groups, when ionized, would repel the predominately negatively charged colloidal species responsible for the clogging. Until such specially treated membranes become available, the only recourse to counter membrane clogging seems to be to locate the sensor in a turbulent region of the sample, where "sticking" of deposits becomes less likely.

3.3 Osmotic, Gravitational, Pressure or Vibrational Effects

The role of the electrolyte layer intervening between the membrane and the cathode, in determining the sensitivity of a POS, was detailed above. Hence physical phenomena which influence directly the thickness of the electrolyte layer can give rise to instability of the response from a sensor.

The membrane which encloses a POS is semipermeable in the sense employed in discussions of osmotic effects, that is, it allows the passage of water vapor but blocks the passage of ions. When a POS is immersed in pure water, therefore, water vapor enters the sensor, diluting the electrolyte. If the walls of the electrolyte chamber were completely rigid, the internal pressure would rise concomitantly, until the water vapor pressure in the electrolytic solution equals the water vapor pressure in the water outside the sensor. Equilibrium would then be attained. In practice, however, the hydrostatic pressure inside the sensor at equilibrium, known as the osmotic pressure, can be as high as 10^4 kPa at equilibrium [20], and could not be restrained by the tension in the membrane. Distillation of water into the sensor continues, therefore, with movement of the membrane to adapt to the increased volume of electrolyte.

This entry of water into the sensor causes an increase in the electrolyte layer thickness, z_e, and a consequent decrease in the output from the sensor.

As an example, we calculate the rate of decrease of the signal delivered by a POS containing 2 mol dm^{-3} KCl as electrolyte, and provided with a 25 μm thick Tefzel membrane, assuming the sensor to be immersed in pure water at 25°C. The vapor pressures of water inside and outside the sensor are 2.96 kPa and 3.03 kPa respectively [12]. From the water permeability of Tefzel, it then follows that the rate of increase of thickness of the electrolyte layer dz_e/dt, is about 1 μm/month. From Eq. (6) we derive:

$$\frac{d}{dz_e}(I_l/I_m) = -\frac{D_m S_m}{z_m D_e S_e} \sim -10^{-3} \ \mu m^{-1}. \tag{11}$$

Hence, the downward drift of the signal due to this cause amounts to about 0.1% per month.

This effect increases in significance as the oxygen and or water permeability of the membrane increases.

A hydrostatic pressure in the sample influences the signal from a POS because the activity of dissolved oxygen at constant concentration is changed by this applied pressure. Assuming that the partial molar volume of oxygen V_m (0.032 dm^3 mol^{-1}) is independent of applied pressure in the range of interest, the activity of oxygen and the signal from the sensor are *increased* by an *overpressure* $_wp$ by the factor ([20], p. 318)

$$\exp \ (V_m \cdot \ _wp/RT).$$

The effect is not large: at 2000 kPa overpressure or 200 m water head, for example, the signal is 2.5% higher than at atmospheric pressure (App. A).

Applied pressure can have a substantial influence upon sensor response if gas bubbles are enclosed in the electrolyte by the membrane, for compression or expansion of these bubbles lead to large changes in the thickness of the electrolyte layer. Such problems are minimized by choosing sensors having uncluttered electrolyte reservoirs which present a minimal risk of trapping of air bubbles.

Gravitational effects upon the electrolyte layer are conceptually simpler to understand, for the membrane may sag to some extent if it supports the whole weight of the electrolyte. The effect is most significant if the electrolyte volume and the membrane area are large, and if the membrane is poorly tensioned and supported. It can be minimized by clamping the sensor in a fixed orientation.

Vibrations of the sensor or turbulence in the sample sometimes causes the sensor to deliver a noisy signal. This problem is particularly acute if either the membrane is poorly tensioned, or if air bubbles are enclosed in the sensor. The explanation is thought to be that convection currents are caused in the electrolyte, thus transporting oxygen from the electrolyte reservoir to the cathode.

3.4 Flow Rate Variations of the Sample

Variations of the flow rate of the sample cause variations of the thickness of the boundary layer z_s. A specific example of a "rotating disk" was presented above, but similar conclusions may be drawn for other modes of convection of the sample fluid. It was pointed out that the sample agitation must exceed a certain critical magnitude, characteristic of the particular membrane used, in order that the sensor output be independent of flow rate. Estimates of the critical velocity for a flowing sample stream are usually given in the published specifications of commercially available POS.

3.5 Plating of Anode Metal on Cathode

Ideally, the electrochemical reaction at the anode of a POS should result in the creation of an insoluble porous solid phase which adheres firmly to the anode metal. In practice, the products generated at most of the commonly chosen anode metals, such as silver, lead, cadmium, thallium, or zinc, are soluble in the electrolyte to some extent because of the formation of complexes between the metal ions and the anions of the electrolyte. Diffusion of these ions and subsequent plating of the anode metal on the cathode is therefore a common phenomenon.

The plating is less rapid when the membrane is well tensioned, and when the zone surrounding the cathode, through which diffusion of the metal ions occurs, is long and narrow. Also the useful life of the sensor is longest, for any particular sample, for membranes of low permeability, since the initial rate of formation of the metal ions is proportional to the membrane permeability.

Sometimes the plating occurs over the whole of the area of the cathode, an effect which is most easily noticeable at gold cathodes because of the change of coloration. This is not too troublesome if the anode is silver, but can lead to a decrease in cathode activity when the anode metal is more basic. The cathode should be polished in these circumstances to restore its activity.

Otherwise the plating occurs at the periphery of the cathode, as tree-like dendritic formations, thus extending the sensing area and leading to an increase in displayed concentration at constant sample concentration. Again, the cathode must be polished at regular intervals.

3.6 Anode Blocking

The continuous operating lifetime of a POS is often limited by the blocking of its anode by nonporous reaction products.

It usually occurs when a particular quantity of anode product has been deposited on each unit area of the anode, and therefore the lifetime is longer the larger the area of the anode, the lower the permeability of the membrane, and the lower the concentration of oxygen to which the sensor is exposed. Activity is easily restored by stripping the product off the anode with a suitable cleaning solution. A silver anode, for example, may be cleaned with a commercial detarnishing fluid, or with an ammoniacal solution.

3.7 Drying Out of the Sensor or Dilution of the Electrolyte with Sample Water

In order to function properly, a POS must be filled with an electrolyte having at least a minimum conductivity for current flow. This conductivity is decreased if either the sensor dries out, or if the electrolyte becomes diluted by sample water entering through a "leak" in the membrane sealing system. The symptom manifested by the sensor in this case is a gradual decrease of signal at constant oxygen concentration. To cure the problem, it is only necessary to renew the electrolyte, and to ensure adequate sealing of the membrane to the sensor body.

3.8 Poorly Stabilized Applied Voltage

Theoretically, the output current from a POS should be independent of small ($\leqslant 0.2$ V) variations of the applied voltage, since the slope of the current voltage curve should be zero at the chosen operating voltage. In practice a sudden change in applied voltage usually has two effects: (1) it triggers a transient "kick" of the current — and hence in the displayed oxygen concentration — which can be of substantial proportions, e.g., equivalent in order of magnitude to the signal in air saturated water, followed by an exponential decay to the steady state, lasting some minutes; (2) the steady-state sensitivity of the sensor changes in some degree, depending upon the design of the sensor and the choices of cathode metal and electrolyte.

The best cure for this problem, of course, is to eliminate the instabilities in the applied voltage, using filtering techniques well known in the electronic art.

Particular attention should be paid to this phenomenon when current measurements are made by following the voltage drop across a resistor in series with the sensor. With this method, the voltage across the sensor changes linearly with the steady-state current, and the system is potentially "noisy". "Galvanic" sensors having an applied voltage of 0 V are usually operated in this fashion. Details of better methods of current measurement are to be found in Chapter I.10.

3.9 Temperature Variations

The current delivered by a POS changes by about 3% to 4% per $^\circ$C, at constant oxygen activity. This temperature coefficient is greatest for the least permeable membranes [15].

When temperature changes occur during the course of measurements, as they inevitably do during field work, for example, this temperature coefficient is recalled from a previous determination in order to interpret the measured current in terms of oxygen activity. Hence one is concerned not only with the stability of sensor output at constant temperature, but also with the stability of the temperature coefficient of oxygen solubility and diffusivity in the membrane, and perhaps also, of properties of the electrolyte layer and diffusion boundary layer.

Extra care is necessary when a POS is to be operated well above room temperature, to avoid enclosing an air bubble in the sensor. The reason is that this bubble expands more than the displaced electrolyte would have done, and the additional volume might be created by "inflation" of the membrane. This, of course, leads to a diminished output from the sensor, and sometimes to an irreversible stretching of the membrane.

3.10 Residual Currents

The residual current of a POS is the current produced by the sensor when exposed to a medium containing no oxygen. It is never zero, although the sensor should be selected such that it is negligible in comparison with the signals of interest in any particular study. Commercial sensors are available having residuals in the range from 0.01% to 1% of the signal generated in air-saturated water. Hence the stability of residual currents is of little interest when measurements are to be made in well-aerated clean water, but becomes of crucial importance when measurements are made in anaerobic media.

Even a cursory study of residual currents reveals that they are rarely constant in time (Chap. I.2). For example, a current spike followed by an exponentially decaying current tail is observed whenever a sensor is switched on after being exposed to air whilst off circuit. The time constant for this decay can be minutes or even hours, depending upon the particular sensor, and upon the state of cleanliness of the cathode.

Because of the transient nature of the residual current, the practice of electronically adjusting the zero point of a sensor by subtraction of a constant current is of limited usefulness.

The residual current is completely internal in origin, and so becomes of less relative importance as the oxygen permeability of the membrane increases, because the signal level increases in comparison with the constant error level.

4 The Compromise Between Stability, Speed of Response, Residual Current, and Stirring Requirement

The membrane permeability (Chap. I.3) has been mentioned in several places in the foregoing discussion, in the context of POS stability. This is because the oxygen must pass through the diffusion boundary layer and the electrolyte layer, as well as the membrane, and these two additional impedances are likely to vary during operation; therefore stability can be expected only when the membrane impedance exceeds those of the other two by a safe margin. This reasoning alone would dictate a choice of a low permeability membrane for a POS, whenever stable, long-term monitoring of oxygen is of primary importance.

Another advantage also accrues from the use of a low permeability membrane, namely that the rate of stirring required to reduce the impedance of the diffusion boundary layer to a negligible value in comparison with that of the membrane, is least for these membranes. Indeed, as has been seen, very high stirring rates are needed to supply oxygen at the rate required by thin PTFE, PFA, or FEP membranes.

Why, then, are highly permeable membranes most commonly chosen for POS?

The reason is that, in some applications, speed of response and the ability to measure traces of dissolved oxygen are of predominating importance.

The response time of a POS for 99% of a signal change is [4]

$$\tau_{99} \cong 0.53 \, z_m^2/D_m, \tag{12}$$

assuming that the transient is controlled by the membrane alone. This time is shortest for thin permeable membranes. Relevant values are recorded in Table 1, and it is clear that the lower permeability membranes would be totally unsuitable for any application where speed of reaction is necessary.

Secondly, the lower limit of oxygen concentration c_{lim} which can be measured with any particular sensor is:

$$c_{lim} = I_r/\phi, \tag{13}$$

where I_r is the residual current of the sensor and ϕ is the sensitivity of the sensor expressed as current per unit concentration. Since the residual current is independent of the membrane chosen, while the sensitivity is proportional to the permeability of the membrane toward oxygen, the concentration limit is inversely proportional to the permeability. Hence, any sensor is more capable of measuring to lower concentration limits, when it is fitted with a permeable membrane, than when fitted with an impermeable one.

A compromise between speed of response and high sensitivity, on the one hand, and stability and low stirring requirement on the other is therefore necessary when selecting a sensor for any particular application. When rapid response is indispensible, frequent recalibration is inevitable, and when stability is of prime importance, a sluggish response must be tolerated.

References

1. Aiba S, Huang SY (1969) Oxygen permeability and diffusivity in polymer membranes immersed in liquids. Chem Eng Sci 24:1149–1159
2. Aiba S, Ohashi M, Huang SY (1968) Rapid determination of oxygen permeability of polymer membranes. I + EC Fundam 7:479–502
3. Benedek AA, Heideger WJ (1970) Polarographic oxygen analyzer response: The effect of instrument lag in the non-steady state reaction test. Water Res 4:627–640
4. Berkenbosch A (1967) Time course of response of the membrane-covered oxygen electrode. Acta Physiol Pharmacol Neerl 14:300–316
5. Berkenbosch A, Riedstra JW (1963) Temperature effects in amperometric oxygen determinations with the Clark electrode. Acta Physiol Pharmacol Neerl 12:131–143, 144–156
6. Borkowski JD, Johnson MJ (1967) Long-lived steam sterilizable membrane probes for dissolved oxygen measurement. Biotech Bioeng IX:635–639
7. Corrieu G, Touzel JP (1978) Comparaison de sonde de mesure de la concentration en oxygène dissous: Essais au laboratoire. Tech Sci Munic 73:349–356
8. Gregor HP, Gregor CD (1978) Synthetic-membrane technology. Sci Am 239:88–101
9. Hale JM, Hitchman ML (1980) Some considerations of the steady state and transient behavior of membrane covered dissolved oxygen detectors. J Electroanal Chem 107:281–294
10. Hitchman ML (1978) Measurement of dissolved oxygen. Orbisphere Laboratories, Geneva, p 7
11. Jensen OJ, Jacobsen T, Thomsen J (1978) Membrane covered oxygen electrodes. I. Electrode dimensions and electrode sensitivity. J Electroanal Chem 87:203–211
12. Kaye GWC, Laby TH (1972) Tables of physical and chemical constants, 14th edn. Longman, London, p 222
13. Kok R, Zajik JE (1973) Transient measurement of low dissolved oxygen concentrations. Can J Chem Eng 51:782–787
14. Kolthoff IM, Miller CS (1941) The reduction of oxygen at the dropping mercury electrode. J Am Chem Soc 63:1013–1017
15. Krevelen van DW (1972) Properties of polymers. Elsevier, Amsterdam, p 286
16. Levich VG (1962) Physiochemical hydrodynamics. Prentice Hall, Englewood Cliffs NJ, p 69
17. Linke WF (1965) Solubilities of inorganic and metal-organic compounds, vol II. Am Chem Soc, Washington DC
18. Mancy KH, Okun DA, Reilley CN (1962) A galvanic cell oxygen analyzer. J Electroanal Chem 4:65–92
19. McKeown JJ, Brown LC, Gove GW (1967) Comparative studies of dissolved oxygen analysis methods. J Water Pollut Control Fed 39:1323–1336
20. Moore WJ (1972) Physical chemistry, 5th edn. Longman, London, p 250, 300
21. Pijanowski BJ (1971) A quantitative evaluation of dissolved oxygen instrumentation. Joint conference on sensing of environmental pollutants, Palo Alto, Calif
22. Randles JEB (1952) Kinetics of rapid electrode reactions, part 2. Rate constants and activation energies of electrode reactions. Trans Faraday Soc 48:828–832
23. Schuler R, Kreuzer F (1967) Rapid polarographic in vivo oxygen catheter electrodes. Respir Physiol 3:90–110
24. Schumpe A, Deckwer WD (1978) Oxygen solubilities in synthetic fermentation media. Preprints, 1st Eur Cong Biotechnol, Interlaken, pp 154–155

25. Setschenow J (1889) Concerning the constitution of salt solutions on the basis of their behavior to carbonic acid. Z Phys Chem 4:117–125
26. Winkler LW (1889) The determination of dissolved oxygen in water. Ber Dtsch Chem Ges 22: 1764–1774
27. Yatskovski AM, Fedotov AN (1969) Solubility and diffusion of oxygen in solutions of potassium hydroxide and phosphoric acid. Electrokhimiya 5:1052–1053

Chapter I.2 Calibration and Accuracy of Polarographic Oxygen Sensors

M.L. Hitchman[1]

1 Introduction

A membrane-covered polarographic oxygen detector, or polarographic oxygen sensor (POS) as it is called here, is an electrochemical device in which the steady-state current at the working electrode (the cathode) is, in the ideal case, linearly proportional to the concentration, or more strictly the activity, of oxygen in contact with the external surface of the membrane. This proportionality is most simply expressed by the equation originally derived by Mancy et al. [15]:

$$I_1 = \frac{nFAP_m c_s}{z_m} ,$$

(1)

where the symbols are defined in the discussion below. According to this equation, a plot of detector current, I_1, against oxygen concentration, c_s, will be linear with a slope of $nFAP_m/z_m$. This slope can be regarded as the calibration factor for the sensor.

However, of the terms in this factor only F, the Faraday, is truly constant. All the other terms can be variable. Clearly, A, the cathode area, can be different for different sensors, but even for a given sensor it can change during use. For example, in a polluted environment there may be partial poisoning of the cathode, leading to a reduction in the active area, while in a dirty environment fouling of the membrane could occur, causing a reduction in the effective area through which the oxygen flux can reach the cathode by diffusion normal to the electrode surface. The membrane thickness, z_m, may vary for membrane samples of the same nominal thickness, while P_m, the permeability coefficient (Chap. I.3), can be very different for membranes of the same plastic but from different sources; different techniques of film preparation can give rise to different degrees of crystallinity and hence membrane permeability [10]. Once a given membrane is mounted on the sensor there is still no guarantee that z_m and P_m will not vary with time. Plastic films used for POS membranes are elastic to some extent and so can be readily stretched when they are mounted on the sensor. Gradual relief of the applied stresses will cause z_m to change slowly, and may also be accompanied by internal restructuring, giving crystallinity changes and concomitant permeabil-

---

1 Department of Chemistry and Applied Chemistry, University of Salford, Salford, M5 4WT, Great Britain

Polarographic Oxygen Sensors (ed. by Gnaiger/Forstner)
© Springer-Verlag Berlin Heidelberg 1983

ity changes. Finally, n, the number of electrons associated with the oxygen reduction, need not be equal to the maximum value of 4. Competing electrode processes and the nature of the electrolyte can cause the value of n to be less than 4 and to be time-dependent [10].

In view of the uncertainties associated with the values of the terms in the calibration factor it is apparent that one cannot calculate with any degree of confidence the value of this factor. Therefore one has to rely upon a calibration procedure based on standard oxygen-containing test samples. Furthermore, since there may also be changes with time of the calibration factor any such procedure should be straightforward and easy to carry out. Some methods that are commonly used are described later, but first we consider calibration standards that can be used in these methods.

2 Calibration Standards

The basic requirement of a calibration standard is that it should have a known oxygen activity. This is because the current of the sensor is dependent on the oxygen flux through the membrane, and this flux is, in turn, determined by the difference in activity or chemical potential of oxygen across the membrane [10]. A practical measure of activity is the partial pressure, p_{O_2} — more strictly the fugacity, f_{O_2}, should be used (Chap. I.1) — and Eq. (1) in terms of p_{O_2} is

$$I_1 = \frac{nFAS_m D_m p_{O_2}}{z_m},\qquad(2)$$

where S_m and D_m are the solubility coefficient and diffusion coefficient, respectively, of oxygen in the membrane. Thus the simplest calibration standard is a gas mixture of known oxygen partial pressure, and the most readily available such mixture is air. Dry air contains 20.95% of oxygen [9] and the oxygen partial pressure in this case will clearly be 20.95% of the total air pressure. Air is, however, very rarely completely dry but has a relative humidity associated with it. This humidity will contribute toward the total pressure, and unless the partial pressure arising from the water vapor is subtracted before the oxygen partial pressure is calculated an erroneous calibration will be made. Unfortunately, the determination of relative humidity is difficult to do with any accuracy, and so in order to use air as a calibration standard it should either be dried before use or brought to 100% relative humidity.

If a calibration check at other partial pressures is required, then this can be done with suitable gas mixtures of known oxygen partial pressure (App. B). In practice linearity is found (Fig. 1), as would be expected from Eq. (2). For gas phase calibration at zero oxygen level the detector can be simply placed in an inert gas; e.g., N_2, Ar, He, etc.

Gaseous calibration standards are the obvious choice if subsequent measurements are to be in the gas phase. For measurements of dissolved oxygen, whether in aqueous or nonaqueous solvents, the gaseous standards discussed above can also be used. Since, as has been mentioned, a POS responds to activity or partial pressure, then such a calibration will allow the direct measurement of dissolved oxygen in terms of oxygen

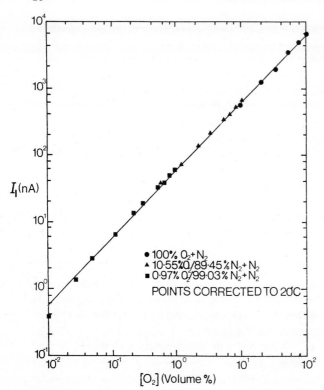

Fig. 1. Calibration curve for a POS. The different oxygen concentrations were obtained with the gas mixtures shown [10]. (Reproduced by permission of John Wiley and Sons, Inc.)

activity. However, if it is necessary to have a measure of dissolved oxygen in terms of concentration, then the calibration is somewhat more complicated since the Henry's law proportionality constant relating activity and concentration at the given temperature and liquid composition must be known (Chap. I.3). The value of this proportionality constant cannot, in general, be predicted, and so one has to rely upon the empirical determinations of solubilities in order to obtain it. For pure, nonaqueous liquids the solubility of oxygen is given in a number of compilations (e.g., [12]), but the data are often only at one temperature and it is difficult to evaluate the accuracy of the data. Hopefully the volume for oxygen in the new Solubility Data Series [1] will remedy this. For water, on the other hand, extensive investigations have been made of the solubility of dissolved oxygen as a function of temperature and pressure (App. A). Using the appropriate equations allows a direct correlation between a calibration made, in terms of partial pressure, with a gaseous standard and the dissolved oxygen concentration at a given temperature. Some caution is needed, however, with subsequent measurements, since the POS will always be measuring oxygen activity and not concentration and the relationship between these two may change; for example, when salting-out occurs (App. A).

For dissolved oxygen measurements it is sometimes more convenient to make the calibration with a dissolved oxygen standard. The standards to be used would then be samples of the pure liquid saturated with oxygen at a known partial pressure. The most common sample of this kind is air-saturated water. Calibration in terms of partial pressure is straightforward, once allowance is made for the vapor pressure of the water

(App. A) at the calibration temperature, while calibration in terms of oxygen concentration has again to rely on solubility tables. The zero oxygen level for liquid phase calibration can be checked either with a deaerated sample of the liquid or with a sodium sulfite solution.

3 Methods of Calibration

3.1 Preparation of Calibration Standards

If air is to be used as a calibration standard it should, as we have pointed out, be either completely dry or brought to 100% relative humidity. Drying of air can be done by several methods [5]. One technique is to pass the moist air along a toruous path through a column containing a solid drying agent; e.g., silica gel or calcium chloride. Another method is to bubble the air slowly through a hygroscopic liquid; e.g., concentrated sulfuric acid or concentrated phosphoric acid. A third method of drying is to freeze out the water by passing the air through a tubular metal coil immersed in a cold bath. The most convenient freezing agent is dry ice ($-56.6°C$) which will give air with a partial pressure of water vapor of < 0.004 kPa and will not freeze out any other common constituent of air. After freezing out the water the dry air should be brought back to ambient temperature by passing it through a second coil in the ambient.

Saturating the air with water vapor is not quite as straightforward as might be imagined if one wishes to humidify a given volume of gas, since it involves not only the mass transfer of water from the liquid to the gaseous phase but also heat transfer, arising from the vaporization of the water, and its interaction with the mass transfer. One should use a device such as an adiabatic saturator [16], where gas passing through an adiabatic spray chamber is cooled and humidified. The exiting gas is, however, only fully saturated at a temperature equal to or less than the water temperature in the saturator, and so must not be allowed to warm up. A far simpler method of obtaining an air calibration standard with 100% relative humidity is to use the air a short distance (a few mm) above the surface of a clean sample of water; this is, in fact, the most commonly used gaseous calibration standard for POS. It is, of course, always necessary to know the temperature of the water used for saturation in order to obtain the appropriate water vapor pressure to subtract from the total air pressure.

For checking of the zero oxygen level of a POS with a gas it is merely necessary to flush out a closed volume with an inert gas (e.g., N_2, Ar, He). Since one is supposed to be measuring zero oxygen partial pressure it is not essential that the inert gas be either at a known total pressure or have a known water vapor content. Gas mixtures with intermediate oxygen partial pressures are available commercially or are readily made up with a high degree of accuracy from pure oxygen and an inert gas using calibrated flow meters or mass flow controllers.

Liquid phase calibration standards are prepared by saturating a sample of pure liquid with oxygen at a known partial pressure. For air-saturated water, a short period (some minutes) of bubbling air through a gas tube dipping into a volume (typically several hundred cm^3) of clean water is often considered to be adequate. Since the water may very well be close to being saturated with air beforehand this preparation is probably enough for the air standard, but with other oxygen partial pressures considerably greater care is needed. This is largely because of the very limited contact between gas and liquid that a simple bubbler produces. In general the total mass transfer to and from a bubble takes place at a rate governed by the difference in partial pressure of the gas in the bubble and the liquid, the interfacial area for transfer, and the mass transfer coefficient in the liquid phase. For a given gas/liquid system it can be shown [16] that this transfer rate (J_o) is fairly well represented by

$$J_o = A k_l P^{0.4} (V_g/u_b)^{1/2}, \tag{3}$$

where A is a constant for the system, k_l is the mass transfer coefficient in the liquid, P is the power input/unit volume to an agitator or stirrer, V_g is the volumetric gas feed rate/cross-sectional area of the containing vessel, and u_b is the bubble rise velocity. It is found that k_l is approximately independent of the power input and impeller characteristics and so, for a given system, can be taken as a constant. Thus for efficient mass transfer to and from the gas bubbles it is necessary to fulfil three requirements. First, to have a large power input to the stirrer to give a high agitation rate and ensure a thorough dispersion of the bubbles throughout the liquid. Second, to have a large volumetric feed of the gas into the liquid to give a high volume fraction of bubbles in the system and so promote a large interfacial area across which mass transfer can occur. Third, to have a slow rise rate for the bubbles in order to provide adequate opportunity for mass transfer to take place during the passage of the bubbles through the liquid.

Apart from bubbling, other means of gas exchange [2] can also be employed to prepare calibration samples with various oxygen concentrations. Electrolytic generation of oxygen [14] and the catalase catalyzed decomposition of hydrogen peroxide [17] have been suggested for the preparation of dissolved oxygen calibration standards, but these techniques would seem to have no real advantages over the more straightforward gas-bubbling method. Indeed, with both electrolytic oxygen generation and hydrogen peroxide decomposition there is the danger of a wrong calibration due to less than 100% efficiency in oxygen production.

Water samples for zero point calibration can be produced by de-aerating with an inert gas. However, because the purging is essentially a first-order rate process the rate of de-aeration will fall off with time as the oxygen partial pressure gradient between the liquid and the bubbles decreases. Therefore it is especially important here to make the rate constant for the process as large as possible by ensuring efficient mass transfer between the two phases. This is particularly necessary if measurements are to be made at the ppb level. An alternative method of producing de-aerated water is simply to boil off the dissolved oxygen and let the water cool down under an inert atmosphere. Or one can take advantage of the thermodynamically, strongly downhill reaction

$$2 \, SO_3^{2-} + O_2 \leftrightharpoons 2 \, SO_4^{2-} \qquad E_{298}^{\circ} = + 1.32 \, V$$

and use a 2%–5% solution of Na_2SO_3 which, as a simple calculation shows, should, and in practice is found to, contain infinitesimally small amounts of dissolved oxygen.

The above discussion of the preparation of liquid phase calibration standards has been in terms of aqueous standards, which are the most common form. For dissolved oxygen analysis in nonaqueous solutions then, provided that measurements are to be made in terms of oxygen partial pressure, there is no fundamental reason why aqueous calibration standards should not be used since the oxygen activity in an aqueous and nonaqueous environment will be the same for a given oxygen partial pressure in the gas phase above the liquids. Even for analysis in nonaqueous solutions in terms of oxygen concentration, aqueous calibration standards can still be used if a partial pressure calibration is made since one can then convert the pressure to a concentration knowing the Henry's Law constant for the pure, nonaqueous liquid at the calibration temperature. This constant will not, however, be the same as that for water, and so a calibration with aqueous standards for dissolved oxygen concentration cannot be used for nonaqueous samples. If a direct calibration of oxygen concentration in a nonaqueous standard is required then, provided that the Henry's Law constant is known, such a standard can be prepared using the gas exchange methods described earlier; direct nonaqueous calibration may be demanded by the need, for example, to avoid contamination of the nonaqueous medium with traces of moisture that could be left on the surfaces of the POS from an aqueous calibration.

3.2 Use of Calibration Standards

Whichever form of standard is used, it is obvious that adequate precautions must be taken to prevent contamination of the standard sample by extraneous gases. In the case where air is the basis of the standard this should not present any real problem, but where the standard has an oxygen content different to that of air, particularly when a zero point calibration is being made, then leakage of air has to be avoided and a suitable calibration chamber must be used. For both gaseous and liquid calibrations continuous flushing with gas is recommended and the outlet from the chamber should be designed so that back-diffusion of air against the gas flow is negligible. Some idea of the conditions needed to fulfil this can be obtained by using the equation [3].

$$\frac{c}{c_\infty} = \exp\left(-<v>\ l/D_a\right),\tag{4}$$

where c_∞ is the gas concentration in the atmosphere at the exit of a flushing tube, c is the concentration at a distance l from the exit inside the tube, D_a is the diffusion coefficient of the gas, and $<v>$ is the mean velocity of the flushing gas. The dimensionless term $<v>\ l/D_a$ is known as the Peclet number, Pe, and it is a measure of the relative importance of forced convection and diffusion for the transport of matter. When Pe $\gg$ 1 the flow is predominantly convective, whereas when Pe $\ll$ 1 it is mainly diffusive. Therefore in order that back diffusion of air against the gas stream can be neglected we can say that it is necessary for

$$D_a < <v>\ l/7,\tag{5}$$

if we take $c \approx 0.001\ c_\infty$ as being negligible. At atmospheric pressure and normal operating temperatures ($\approx 20°C$) for air $D_a \approx 0.3$ cm^2 s^{-1} and so values of $<v>$ and l required to fulfil the criterion can be calculated. For example, with a gas flow 1 cm^3 s^{-1} and an exit tube cross-sectional area of 1 cm^2 the tube length required is ≈ 2 cm. A flow of 1 cm^3 s^{-1} is not unreasonable for flushing a liquid or a gas chamber, but if for a liquid calibration standard a lower rate of flushing is desired then a rather longer and/or narrower exit tube should be used, or gas should be simultaneously flushed over the surface of the liquid but at a higher rate than the flushing through the liquid.

In addition to ensuring that the calibration standard has the oxygen partial pressure expected it is also essential to make sure that this partial pressure prevails right up to the outer surface of the membrane of the POS. In other words, that there is no depletion of the oxygen close to the membrane due to mass transfer limitations. A very simple criterion for this is that the rate of mass transfer through the membrane must be much slower than the transfer through the diffusion layer at the outer face of the membrane in the calibration medium. This criterion for diffusion in the membrane to be rate-determining can be expressed by

$$\frac{\delta_s}{D_s} \ll \frac{z_m}{P_m} , \tag{6}$$

where δ_s is thickness of the diffusion layer in the test environment, D_s is the diffusion coefficient for O_2 in this environment, and z_m and P_m have the same meaning as in Eq. (1) (cf. Chap. I.4). Considering first the case of gaseous calibration then, assuming that the gas flows tangentially past the membrane, it can be shown [11] that

$$\delta_s \sim \left(\frac{l D_s}{<v>} \right)^{1/2} , \tag{7}$$

where l is a length along the membrane surface and $<v>$ is the mean gas velocity. Condition (6) then becomes

$$<v> \gg \frac{l}{D_s} \left(\frac{P_m}{z_m} \right)^2 . \tag{8}$$

Taking typical values of $l \approx 1$ cm, $D_s \approx 0.2$ cm^2 s^{-1}, $P_m \approx 10^{-6}$ cm^2 s^{-1} and $z_m \approx 25$ μm, then it is only necessary for $<v>$ to be much greater than $\approx 10^{-6}$ cm s^{-1}, and this will always be true since even natural convection will provide a flow greater than this. With very thin and highly permeable membranes it is still unlikely that the criterion will not be valid. So with gaseous calibration there is no need to be concerned about oxygen depletion near the outer surface of the membrane.

With liquid calibration there is more of a problem. The diffusion layer thickness is now given by ([10]; cf. Chap. I.1)

$$\delta_s \sim D_s^{1/3} v_s^{1/6} \left(\frac{l}{<v>} \right)^{1/2} . \tag{9}$$

where v_s is the kinematic viscosity of the liquid. Condition (6) now becomes

$$<v> \gg \frac{l}{D_s} \left(\frac{P_m}{z_m} \right)^2 \left(\frac{v_s}{D_s} \right)^{1/3} . \tag{10}$$

Taking values of l, P_m and z_m as before, and with typical values of v_s and D_s of 10^{-2} cm^2 s^{-1} and 10^{-5} cm^2 s^{-1} respectively, the right hand side of the inequality is ≈ 0.2 cm s^{-1}, which is a much more demanding criterion. For example, if we take the inequality as being valid only at the 1% level then a liquid flow rate of 20 cm s^{-1} is required. In practice it is found that the sensitivity of a POS is indeed independent of test solution flow rates only at higher levels than this [10, 15]. Thus the great care in having adequate agitation when making calibration with liquids is emphasized. Also highlighted is a problem often encountered when calibrating in a gaseous environment for measurements eventually to be made in a liquid. Because of the difficulty of providing sufficient agitation to eliminate the effects of the diffusion layer in a liquid, a gas phase calibration (e.g., with air a short distance above an air-saturated water) can give a higher reading than a liquid phase calibration (e.g., with the water). If there is any possibility of the liquid flow rate being inadequate then it is probably better to calibrate in the liquid with the same stirring as will be used for the measurement. In this way calibration and measurement will at least be done under equivalent conditions.

We have already pointed out in the previous section the need for temperature control when using for calibration a gas sample with 100% relative humidity in order that the appropriate water vapor correction can be made. In fact, some form of temperature control is necessary with any form of calibration standard because of the large temperature coefficient of the current of a POS [10]. Automatic compensation can help to alleviate this problem. Typical errors quoted for automatic correction of commercial system are ± 1% for temperatures within ± 5°C of the calibration temperature and ± 5% over the complete temperature range of 0°−50°C usually encountered (cf. Chaps. I.10, III.1). However, membrane characteristics often show a strong hysteresis when subject to large temperature changes and so some caution is necessary in using compensating circuits. For the most accurate work automatic compensation should not be relied upon and a calibration should always be made either at the same temperature at which the measurement is to be made, or as close as possible in order to allow a correction [10] to be made with confidence.

The calibration of a POS can change with time as a result of variations in the terms of the calibration factor. There may also be in addition external factors causing instabilities or changes in the sensor current (Chaps. I.1, I.6). Because of the very nature of many of these influences it is not possible to give a frequency for calibration checks categorically. All that can be said is that certainly a check should be made whenever the membrane is changed and when the sensor has not been used or checked for some time. For the rest one must rely upon both scientific and basic common sense.

As far as the number of calibration points is concerned, according to Eq. (2) only a single point is needed. In practice, because of residual currents an additional zero check has generally to be made in order to back-off any such residuals. It is probably also worthwhile to check the linearity of response from time to time with intermediate calibration standards. This is especially so at low oxygen partial pressures where deviations are most likely to occur as a result of the residuals.

3.3 Minimizing Residual Currents

Because residual currents are rarely constant in time frequent zero checks and/or calibration at low levels must be made, and so a better solution would be to attempt to eliminate the residuals altogether. But any attempt at eliminating residuals presupposes that the cause of such currents is known. Therein lies a problem, in that there is very often more than one cause and it is not always obvious which one is predominant.

Two sources that can give rise to varying current residuals are the reduction of surface oxides on the cathode [10] and the charging of a pseudocapacitance arising from the formation of adsorbed intermediates at the electrode surface. Both these phenomena are, however, only likely to contribute significantly to the residual current during the settling down of the detector when it is switched on, and once a true steady-state situation has been reached they will probably only make a small contribution to the overall residual.

Residual currents can also arise from reduction of species other than dissolved oxygen in solution. For example, if the electrolyte is not completely stable to oxidation then reduction of the oxidation products can occur at the cathode (Chap. I.5). Some detectors use an iodide electrolyte and the I^- anion, as is well known, is readily oxidized to I_2 in the presence of light. The I_2/I^- couple has a high, positive standard potential ($E^o_{298} = 0.535$ V) and so I_2 is easily reduced at the cathode. Electrolytes not resistant to oxidation, or indeed showing any form of instability, are to be avoided.

Other reducible species that can appear in the detector electrolyte are heavy metal ions. A POS that has been running for some time will have an electrolyte containing such ions as a result of the anodic dissolution of the auxiliary/reference electrode. The concentration of these cations and the resulting residual current from their reduction will depend very markedly on the anode couple used. For a POS with a Ag/AgCl anode in an electrolyte of 2 mol dm^{-3} KCl the Ag^+ concentration will be $\approx 10^{-10}$ mol dm^{-3} and this concentration could be expected to produce a maximum current density $\approx$ 1% of that generated by 1 ppb of dissolved oxygen. On the other hand for a Ag/Ag$_2$O anode with an electrolyte of 2 mol dm^{-3} KOH the Ag^+ concentration will be $\approx 10^{-8}$ mol dm^{-3} (the solubility product of Ag$_2$O being about 100 times greater than that of AgCl) and the maximum current from the reduction of this concentration will be comparable to that from 1 ppb of dissolved oxygen. Therefore some care is needed in the choice of anode couple if low level oxygen calibrations are to be made. Here it is worth noting that any electrolyte which does not initially have a low pH or which is not strongly buffered will ultimately become alkaline anyway [8]. This change of pH will, of course, take some time to establish and during this period one may expect to observe a drift in the detector sensitivity. It is therefore better to start off with a high pH electrolyte, but at the same time to have a high concentration of an anion which has a low solubility product for the cations formed by the anodic dissolution of the anode.

Another reason for using a high pH electrolyte is to discourage, right from the moment of switching on the detector, the formation of hydrogen peroxide at the cathode. This is produced by incomplete reduction of oxygen at low pH values and it can build up to a steady concentration in the bulk of the electrolyte [6]. Calibrations made at low oxygen levels with a detector which initially had a low pH electrolyte and which has previously been used at high oxygen partial pressures can show residual currents arising from either the direct reduction of the H_2O_2 or from reduction of oxygen produced by the catalytic decomposition of the H_2O_2 on the cathode. Hydrogen peroxide formation is also minimized by using a high reduction potential [6], but too negative a potential can lead to a current residual from another source — the reduction of the solvent. Thus a compromise must be achieved. In this context it should be noted that Pt is a significantly better catalyst for hydrogen evolution than Au [4], so that Au is to be preferred to Pt from this point of view.

All that we have said so far has concerned residual currents arising from processes other than the reduction of oxygen. In order to minimize these residuals it can be seen that one must simultaneously satisfy a number of requirements: anode couple producing low concentration of soluble cations; stable electrolyte of high pH and with anions to precipitate cations from the anode dissolution; cathode with high overpotential for water reduction; high applied potential. These conditions are probably most satisfactorily fulfilled with a POS having an Au cathode, a concentrated KCl electrolyte with a pH $\approx$ 13, an Ag/AgCl anode, and an applied potential of -0.8 V.

In addition, however, to residuals from these various sources there is also the possibility of residual currents from oxygen reduction itself. This oxygen is in excess to that in the test or calibration environment and is, in fact, residual oxygen left behind in the detector from a time when the sensor was in an oxygen-rich environment. It can accumulate in various places, such as in the electrolyte reservoir, in electrolyte in microcracks in between the cathode and the insulating region surrounding it, and even in the body of the sensor itself. But whatever the final source, it is clear that this oxygen reaches the cathode by radial diffusion in contrast to the diffusion normal to the cathode of the oxygen from outside the membrane of the POS. The problem of the contribution to the current of a POS from oxygen diffusing to the cathode radially from the electrolyte reservoir has been examined in some detail [13]. This solution is complex, involving modified Bessel functions, but a simpler, approximate solution can be obtained which allows one to obtain a good estimate of the current from radial diffusion of oxygen [7]. We consider the differential equation for radial diffusion:

$$\frac{\partial}{\partial r}\left(\frac{r\,\partial c}{\partial r}\right) = 0 \tag{11}$$

with the boundary conditions

$$r = r_0 \qquad c = 0 \tag{12}$$

$$r \geqslant r_E \qquad c = c_e. \tag{13}$$

The boundary conditions apply to a detector design with a central disk cathode of radius r_0. The distance from the center of the electrode to the edge of the insulating land between the cathode and the electrolyte reservoir is r_E, and in the reservoir it is assumed there is a constant oxygen concentration, c_e; since this will, in practice, clearly not be the case, because of loss of oxygen through the membrane and also removal by the radial diffusion process, any contribution to the current from radial diffusion will be less than what we shall estimate here. It is also assumed that the cathode is a cylinder with a radius equal to that of the actual disk cathode; this is a good approximation since the thickness of the electrolyte layer is, in general, much less than r_E. Solving Eq. (11) subject to Eqs. (12) and (13) gives

$$c = c_e\,\frac{\ln r/r_0}{\ln r_E/r_0} \tag{14}$$

and so the radial diffusion current I_r is

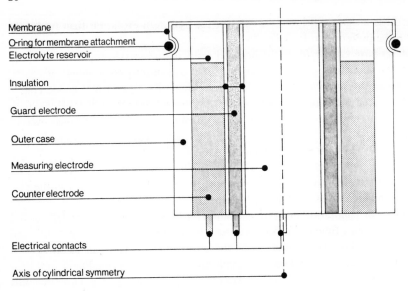

Membrane
O-ring for membrane attachment
Electrolyte reservoir

Insulation

Guard electrode

Outer case

Measuring electrode

Counter electrode

Electrical contacts

Axis of cylindrical symmetry

Fig. 2. A POS with a guard ring. (Reproduced by permission of Orbisphere Corporation)

$$I_r = nFAD_e \left(\frac{\partial c}{\partial r}\right)_{r=r_0} = \frac{8\pi z_e FD_e c_e}{\ln r_E/r_0} , \tag{15}$$

where z_e is the thickness of the electrolyte film over the cathode and D_e is the diffusion coefficient for oxygen in the electrolyte solution. The concentration of oxygen in the electrolyte reservoir will be less than that in air-saturated water because of the salting-out effect, and might typically be 125 μmol dm^{-3} at 25°C [10]. Then with $z_e \approx$ 5 μm and with $r_E/r_0 \approx 1.2$, which is typical for many POS, the value of $I_r \approx 9$ nA. With a POS having a cathode area ≈ 0.3 cm^2 the corresponding current density is ≈ 30 nAcm^{-2} and current residuals are often at this level; i.e., equivalent to ≈ 2 ppb of dissolved oxygen. As we have pointed out, the value of I_r we have calculated will be greater than any actual value, but clearly radial diffusion of oxygen dissolved in the electrolyte reservoir could be a cause of the residual currents commonly found. Equation (15) shows that to reduce the current from radial oxygen diffusion one requires a thin electrolyte layer and as long a length as possible of this layer between the edge of the cathode and the electrolyte reservoir. In practice, one is limited by the requirement that the ohmic drop between cathode and anode should not be too high, and additional precautions may need to be taken. For example, by designing the sensor so that there is only a small area of the electrolyte reservoir exposed to the ambient atmosphere, or by incorporating a second cathode into the sensor which guards the current measuring cathode from oxygen diffusing radially from the reservoir. Figure 2 shows schematically the design of a three-electrode sensor. The guard electrode and measuring electrode both function as cathodes and reduce oxygen, but only the current from the central electrode is used as a measure of the oxygen in the test or calibration environment. Calculations show that the guard electrode will remove all but 0.001% of the oxygen diffusing radially from the reservoir.

One final point about residual currents should be made. The term polarographic oxygen sensor is taken to include both amperometric and galvanometric POS. The residual currents in all cases arise from processes occurring at the cathode and there is no fundamental electrochemical difference between these processes in the two types of cell. The interfacial potential at the cathode/electrolyte interface, which is the potential difference that governs any current, is the same whether the cell potential is being applied from an external source (amperometric POS) or arises from the intrinsic relative potential difference between cathode and anode (galvanometric POS). The methods suggested for minimizing residual currents therefore apply to both types of sensor.

In conclusion, we can say that while it may be a relatively easy matter to produce a calibration standard of zero or low oxygen content using such a standard to give an accurate calibration is by no means trivial. Indeed, if one wants to do more than just make an electrical adjustment to back-off any residual current considerable problems can arise.

4 Accuracy of Calibration

Having decided on the calibration standards to be used, and having taken all the necessary precautions to make sure that they have been used correctly, the final question to be discussed is the following: How accurate (i.e., how close to the true value) will a measurement with a POS be? Before we can answer this question it is necessary to ask some further questions in order to define the boundary conditions within which we are going to operate. So, for example, are we going to assume that the sensor is stable and that there is no long-term drift (Chap. I.1)? Is there going to be no poisoning of the detector by gaseous pollutants (Chap. I.6)? Is the measurement going to be made in an identical environment to that in which the calibration was made (Chaps. II.10, III.1)? As indicated, the problems of signal instability and electrode poisoning are dealt with in other sections, and here we just briefly consider the question concerning "identical environment". By this term we mean not only the same form of environment, but also all the same conditions except, of course, for the oxygen partial pressure. Thus, as we mentioned earlier, gaseous calibration standards are the logical choice for gaseous measurements, but, as we have also discussed, the temperature of the measurement and the calibration should ideally be the same because of the large temperature coefficient of the sensor current. If the read-out is to be in terms of volume percent rather than oxygen partial pressure, then there is the further constraint that there must be the same conditions of total pressure and relative humidity for both measurement and calibration.

For measurements in solution, gaseous calibration standards can be used, but there is then the problem of providing adequate agitation of the test solution to allow the calibration to be valid. For this reason it is probably better to calibrate with dissolved oxygen standards for dissolved oxygen measurements. As with gaseous phase measurements, calibration and measurement should preferably be done at the same temperature, and even with automatic compensation the difference in temperature should be

± 5°C at the most if an accuracy of ± 1% or better is required. Read-out of dissolved oxygen can be as partial oxygen pressure or oxygen tension, p_{O_2} [kPa], as concentration [mmol dm^{-3}], or percent saturation. With partial pressure no particular problems arise, but with concentration and percent saturation a number of factors can lead to errors. As has been emphasized several times, a POS measures oxygen activity rather than concentration (Chap. I.3), and so any deviation from the expected relationship between the two — when, for example, measurements are made in saline solution or at great depths — will give a wrong concentration reading.

So in answer to our original question "How accurate will a measurement with a POS be?" we can say that if the answer to each of the supplementary questions is "yes", then there is no reason why a measurement should not be made with an extremely high degree of accuracy. If the answer, on the other hand, to any of the supplementary questions is "no", then the measurement will be inaccurate to a greater or lesser extent. But to put a general figure to the level of inaccuracy is not possible — in spite of what many manufacturers of POS would like us to believe.

References

1. Battino R (1979) Solubility data series, vol V. Oxygen and ozone — gas solubilities. Pergamon Press, Oxford
2. Battino R, Clever HL (1966) The solubility of gases in liquids. Chem Rev 66:395–463
3. Beek WJ, Muttzall KMK (1975) Transport phenomena, ch 1. John Wiley and Sons, New York, x + 298 pp
4. Bockris JO'M, Reddy AKN (1970) Modern electrochemistry, vol II, ch 10. Plenum Press, New York, lvi + 709 pp
5. Coulson JM, Richardson JF (1968) Chemical engineering, vol II, ch 15. Pergamon Press, Oxford, xvii + 790 pp
6. Hahn CEW, Davis AH, Albery WJ (1975) Electrochemical improvement of the performance of pO$_2$ electrodes. Respir Physiol 25:109–133
7. Hale JM, Hitchman ML (1979) Some considerations of the steady-state and transient behaviour of membrane covered dissolved oxygen detectors. RCA Tech Rep ZRRL-79-TR-002, 52 pp
8. Hale JM, Hitchman ML (1980) Some considerations of the steady-state and transient behavior of membrane covered dissolved oxygen detectors. J Electroanal Chem 107:281–294
9. Handbook of chemistry and physics (1969), 50th ed. Weast RC (ed) Chemical Rubber Co, Cleveland, p 151
10. Hitchman ML (1978) Meaurement of dissolved oxygen. John Wiley and Sons, New York, xvi + 255 pp
11. Hitchman ML (1980) A consideration of the effect of the thermal boundary layer on CVD growth rates. J Cryst Growth 48:394–402
12. Washburn EW (ed) (1928) International critical tables, vol III. McGraw Hill, New York, xiv + 444 pp
13. Jenson OJ, Jacobsen T, Thomsen K (1978) Membrane covered oxygen electrodes. I Electrode dimensions and electrode sensitivity. J Electroanal Chem 87:203–211
14. Keidel FA (1960) Coulometric analyser for trace quantities of oxygen. Ind Eng Chem 52:490–493
15. Mancy KH, Okun DA, Reilley CN (1962) A galvanic cell oxygen analyser. J Electroanal Chem 4:65–92
16. McCabe WL, Smith JC (1976) Unit operations of chemical engineering, 3rd edn, ch 22, 24. McGraw-Hill, New York, ix + 1028 pp
17. Wingo WJ, Emerson GM (1975) Calibration of oxygen polarographs by catalase catalysed decomposition of hydrogen peroxide. Anal Chem 47:351–352

Chapter I.3   A Thermodynamic Consideration of Permeability Coefficients of Membranes

M.L. Hitchman[1] and E. Gnaiger[2]

1 Introduction

The permeability coefficient of the membrane appears in various places in this volume either as P_m with units of $[m^2\ s^{-1}]$ or as P_m with units of $[mol\ m^{-1}\ s^{-1}\ kPa^{-1}]$. Here we show how these permeability coefficients arise and how they are related.

2 Oxygen Distribution Between a Solution and a Membrane

If we consider the distribution of oxygen between a test solution (s) and a membrane (m) then we can write an equilibrium

$$O_2(s) \rightleftharpoons O_2(m).$$

From basic thermodynamics we can write for the test solution

$$\mu_s = \mu_s^{\circ} + RT \ln a_s \tag{1}$$

and for the membrane

$$\mu_m = \mu_m^{\circ} + RT \ln a_m, \tag{2}$$

where μ_s and μ_m are the chemical potentials of the oxygen in the two phases, μ_s° and μ_m° are the corresponding chemical potentials in the standard state, and a_s and a_m are relative activities [4]. The standard state for defining μ_s° and μ_m° is taken as the hypothetical state at unit molarity of dissolved oxygen, but in which the environment of each molecule is the same as at infinite dilution of oxygen. Since these environments will clearly not be the same in the case of the membrane and the test solution then the chemical potentials in each phase under standard-state conditions will be different. The dimensionless relative activities are defined by:

---

1 Department of Chemistry and Applied Chemistry, University of Salford, Salford, M5 4WT, Great Britain
2 Institut für Zoologie, Abteilung Zoophysiologie, Universität Innsbruck, Peter-Mayr-Str. 1A, A–6020 Innsbruck, Austria

Polarographic Oxygen Sensors (ed. by Gnaiger/Forstner)
© Springer-Verlag Berlin Heidelberg 1983

$$a_s = y_s \frac{c_s}{c_s^o} \tag{3}$$

and

$$a_m = y_m \frac{c_m}{c_m^o}, \tag{4}$$

where y_s and y_m are the activity coefficients for the two phases, c_s and c_m are the concentrations in terms of molarities, and c_s^o and c_m^o are standard concentrations, which are usually chosen to be 1 mol dm^{-3}. Since for most practical situations that one is concerned with in using a POS the interaction between oxygen molecules is negligible [1], then, on the basis of the standard state that has been chosen for defining the μ^o values, the activity coefficients in Eqs. (3) and (4) will only deviate from unity if there are any other solutes which change the environment of the dissolved oxygen; e.g., ionized salts (App. A).

For equilibrium at the test solution/membrane interface

$$\mu_s = \mu_m \tag{5}$$

and hence

$$\ln \frac{a_m}{a_s} = \frac{\mu_s^o - \mu_m^o}{RT} = -\frac{\Delta G^o}{RT}, \tag{6}$$

where ΔG^o is the difference in molar standard free energy between oxygen in the membrane and the test solution. However, since ΔG^o is simply related to the thermodynamic equilibrium constant K then we can write

$$K = \frac{a_m}{a_s}. \tag{7}$$

This equation shows that K is the dimensionless distribution or partition coefficient in terms of activities [3].

3 The Current at a POS and the Permeability Coefficient P_m

The signal normally obtained with a POS is a function of the flux of oxygen to the cathode. Figure 1 shows schematically the situation at the electrode under steady-state conditions. It is assumed that there is an equilibrium distribution of oxygen between the test solution and the membrane, that there is no limitation on transport of oxygen to the cathode through the electrolyte layer between the membrane and the cathode (cf. Chap. I.1), and that the transport is completely one-dimensional (cf. Chap. I.2). The flux, J_{O_2} [mol m^{-2} s^{-1}], of oxygen to the cathode is given by Fick's Law:

$$J_{O_2} = D_m \left(\frac{dc_{m,z}}{dz} \right)_{z=o}, \tag{8}$$

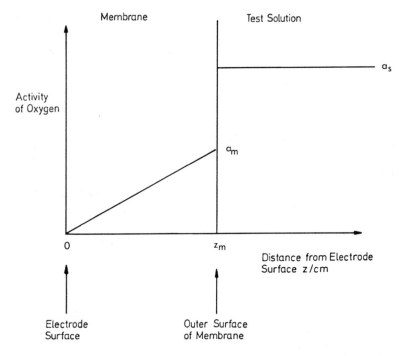

Fig. 1. Scheme of the situation at the electrode under steady-state conditions

where $(dc_{m,z}/dz)_{z=0}$ is the concentration gradient $[\text{mol m}^{-3}\,\text{m}^{-1}]$ in the membrane at the electrode surface and D_m $[\text{m}^2\,\text{s}^{-1}]$ is the diffusion coefficient for the oxygen in the membrane. Assuming that the concentration gradient is linear over the total membrane thickness (z_m), then with $c_{m,0} = 0$ Eq. (8) becomes

$$J_{O_2} = D_m\, \frac{c_m}{z_m} \,,\tag{9}$$

where c_m is simply written for the dissolved oxygen at the outer edge of the membrane. Using Eqs. (4) and (7), this equation becomes

$$J_{O_2} = \frac{c_m^{\circ}}{y_m}\, \frac{KD_m}{z_m}\, a_s.\tag{10}$$

The term c_m°/y_m will be unity provided that there is, for example, no salting out of oxygen in the membrane; i.e., $y_m = 1$. This is likely to be the case since membranes used in practice for POS are impermeable to electrolytes. The product of K and D_m we can write as

$$P_m = KD_m,\tag{11}$$

where P_m is the permeability coefficient with the same units as D_m i.e. $[\text{m}^2\,\text{s}^{-1}]$, since K is dimensionless. It is a constant at a given temperature and, since K is defined in terms of activities, it is independent of any salt effects in the test solution. Eq. (10) can now be written as

$$J_{O_2} = \frac{P_m}{z_m} a_s.$$ (12)

Although a_s is dimensionless [Eq. (3)], the units of J_{O_2} are [mol m^{-2} s^{-1}], as is requir-ed for a flux, since the equation contains the "hidden" concentration c_m^o.

The transport limited current, I_1, at a POS will be given by

$$I_1 = nFAJ_{O_2},$$ (13)

where n is the number of electrons associated with the oxygen reduction, F is the Faraday, and A is the cathode area. Hence

$$I_1 = nFA \frac{P_m}{z_m} a_s.$$ (14)

In the case where there is no salting-out in the test solution, y_s is unity in Eq. (3) and

$$a_s = \frac{c_s}{c_s^o}.$$ (15)

Equation (14) can thus be written as

$$I_1 = nFA \frac{P_m}{z_m} \frac{c_m^o}{c_s^o} c_s,$$ (16)

where we have reintroduced c_m^o to show that the right hand side of the equation is still dimensionally correct. Since $c_m^o = c_s^o = 1$ mol dm^{-3} then the equation becomes

$$I_1 = nFA \frac{P_m}{z_m} c_s.$$ (17)

This is the equation as originally obtained by Mancy et al. [5] and as given by Eq. (1) in Chap. I.2. For some applications this equation can be used to describe the current obtained with a POS, but in many cases the activity coefficient y_s has to be considered and then Eq. (14) must be used.

4 The Current at a POS and the Permeability Coefficient P_m

In order to obtain an expression for the flux of oxygen to the cathode of a POS in terms of the permeability coefficient P_m with units of [mol m^{-1} s^{-1} kPa^{-1}], we define a solubility, S_s

$$S_s = \frac{c_s}{p_s} \quad \text{[mol dm}^{-3} \text{ kPa}^{-1}],$$ (18)

where c_s is concentration in terms of molarity and p_s is the partial pressure (kPa) of oxygen in the test environment. With the definition of a_s [Eq. (3)] we obtain

$$S_s = \frac{a_s c_s^\circ}{y_s p_s} \; . \tag{19}$$

Substituting from this equation for a_s into Eq. (10) for the oxygen flux to the cathode we then have

$$J_{O_2} = \frac{y_s c_m^\circ}{y_m c_s^\circ} \; \frac{KD_m}{z_m} \; S_s p_s. \tag{20}$$

If we now write the flux in the general form [cf. Eq. (12)]

$$J_{O_2} = \frac{1}{z_m} \times (\text{Permeability Coefficient}) \times (\text{Measure of Oxygen ``Concentration''}) \tag{21}$$

then with the partial pressure being a measure of the oxygen concentration, we can write

$$J_{O_2} = \frac{P_m}{z_m} \; p_s, \tag{22}$$

where the permeability coefficient is now defined by

$$P_m = KD_m S_s \frac{y_s c_m^\circ}{y_m c_s^\circ} \quad [\text{mol m}^{-1} \text{ s}^{-1} \text{ kPa}^{-1}]. \tag{23}$$

This permeability coefficient, like P_m, is a constant at a given temperature. Since S_s is in terms of concentration [Eq. (18)], and thus inversely proportional to y_s, then the presence of the product $S_s y_s$ means that P_m is also independent of salt effects in the test solution.

5 The Relationship Between P_m and $\mathbf{P}_m$

From Eqs. (11) and (23) it is obvious that

$$\mathbf{P}_m = P_m S_s \frac{y_s c_m^\circ}{y_m c_s^\circ} \; . \tag{24}$$

It is also interesting to note that from the two equations for the oxygen flux [Eqs. (10) and (20)], which of necessity must be equal, we get the relationship

$$P_s = \frac{c_s^\circ}{S_s y_s} \; a_s. \tag{25}$$

This equation is simply a form of Henry's Law in which the proportionality constant k_s between partial pressure (or more strictly fugacity) and activity is given by

$$k_s = \frac{c_s^\circ}{S_s y_s} \quad [\text{kPa}] . \tag{26}$$

Now since the oxygen flux at the cathode can be expressed in terms of partial pressure as well as in terms of activity, then the chemical potentials of oxygen in the test solution and membrane can also be expressed in terms of partial pressures. In other words, instead of Eqs. (1) and (2) we can write

$$\mu_s = \mu_s^* + RT \ln \frac{p_s}{p_s^\circ} \tag{27}$$

and

$$\mu_m = \mu_m^* + RT \ln \frac{p_m}{p_m^\circ} , \tag{28}$$

where the μ^* can be regarded as the chemical potential of the gas in each phase when the partial pressure is unity. From Eqs. (1) and (2) we know that the μ° are the chemical potentials at unit activity and that the state of unit activity is the standard state. If this standard state is chosen as that in which the partial pressure of the gas is also unity, then this choice makes the μ° in Eqs. (1) and (2) numerically equal to the corresponding μ^* in Eqs. (27) and (28). Or, looking at it another way, the Henry's Lay constant k_s [Eq. (26)] has a value of unity. We can then write instead of Eq. (25)

$$\frac{p_s}{p_s^\circ} = a_s , \tag{29}$$

where $p_s^\circ = p_m^\circ$ represents unit partial pressure ($k_s = k_m = 1$ atm $= 101.325$ kPa).

For a gaseous system it is often advantageous to choose the standard state of unit activity as that in which the partial pressure of the gas is unity at a given temperature [2]. With this choice of standard state, from the equation relating the two permeability coefficients [Eq. (24)], it can be seen that P_m and P_m become numerically identical.

References

1. Battino R, Clever HL (1966) The solubility of gases in liquids. Chem Rev 66:395–463
2. Glasstone S (1947) Thermodynamics for chemists, ch XII. Van Nostrand, Princeton NJ, viii + 522 pp
3. Glasstone S (1948) Textbook of physical chemistry, ch X. McMillan, London, xiii + 1230 pp
4. IUPAC (1979) Manual of symbols and terminology for physicochemical quantities and units. Pergamon Press, Oxford, App I
5. Mancy KH, Okun DA, Reilley CN (1962) A galvanic cell oxygen analyser. J Electroanal Chem 4: 65–92

Chapter I.4 Microcoaxial Needle Sensor for Polarographic Measurement of Local O_2 Pressure in the Cellular Range of Living Tissue. Its Construction and Properties

H. Baumgärtl and D.W. Lübbers[1]

1 Introduction

With the progress in polarographic analysis and with the development of various types of needle sensors, in vivo determination of oxygen supply to tissue has become feasible [10–15, 19–21, 23, 43, 56, 57, 68–70, 75, 77]. The smallest types of these polarographic oxygen sensors (POS) have tip diameters of $1-5$ μm. Their spatial and temporal resolutions are sufficiently high for satisfactory measurement of intercapillary oxygen partial pressure p_{O_2} [4, 11, 24, 29–32, 40, 49, 52, 67]. However, POS of this size may still cause considerable damage to the tissue. To avoid structural damages and to allow almost punctiform measurements of p_{O_2} in the cellular range, further miniaturizing has been desirable. For this purpose, the techniques and materials used so far have proved insufficient. Aside from poor polarogram characteristics, increasing instability of the measuring signal, sensitivity to mechanical pressure, high residual currents, intolerable drift of calibration curves, reduced stability, as well as insufficient insulation resistance and relatively high ionic sensitivity of the glass shaft of the sensor were observed.

Investigation of the technical details which influence the function of the POS has shown that the measuring properties essentially depend on the quality of the cathode material, chemical composition of the glass, features of the reference electrode, quality of the membrane, and special techniques of treatment. By choice of the proper raw material and suitable techniques – among others vacuum coating – we succeeded in eliminating the main sources of error and have been able to produce a small and reliable needle sensor of mean tip diameter of 0.6 μm, the so-called microcoaxial sensor.

2 Construction of the p_{O_2} Microcoaxial Needle Sensor

Figure 1 shows a schematic drawing of the microcoaxial needle sensor used for polarographic p_{O_2} measurements in the cellular range of living organisms. It is an improved

1 Max-Planck-Institut für Systemphysiologie, Rheinlanddamm 201, D–4600 Dortmund, FRG

Polarographic Oxygen Sensors (ed. by Gnaiger/Forstner)
© Springer-Verlag Berlin Heidelberg 1983

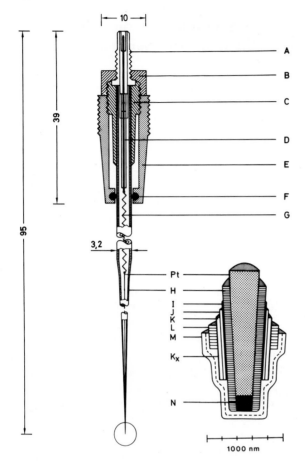

Fig. 1. Schematic drawing of the p_{O_2} microcoaxial needle sensor. *A* chassis socket; *B* nut; *C* metallic contact; *D* polyethylene-coated Cu wire; *E* readily exchangeable Plexiglas housing ground conically and fitting into screw thread; *F* siliconized O-ring; *G* metallic spiral; *H* glass shaft; *I* sputtered Ta-layer, ca. 8 nm; *J* sputtered Pt-layer, ca. 30 nm; *K* sputtered Ag-layer, ca. 62 nm; K_x electrochemically produced AgCl-layer; *L* sputtered dielectric layer such as SiO_2, Al_2O_3, Si_3N_4, Ta_2O_5, $BaTiO_3$; *M* adhesive, nontoxic O_2-permeable double-membrane (collodion, polystyrene, silicone, acrylic-polymere); *N* electrochemically polished cathode surface with recess, electrodeposited gold-layer and electropolymerized membrane [polystyrene or poly-diacetone-acrylamide (PDAA)]

modification of the sensors described by Baumgärtl and Lübbers [13, 14] and Baumgärtl [10] where platinum cathode and reference electrode are situated behind one and the same membrane. The characteristics of this POS are as follows:

1. The active measuring element, the cathode (Pt), is made of a physically or spectrally pure platinum wire (W.C. Heraeus GmbH, D–6450 Hanau) that is 0.2 mm thick and approx. 50 mm long for reasons of mechanical stability. On one end the wire is conically etched down to tip diameters between 0.1 and 0.5 μm over a length of 15–25 mm, using alternating current of controlled voltage.

2. The electrode shaft (H) consists of melting glass free from lead, (GW-Glas, Glaswerk Wertheim, D–6980 Wertheim; or melting glass – 8510, Jenaer Glaswerk Schott & Gen., D–6500 Mainz). The glasses have been selected with regard to their electrical insulation properties, viscosity, wettability, and chemical resistance. Their coefficient of thermal expansion corresponds to that of platinum. Before melting, the glass is thoroughly cleaned in aqueous watery detergent with ultrasound during the organic solvent vapor phase, as well as by heat treatment and ionic bombardment in glow discharge.

3. Under microscopic control the tip of the platinum cathode is fused to the glass capillary (H) over a length of ca. 25 mm with an electrically heated loop. Temperature is continuously controlled. After finishing the tip, the electrode is re-fused in the tip area over about 20 μm with a micro heating loop in order to ensure high stability of the fusion zone platinum/glass and to remove splits and microcracks.

4. Mechanical stability and rigidity, chemical stability and electrical insulation resistance of the glass shaft (H), which is only 100 nm thick in the tip region, are essentially increased by further treatment, such as annealing, chemical heat treatment through ionic exchange in a KNO_3-melt at $350°–400°C$, applying electrical tension, ionic bombardment in a glow discharge, or by sputtering thin dielectric layers, such as Ta_2O_5, SiO_2, Al_2O_3, Si_3N_4, $BaTiO_3$, etc. Best results have been obtained with layers of Al_2O_3 or SiO_2, as well as with double layers of Al_2O_3/SiO_2.

5. The form of the electrode tip is determined either by etching and grinding processes or by "pushing" to break off the projecting glass of the tip. The following tip forms are produced: plane (Fig. 1, enlargement of cutout), obliquely ground at an angle of $20°–30°$ (Fig. 2a) and cone-shaped (Fig. 2b) ones. These procedures result in circular, elliptic, and conic cathode surfaces. In general, the smallest circular cathodes

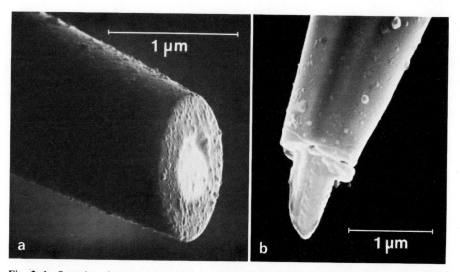

Fig. 2a,b. Scanning electron micrographs of tips of two needle electrodes: a obliquely ground, b etched with hydrofluoric acid

have diameters between 0.2 and 0.5 μm. This corresponds to surfaces of ca. 0.03–0.2 μm^2. In the polarographic circuit the small cathode surfaces produce low O_2 reduction currents ranging from 1×10^{-10} to 2.2×10^{-10} A in air-saturated solutions. The oxygen consumption of the cathodes themselves of 3.5×10^{-7}–7.7×10^{-7} mm^3 O_2/min ($= 1.5 \times 10^{-8}$–3.4×10^{-8} μmol O_2/min; or 5.0×10^{-7}–1.1×10^{-6} μg O_2/min) is negligibly small and does not essentially influence the p_{O_2} measurements. Under these conditions, the theoretically expected decrease in O_2 pressure of a fluid sample of 1 mm^3 with solubility coefficient $\alpha = 0.01$ (cm^3 O_2/cm^3 atm^{-1}) is ca. 3.6–7.7 Pa/min ($= 2.7 \times 10^{-2}$–5.8×10^{-2} mm Hg/min). The signal obtained with an obliquely ground electrode tip (angle of 30°) is about 10% higher than that of the plane electrode of identical total diameter. A cathode obliquely ground at 45° augments the signal by about 20%. Cone-shaped cathode surfaces provide an increased signal as well. This is of advantage for measurements requiring an increased signal which is then achieved without increasing the outer diameter of the electrode.

6. A sputtered thin-layer package of Ta/Pt/Ag/AgCl serves as the reference electrode (I-K$_x$) located at short distance from the tip. It is covered by a thin, sputtered, high-ohmic insulating layer (L) such as to from a small ring. Due to this arrangement, artifacts caused by dc-voltage shifts are rather ineffective. When using rf-sputtering, the very thin metallic reference element is produced with high accuracy. The total thickness of the multicomponent layer is ca. 100 nm (Ta-layer = 8 nm $\triangleq$ 80 Å; Pt-layer = ca. 28 nm; Ag-layer 2 ca. 64 nm). The Ta-layer (I) sticks well to the glass surface, while the intermediate platinum layer (J) guarantees a good electrical conductivity even if the Ag-layer (K) has chlorinated. Usually, a thin AgCl-layer (K$_x$) develops during routine tests of the measuring properties in the polarographic circuit at ca. -800 mV. The distance of 1–5 μm between platinum cathode and metallic layer serving as the reference anode is produced by etching under the microscope. In addition, the metallic layer is a good electrical shield. The electrical stability of such a reference electrode is excellent.

7. The surface of the platinum cathode may be polluted by foreign metallic influence or abrasives etc. and therefore it is electrochemically polished. The surface condition reached this way favorably influences the measuring properties of the p_{O_2} needle sensor. In some cases, the small recess (N) additionally developing through electropolishing, is also electroplated with a gold layer and thus improves the quality of the cathode surface. Moreover, the recess may be used to reinforce membrane (M) on the cathode and to increase its adhesiveness.

The polarograms of the electrochemically polished platinum cathode and the gold-tipped electrode usually show a clear plateau between -550 and -850 mV and between ca. -500 and -1200 mV, respectively, after intensive rinsing. The electrodeposited gold layer enlarges the cathode surface by its fine granular structure and produces a signal distinctly higher than that of the polished platinum surface. In a p_{O_2} range of 1.3–18.0 kPa ($= 10$–135 mmHg) the signal drifts $\pm 9.3\%$ in 30 days with either type. At high O_2 concentration (above 85%) the value is less constant. The small recess (N) can also be filled with fine-spread palladium. This increases the sensitivity to hydrogen so that the sensor is applicable to p_{H_2} measurements after adequate polarization. p_{Cl^-} sensors with good selectivity are produced by coating the platinum surface with silver and chlorinating it.

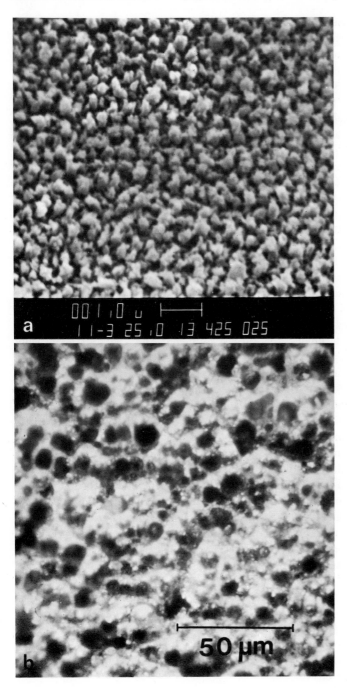

Fig. 3. a Scanning electron micrograph of a polystyrene membrane produced by gas phase polymerization in glow discharge. Nontempered membranes show a coarse-grained structure (size of grain 0.1–0.3 μm), while layers tempered at 200°C are characterized by a fine-grained surface with good sticking properties (bar = 1 μm). **b** Light micrograph of an electropolymerized polydiacetonacrylamid layer (PDAA) on a round, plane macro-Pt cathode

8. The sensor is covered with nontoxic, O_2-permeable membranes (M) made of hydrophilic and hydrophobic materials such as collodion, polystyrene, zapon lacquer, silicone, acrylic polymeres. The filmlike layers are single or multiple, and are produced either in a dipping process or by electropolymerization or by gas phase polymerization in glow discharge (Fig. 3a,b); they adhere to cathode and anode. They produce a defined diffusion field in front of the sensor. This is necessary for absolute p_{O_2} measurements; only a defined p_{O_2} field allows p_{O_2} calibration of the POS. The membrane thickness of our POS approximately corresponds to the dimension of the cathode diameter. For a double membrane of collodion and polystyrene, the diffusion error is $< 5\%$ and the response time $1-3$ s. In most cases, this is sufficient to measure the kinetics of the p_{O_2} field. The quality of the membrane must be controlled in a special test.

9. The tip diameter of the p_{O_2} microcoaxial sensor, visualized under the scanning electron microscope, is about 0.6 μm when measured about 0.2 μm from the extreme tip, while the diameter of the platinum cathode, the glass shaft and the membrane each account for about 0.2 μm (see enlargement in Fig. 1). At $1-5$ μm from the tip the thin-layer reference element (I-K$_x$) thickens the sensor for another 0.2 μm. A total diameter of about 15 μm is measured about 150 μm from the tip. Due to these dimensions, compression of the vessels is avoided in most cases. Neither microcirculation nor O_2 diffusion are influenced so that almost punctiform measurements are feasible. Further improvement is attained with optimum membranization as achieved by electropolymerization. With this technique the surface is membranized without appreciably thickening the electrode shaft. The diameter of the POS is reduced by about 0.2 μm and so an actual diameter of $0.3-0.4$ μm is obtained.

10. A connecting system for the small measuring current (A-G) protected against wetness also serves as the sensor housing (E) with ground and screw thread and makes the sensor handy and easy to exchange by a simple plug (A,C,D,G). The complete POS is sterilizable and may also be used for chronic implantations.

The production of this type of POS requires a relatively high technical investment in a specially equipped laboratory. The laboratory is under slight excess pressure and can only be entered via several dust-binding mats (Unichem GmbH, D–6000 Frankfurt 90). Here, three Laminar-flow clean work benches are combined with two clean work cabins with vertical current (Ceag-Schirp-Reinraumtechnik, D–4714 Selm-Bork) in such a way that the critical phases of production can be carried out under the conditions of extreme cleanliness absolutely necessary. The single steps in producing the p_{O_2} [9] microcoaxial needle sensor are shown as a schematic block picture in Fig. 4. A detailed description of production methods will be published elsewhere.

3 Test of Function and Measuring Properties of p_{O_2} Microcoaxial Needle Sensor

3.1 Performance of Bare Electrodes

3.1.1 p_{O_2} Signal of Platinum Cathode

The quality of a p_{O_2} electrode is, apart from the melting process which determines the boundary platinum/glass, mainly influenced by the surface condition of the cathode.

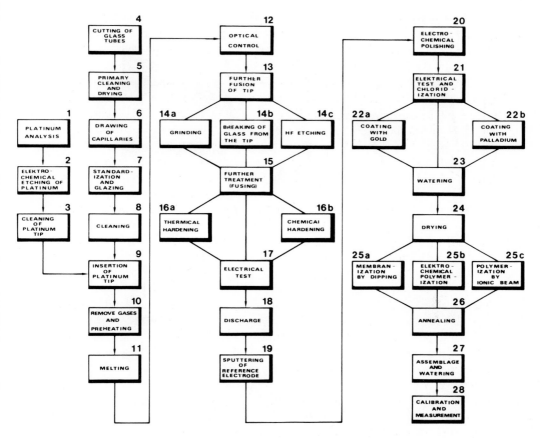

Fig. 4. Schematic drawing of the single steps necessary for producing the p_{O_2} microcoaxial needle sensor

Experience has shown that high cleanliness of the surface and the purity of the platinum used are of extreme importance. Tests in the polarographic measuring circuit have shown that the surface condition is favorably influenced by electrochemical polishing of the Pt-cathode. Figure 5 shows an original registration of the O_2 reduction currents of three needle sensors in an unpolished (I) and electrochemically polished (II) condition. In an untreated condition the three sensors show continuously increasing signals, whereby sensors a and b exhibit an especially unsteady behavior. After electropolishing the same sensors in diluted KCN-solution with ac current, the reproduceability of the signal was improved, the unsteadiness was essentially diminished, and the drift was reduced to a minimum. An effect of electropolishing was observed in 95% of sensors. Only in a few cases was re-etching necessary.

Apart from a clean cathode surface, electropolishing produces a small recess, the size of which is assessed by determining the so-called stirring effect R_e (diffusion error). The stirring effect is determined under defined convection conditions and is calculated according to formula

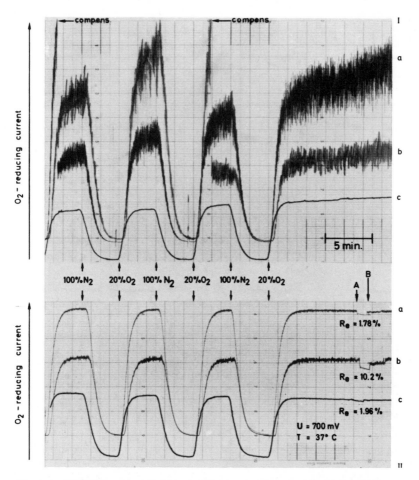

Fig. 5. Behavior of nontreated (I) and electrochemically polished (II) platinum needle sensors. The signals of unpolished sensors (*a, b, c*), recorded in Theorell buffer solution (pH = 7.2) at continual change of calibration gas steadily increase, whereby sensors a and b show considerable unrest. After polishing with ac current in dilute KCN solution, the reproduceability of the signal improves, the high amplitude essentially diminishes and the drift is eliminated. The small stirring effects (R_e) of sensors *a* and *c* indicate recesses greater than that of *b*. A stop of gas supply (unstirred medium), B switching on gas supply (stirred medium)

$$R_e = \frac{I - I_1}{I} \times 100\% \text{ or } \left(1 - \frac{I_1}{I}\right) \times 100\%, \tag{1}$$

where I is the p_{O_2} signal produced in stirred solution and I_1, the p_{O_2} signal in unstirred solution.

A low R_e value suggests a big recess and means that the O_2 particle flow is mainly restricted to the interior of the recess [see Eq. (4) with large z]. Figure 5 shows a big recess in curve a (R_e = 1.78%) and curve c (R_e = 1.96%) and a smaller one in curve b (R_e = 10.2%).

Table 1. Reduction current I_1 in H_2O and diluted aqueous solutions (25°C and 37°C) at air saturation in dependence on the size of platinum cathode surface. r_0 = radius of platinum cathode, A = surface area of platinum cathode

Reduction current (nA)		Dimension of cathode	
25°C	37°C	r_0 (μm)	A (μm^2)
–	0.04	0.05	0.008
0.07	0.09	0.1	0.031
0.1	0.13	0.15	0.071
0.13	0.18	0.2	0.126
0.16	0.22	0.25	0.196
0.19	0.27	0.3	0.283
0.22	0.32	0.35	0.385
0.26	0.36	0.4	0.503
0.29	0.4	0.45	0.636
0.33	0.45	0.5	0.785
0.36	0.5	0.55	0.950
0.4	0.55	0.6	1.131
0.43	0.6	0.65	1.327
0.47	0.64	0.7	1.539
0.5	0.69	0.75	1.767
0.54	0.74	0.8	2.011
0.58	0.78	0.85	2.270
0.61	0.83	0.9	2.545
0.64	0.89	0.95	2.835
0.68	0.95	1.0	3.141

Under conditions in other respects constant, the O_2 reduction current depends on the size of the cathode surface and so this parameter provides information about the platinum surface of the needle electrode. Table 1 shows the data obtained from measurements on bigger and exactly defined platinum surfaces extrapolated to thin cathode diameters. In general, the O_2 reduction currents of our smallest needle electrodes amount to 0.06–0.25 nA in air-saturated, diluted electrolytes (such as 0.1 mol dm^{-3} KCl, NaCl etc.) at 37°C, which corresponds to cathode diameters of 0.15–0.5 μm. This has been confirmed with control measurements made under the scanning electron microscope.

3.1.2 p_{O_2} Signal of the Gold Cathode

In comparison to platinum as a cathode material, gold has a broader plateau in the polarographic measuring circuit. This has the advantage that with p_{O_2} measurements in vivo in various tissues, the effective polarization current is not influenced by small biological dc fluctuations. We prefer electrodeposition of Au on the Pt-cathode surface, as it is extremely difficult to seal thin tips of gold wire into glass because no glass available matches the coefficient for thermal expansion. Satisfactory layers must be

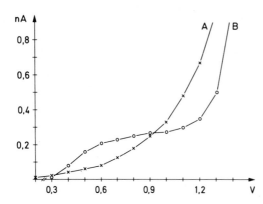

Fig. 6. Polarogram of a gold-plated p_{O_2} microcoaxial needle sensor recorded immediately after gold-plating (A) and after intensive soaking for 10 days (B). The broad polarogram typical for gold running almost in parallel with the abscissa, is distinct only in curve B

very thin and require careful cleaning in bidistilled water. The best measuring properties of galvanically gold-plated electrode tips were seen after watering for 10days.

Figure 6 shows a polarogram of a galvanically gold-plated microcoaxial needle sensor with a layer thickness of about 1 μm. Polarogram (A) was recorded immediately after galvanizing and (B) after 10 days of thorough cleaning by watering. The broad plateau seen in curve B reaches from 600 to 1100 mV which is typical for gold. Gold-plated needle sensors usually have a higher signal, because the fine-grained surface structure enlarges the cathode surface. According to information received from MRC (Materials Research Corporation, Orangeburg/New York), platinum even of the highest degree of purity produceable contains, aside from dissolved gases (14.5 ppm of H_2, O_2, N_2), impurities such as 17.5 ppm nonmetal (C, S), 30 ppm Fe, 15 ppm Rh, 7 ppm Al, 5 ppm W, 2.5 ppm Cr, Ni, Ti, Zr each. It cannot be excluded that some of these impurities unfavorably influence the p_{O_2} measurements. For some applications we therefore use the galvanic Au-layer for further improvement of the platinum micro electrodes.

3.1.3 Insulation Resistance and Ionic Sensitivity of Thin Glass Electrode Shaft

Routine control has shown that the quality of electrical insulation is recognizable from the size of the current in N_2 saturated solution, I_r.

This current should be very low. The ratio I_{air}/I_r is high if

1. the Pt-cathode is especially well fused to the glass,
2. the glass has a high electric insulation resistance and
3. the thin glass wall has not been damaged by microcracks.

In general, a mean quotient of 35 (max: > 100; $n = 923$) is reached with GW-glasses. POS with a quotient < 5 should not be used for p_{O_2} measurements. Such sensors usually have microcracks and splits between platinum and glass wall. In most cases, they also have long response times.

Apart from GW-glass, the lead-free soft glass 8510 has been used for the glass shaft of the sensor. Its coefficient of thermal expansion is about 9.3×10^{-6} in the range

$20°-300°C$. Its insulation resistance ($t_K = 322$) is especially high, and during fusing in it forms very thin and mechanically stable insulation layers on the platinum wire. As compared with other melting glasses, this glass shows little sensitivity to K^+ and Na^+ ions. Layers of untreated and thin GW-glass of 0.1 μm may have a pH sensitivity of 16–20 mV/pH. The ionic sensitivity is strongly decreased by a layer of Si_3N_4 sputtered onto the glass shaft and by the metallic thin-layer package mentioned above, which is used as the reference electrode. Though lead glasses are superior because of their high insulation resistance, they cannot be used for sealing platinum cathodes because their chemical resistance in aqueous solutions is so low that Pb is released.

3.2 Performance of Membranized Sensor

3.2.1 Effect of the Membrane on the O_2 Diffusion Field in Front of the Platinum Cathode

With polarographic p_{O_2} measurement, oxygen molecules are reduced on the cathode surface. In the neutral and alkaline ranges the chemical reaction is as follows:

$$O_2 + 2\,H_2O + 4\,e^- = 4\,OH^-. \tag{2}$$

Electrochemical reduction generates the signal in the polarographic circuit. Since the cathode continuously consumes oxygen, a continuous flux of oxygen molecules toward the electrode develops. Under the assumption that at optimum polarization all of the oxygen molecules are immediately reduced, the number of oxygen molecules on the cathode surface and, hence, p_{O_2} are zero. With this boundary condition, the p_{O_2} in the diffusion field developing in front of the electrode has been calculated. The calculation provides information about the catchment range of the electrode [28, 29, 65]. Using cylindrical coordinates, the stationary state of the O_2 diffusion field of the plane and circular electrode is described by

$$p\,(r_E,z) = p_{O_2}\,(1 - \frac{2}{\pi}) \int_0^\infty \frac{\sin\,(\lambda r_o)}{\lambda}\,J_o\,(r_E\lambda)\exp\,(-\lambda z)d\lambda, \tag{3}$$

where $p(r_E,z)$ is p_{O_2} in the diffusion field
$\quad\quad p_{O_2}$ true p_{O_2} of medium, uninfluenced by O_2 diffusion field
$\quad\quad r_o$ radius of cathode
$\quad\quad r_E$ horizontal distance from the centre of electrode
$\quad\quad z$ vertical distance from that plane
$\quad\quad J_o$ zero order Bessel function
$\quad\quad \lambda$ integration variable

(for solutions of Eq. (3) see [28]).

    Figure 7 shows the p_{O_2} distribution calculated for three different sites (r_E = 0.73 mm, 0.6 mm, 0.01 mm) of the bare, circular cathode surface (r_o = 0.75 mm). Percentage p_{O_2} is plotted on the ordinate and distance (z) from the cathode surface in mm, on the abscissa. The zero point on the abscissa indicates the cathode surface where p_{O_2} is zero. We see that the diffusion field of a bare electrode in any case

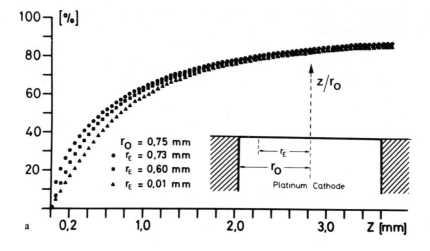

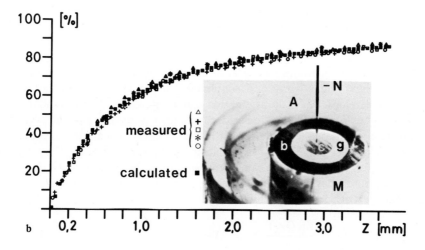

Fig. 7. a Calculated p_{O_2} distributions in the stationary diffusion field of a plane, circular cathode surface (r_0 = 0.75 mm) for three different horizontal distances from the center of the electrode (r_E = 0.73 mm, 0.6 mm, 0.01 mm). *Ordinate:* percentage p_{O_2} of true tissue p_{O_2} of medium, *abscissa:* distance (z) from cathode surface. The zero point on the abscissa indicates the cathode surface where p_{O_2} is zero. With a bare electrode, the diffusion field spreads far into the medium to be measured. 3 mm ($\triangleq$ 4 r_0) distant from the cathode surface, p_{O_2} is about 85% of the true p_{O_2} of the medium. The p_{O_2} distribution is different at different sites. In the center of the cathode, its course is flatter than at the periphery, since the flow density increases toward the border of the cathode. **b** Comparison between measured and calculated p_{O_2} distribution in the diffusion field of a circular Pt-cathode (*c*) (r_0 = 0.75 mm) for a horizontal distance from the electrode, r_E = 0.6 mm. With a macro POS, *M* (*c* plane, circular Pt cathode of 1.5 mm diameter, *g* glass insulation, *b* ring-shaped Ag/AgCl reference electrode) a diffusion field is produced in a defined agarose layer, *A* (d 13.0 mm; z 6 mm). The p_{O_2} distribution satisfactorily agrees with the calculated p_{O_2} course, as measured with a microneedle sensor, *N*, by advancing it continuously from the agarose layer toward the macro POS

spreads asymptotically with increasing distance from the cathode surface, reaching far into the medium to be measured. When defining the catchment range of an electrode as being that area which supplies 90% of the oxygen to the electrode, the approximate value is $z_{90} = 6.3 \, r_0$. Accordingly, in our example, the distance of the z_{90} plane from the plane circular electrode is 4.7 mm, which implies that up to this distance from the cathode surface, the electrode current is influenced by disturbances of the O_2 diffusion field, e.g., by convection or changes of the diffusive properties of the medium. Because of the particularly large catchment range of the bare electrode, such electrodes mostly do not develop a constant diffusion field in front of the platinum wire and therefore are not applicable to absolute p_{O_2} measurements in the tissue. With an ideal membrane the O_2 permeability, P_m of which is very small as compared to that of the sample medium, P_s the p_{O_2} decrease would be entirely restricted to the membrane. In practice, however, the diffusion field, i.e., the p_{O_2} decrease, always begins in the medium. The effect of a p_{O_2} decrease in the medium can be shown by stirring. The difference in the reduction current between stirred and unstirred medium of a membranized electrode directly indicates the diffusion error, i.e., the difference between true p_{O_2} of the medium (stirred) and p_{O_2} which is falsified by the O_2 diffusion within the medium (unstirred). Therefore the stirring effect can be used to test the POS.

The p_{O_2} distribution determined theoretically in the stationary diffusion field of a bare electrode has been experimentally confirmed [11, 13]. The result of such an experiment is shown in Fig. 7b. In that case, an O_2 diffusion field was produced with a macro POS, M, in an agarose layer, A, superimposing the sensor. The p_{O_2} distribution thus produced was measured with a membranized needle sensor, N. Comparison of the measured values (out of five experiments) with the calculated p_{O_2} distributions shows a very good agreement which substantiates the validity of the model mentioned above. Apart from the fact that all of the oxygen molecules are immediately reduced on the cathode surface and so p_{O_2} on the cathode surface is zero, the experiments show that the spatial resolution, i.e., the catchment range, of our micro needle sensor is small enough to allow absolute – almost punctiform – p_{O_2} measurements.

3.2.2 Control of the Membrane

Exact p_{O_2} measurements require covering of the electrode with a membrane which is homogeneous, pinhole-free, adhesive, nontoxic, gas-permeable and water-resistant. The p_{O_2} can be clearly related to the O_2 reduction current only if such a membrane determines the O_2 diffusion toward the electrode. Not being visible on our small needle sensors, the membranes are not optically measurable, and so indirect methods such as determination of the stirring effect must be used.

With the aid of the stirring effect the signal difference between stirred and unstirred solutions is described. According to [28, 29], the stirring effect (diffusion error) for a membranized sensor is:

$$R_e = \frac{1}{1+\delta} \times 100\% \quad \text{with} \quad \delta = \frac{P_s}{P_m} \times \frac{z_m}{r_0} . \tag{4}$$

Table 2. Stirring effect (R_e) in dependence on membrane thickness and radius of cathode (z_m/r_o) under different permeability coefficients (medium/membrane = P_s/P_m). The stirring effect becomes smaller (1) with increasing thickness of membrane, (2) with increasing ratios of permeability, (3) with decreasing radius of the cathode. According to the values reported for oxygen permeability (37°C) in the literature (see Table 3), the ratios for permeability P_s/P_m of an electrode covered with a collodion membrane ($P_m = 1.49 \times 10^{-12}$ mol O_2 cm^{-1} s^{-1} atm^{-1} $\triangleq 1.47 \times 10^{-14}$ mol O_2 cm^{-1} s^{-1} kPa^{-1}) are: 23.5 in water (with $P_s = 3.5 \times 10^{-11}$ mol O_2 cm^{-1} s^{-1} atm^{-1} $\triangleq 3.45 \times 10^{-13}$ mol O_2 cm^{-1} s^{-1} kPa^{-1}); 13.5 in brain cortex (with $P_s = 2.01 \times 10^{-11}$ mol O_2 cm^{-1} s^{-1} atm^{-1} $\triangleq 1.98 \times 10^{-13}$ mol O_2 cm^{-1} s^{-1} kPa^{-1}); 12.5 in lung tissue (with $P_s = 1.86 \times 10^{-11}$ mol O_2 cm^{-1} s^{-1} atm^{-1} $\triangleq 1.82 \times 10^{-13}$ mol O_2 cm^{-1} s^{-1} kPa^{-1}); 12.2 in heart muscle (with $P_s = 1.82 \times 10^{-11}$ mol O_2 cm^{-1} s^{-1} atm^{-1} $\triangleq 1.81 \times 10^{-13}$ mol O_2 cm^{-1} s^{-1} kPa^{-1}); 4.0 in 30% protein solution (with $P_s = 5.95 \times 10^{-12}$ mol O_2 cm^{-1} s^{-1} atm^{-1} $\triangleq 5.87 \times 10^{-14}$ mol O_2 cm^{-1} s^{-1} kPa^{-1}). To obtain a stirring effect lower than 5% e.g. when measuring p_{O_2} in brain tissue, the membrane thickness should be 1.5 times the cathode radius

z_m/r_o	$P_s/P_m = 1$	$P_s/P_m = 2$	$P_s/P_m = 5$	$P_s/P_m = 10$	$P_s/P_m = 15$	$P_s/P_m = 20$	$P_s/P_m = 25$
0.1	90.9	83.3	66.7	50.0	40.0	33.3	28.6
0.5	66.7	50.0	28.6	16.7	11.8	9.1	7.4
1.0	50.0	33.3	16.7	9.1	6.2	4.8	3.8
1.5	40.0	25.0	11.8	6.3	4.3	3.2	2.6
2.0	33.3	20.0	9.1	4.8	3.2	2.4	2.0
2.5	28.6	16.7	7.4	3.8	2.6	2.0	1.6
3.0	25.0	14.3	6.3	3.2	2.2	1.6	1.3
3.5	22.2	12.5	5.4	2.8	1.9	1.4	1.1
4.0	20.0	11.1	4.8	2.4	1.6	1.2	1.0
4.5	18.2	10.0	4.3	2.2	1.5	1.1	0.9
5.0	16.7	9.1	3.8	2.0	1.3	1.0	0.8

Here $P_s = S_s \times D_s$ is the permeability coefficient (Chap. I.3) of oxygen in the sample medium where S_s is the solubility coefficient and D_s the diffusion coefficient (see Tables 2 and 3), P_m is the permeability coefficient of oxygen in the membrane (see Table 2 and Chap. I.1), z_m is the thickness of the membrane, and r_o the radius of the cathode.

Accordingly, diminishing with increasing δ, a small stirring effect is achieved if (1) the ratio of the permeability coefficients P_s/P_m is great, (2) the membrane is thick, (3) the radius of the electrode is small. Table 2 shows the stirring effect in dependence on the membrane thickness under various conditions of diffusion permeability. It follows from Eq. (4) that an insufficient membrane (small z_m, P_m close to P_s) produces a large stirring effect.

However, there may be cases where determination of the stirring effect is not sufficient and so an additional test becomes necessary. Since significant differences exist between bare and membranized electrodes in different solutions, the difference can be used to determine quickly and reliably the quality of a membrane. The p_{O_2} signal of POS with unsatisfactory membranes continuously decreases down to insensitivity to oxygen in calibration solutions containing Ca^{2+} and Mg^{2+} as cations and phosphate, carbonate and/or sulfate as anions. In contrast, optimally membranized sensors show stable measuring properties in the presence of these critical ions. A bicarbonate buffer (pH = 7.4) of the following composition has proved to be a good test solution:

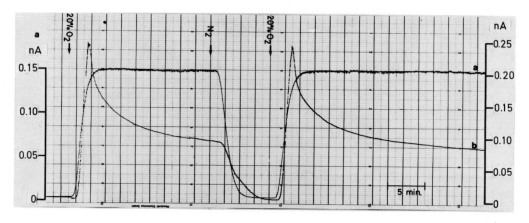

Fig. 8. p_{O_2} signals of (a) an optimally membranized and (b) a bare or too thinly membranized needle sensor in CO_2-bubbled Ringer solution containing calcium and magnesium (37°C). The signals of needle sensors with defect or too thin membranes drift considerably to low values and may sink down to O_2-insensitivity when kept in this solution for some time. In addition, when changing the calibration gas from N_2 to O_2, the signal shows an overshoot. Changing calibration gas from O_2 to N_2 results in a two-phase course of the signal. Optimally membranized sensors are superior due to stable measuring behavior

26 mmol dm^{-3} $NaHCO_3$, 2.6 mmol dm^{-3} $CaCl_2 \times 2 H_2O$, 10 mmol dm^{-3} glucose, 124 mmol dm^{-3} NaCl, 4.9 mmol dm^{-3} KCl, 1.3 mmol dm^{-3} KH_2PO_4 and 1.3 mmol dm^{-3} $MgSO_4 \times 7 H_2O$.

Figure 8 shows an example of an original registration. The p_{O_2} signal of sensor a having an intact membrane always returns to initial values after change of the calibration gas, whereas the O_2 reduction current of sensor b having a defect membrane continuously decreases and, in addition, strongly overshoots in CO_2-bubbled bicarbonate solution when changing from N_2 to air. At change from air to N_2, in most cases the curves of the insufficiently membranized POS show a two-phase course in that test solution (see sensor b). We have not yet investigated which chemical reactions are responsible for this behavior. We assume that insoluble compounds of Ca^{2+} and Mg^{2+} with hydroxides, phosphates, carbonates, and sulfates are formed and precipitate on the defect membrane and perhaps on the bare platinum surface. As a consequence, the diffusion path for O_2 increases continuously, the precipitate also changes the surface conditions of the platinum cathodes and thus, the p_{O_2} signal decreases. This has been substantiated by the fact that the measuring properties of such POS are almost completely normalized by dissolving the precipitates of hydroxide, phosphate, and carbonate by acidification of the calibration medium to pH of approx. 4.5. Here it should be mentioned that salts of carbonate, sulfate, and phosphate, which do not form precipitates, do not affect the polarographic behavior of the platinum cathode.

Organic sulfur compounds especially influence imcompletely membranized sensors. We added 16 mg of cystein to a 200 cm^3 magnesium-phosphate calibration solution and observed extreme changes in the signal. While an optimally membranized sensor scarcely perceives a deviation of the signal, with incompletely membranized sensors the signal decreases in the presence of O_2 and increases in N_2-equilibrated solution. This behavior is reversible by acidifying the medium to pH = 4.0. The above effects are eliminated by repeatedly membranizing the sensor. As the membrane determines the response time, it must not be too thick in order to avoid long response times. For application to tissue, the properties of the membrane were additionally improved by keeping the POS for some time in tissue, tissue homogenates or protein solutions. Occasionally, with too-thick membranes of polystyrene sealed directly onto the cathode surface in a sinter process at $200°-250°C$, the p_{O_2} signals slowly and continuously decreased. Such a decrease of the signal was observed only in neutral or alkaline milieu of the calibration solution. In the acid range, at pH of about 4, a stable signal was attained again. We assume that this behavior is caused by the different chemical reactions during electrochemical reduction, where either the removal of OH^- ions from the cathode surface is hindered by the membrane, or the OH^- ions are immediately neutralized to water by H^+ ions.

3.2.3 Response Time

The time needed by the POS to reach 90% of the new gas pressure value after a sudden change in gas pressure, is called response time, τ_{90}:

$$\tau_{90} \cong c \, \frac{z_m^2}{D_m} \, , \tag{5}$$

where z_m is thickness of membrane; D_m, diffusion coefficient of membrane; c, a constant ([28, 30] Chap. I.1).

The response time can be kept short if the membrane is easily diffusible and its thickness is low. As mentioned above, the membrane characteristics determine the stirring effect as well. Both parameters appear as systematic errors. Since they behave contrarily, a sensible compromise must be made by choosing the proper membrane:

1. a thick membrane for a small stirring effect, and a thin membrane for a short response time. However, the membrane must still be thick enough to prevent the O_2 signal from overshooting;
2. a low diffusion conductivity for a small stirring effect, and a high diffusion coefficient for a short response time.

In practice it is very difficult to fulfill all of these conditions at the same time. It has to be taken into account that the membrane thickness influences the response time quadratically, the stirring effect linearly. A material suitable for a good membrane should have a high O_2 diffusion coefficient, but a small solubility for O_2. So far, various materials have been used, such as collodion [13, 22, 77], polystyrene [8, 12, 68], zapon lacquer [13] as well as other film-forming polymeres [15, 34, 69]. We measured 90% response times of 1–3 s with membranes of collodion, polystyrene, and cellulose derivates whose thicknesses approximately corresponded to the diameter of the cathode. Hydron membranes (Hydron-Polymer Type XE-3, Hydron Laboratories Inc., New Brunswick, N.J. 08902) made of polyhydroxyethylmethaceylates show an interesting behavior. Their response time is short when the stirring effect is reduced. The good compatibility with biological tissue is another advantage of hydron.

3.2.4 Sensitivity to Temperature

We tested the temperature sensitivity in the range between $1°$ and $50°C$ on 36 sensors and found that the mean changes in the signal per $°C$ were 2.1% from $15°-25°C$, 2.3% from $25°-35°C$ and 2.7% from $35°-45°C$. The temperature sensitivities of different sensors differ immensely. Precise measurements require calibration of each POS at the temperature of the object to be measured.

3.2.5 Sensitivity to Light

At certain wavelengths the microcoaxial needle sensor shows sensitivity to light and reacts with an increase in the signal. In some special calibration experimensts we illuminated the shaft of the sensors laterally with a microscopic lamp (Osram 6 V/15 W with heat filter) in Theorell buffer solution (pH = 7.2) and found an immediate increase in the signal by about 1.4%. As soon as illumination was interrupted, the signal decreased to its initial value. At re-illumination, the signal increased again by the same amount. Illumination of the same sensor with light guides (type KL 150 B, Schott/ Mainz) did not change the signal.

3.2.6 Stability and Drift of the Signal

For each POS the signal is different and can change with time. We investigated the stability of the signals of seven microcoaxial needle sensors (0.1–0.15 nA) in air-saturated Theorell buffer solution at $37°C$ over a period of 30 days. During this period we observed a change of $\pm 9.3\%$ in the air value. The change of the zero point (in nitrogen) was only $\pm 1.3\%$. It was interesting that the most constant values were measured between the 11th and 23rd day (during this period the stability of the polarogram was also best).

3.3 Accuracy of Calibration and Measurement of Needle Sensors

To allow an exact quantitative p_{O_2} measurement, needle sensors should be calibrated in a solution whose diffusion conductivity for O_2 resembles that of the medium to be measured. We calibrate in a glass vessel placed in a Faraday cage. The vessel made of highly insulating laboratory glass is earthed by a calomel electrode. Temperature is continuously monitored by a thermosensor. Four sensors can be calibrated at the same time. For reasons of accuracy, calibration curves should be measured in the p_{O_2} range in which the measurements are to be made. Figure 9 shows that linear calibration curves are obtained in both high (A) and low (B) p_{O_2} ranges.

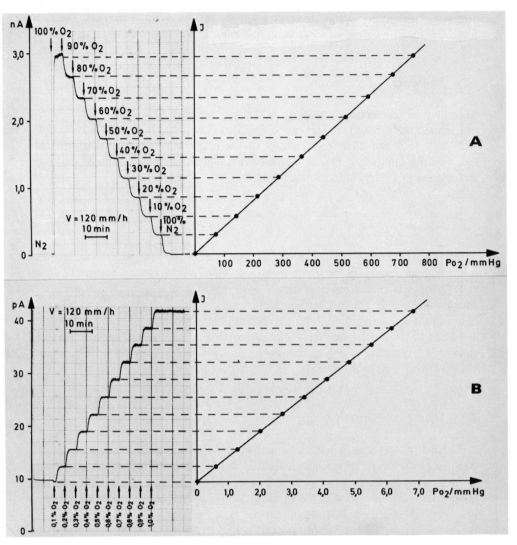

Fig. 9A,B. Calibration curves of a p_{O_2} microcoaxial needle sensor at high (**A** 0–760 mmHg $\triangleq$ 0–101 kPa) and low (**B** 0–7 mmHg $\triangleq$ 0–0.9 kPa) p_{O_2} ranges. The two original registrations show 11 spot calibrations in Theorell buffer (pH = 7.2) at 37°C. Plotting reduction current (*ordinate*) versus O_2 partial pressure (*abscissa*) results in linear calibration curves in both pressure ranges

Fig. 10a. Comparison of the p_{O_2} signals to be theoretically expected (– ● –) with those actually recorded (– * –) using bare platinum cathodes with circular surfaces (r_0 = 10 μm) in different biological media (37°C). *Dotted area* indicates the size of the signals to be theoretically expected in water. The experimentally determined values for both water and diluted aqueous solutions are within these boundaries. When relating the p_{O_2} values measured in tissue to the calibration curve recorded in diluted electrolyte solution, considerable errors of measurement may arise if either bare electrodes or POS with defective membranes were used. These errors are due to the different diffusive properties. The absolute values determined may be too low: in brain by a factor of 1.4, in the lungs of 1.5, in heart of 1.7, in muscle of 2.6, and in connective tissue of 3.2

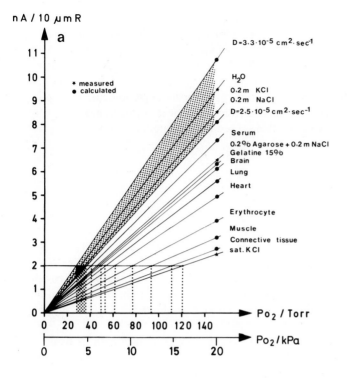

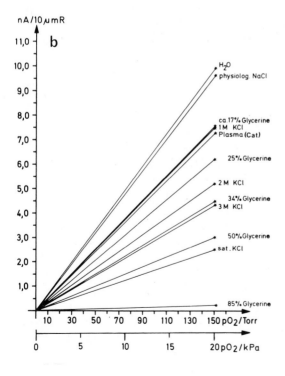

Fig. 10b. p_{O_2} signals experimentally determined for a bare platinum electrode with circular surface (r_O = 10 μm) in media of different concentrations (37°C). Calibration errors can be minimized by adapting the diffusive properties for O_2 of the calibration solutions to those of the media to be measured. For example, with different concentrations of solutions of glycerine or KCl diffusion conditions can be attained which resemble those of tissue

In practice, in most cases a 3- or 4-point calibration is sufficient. We use tanks of "extremely pure" nitrogen and oxygen. The calibration gases are two-component mixtures of oxygen and nitrogen produced by a calibrated gas-mixing pump (Type 1 M-100/a-F, Apparatebau H. Wösthoff, D–4630 Bochum). Zero point can also be determined in a saturated Na_2SO_3 solution.

For determination of the calibration curve p_{O_2} versus reduction current, the respective oxygen partial pressure is calculated from the known oxygen concentration of the calibration gas. The following formula is used to calculate the oxygen partial pressure:

$$p_{O_2} = (_bp - p_{H_2O}) \times \phi_{O_2},$$ (6)

where p_{O_2} is oxygen partial pressure in kPa (or mmHg); $_bp$, temperature-corrected barometer pressure in kPa (or mmHg); p_{H_2O}, water vapor pressure at calibration temperature in kPa (or mmHg); ϕ_{O_2}, volume fraction of oxygen in dry calibration gas (App. B).

Since the slope of the calibration curve can change with time, it is advisable to calibrate the POS before and after each experiment. On an average, the sensitivity of the microcoaxial needle sensor in watery system at $37°C$ is about 1.2×10^{-11} A kPa^{-1} ($\hat{=} 1.6 \times 10^{-12}$ A/mmHg).

Special care is needed to relate the O_2-reduction currents measured in tissue to those of the calibration curves obtained in watery medium. Too thin or defect membranes can produce immense measuring errors if the O_2-diffusion properties of calibration solution and biological media are different. The extent of such errors which might theoretically occur is shown in Fig. 10a. The O_2 reduction current of a bare electrode with circular surface ($r_0 = 10$ μm) at $37°C$ has been calculated for various media with different diffusion properties. The signal is plotted on the ordinate in dependence on O_2 up to a value of 20 kPA = 150 mmHg on the abscissa. The p_{O_2} signal was calculated as follows:

$$I_l = 4 n F r_0 S_s D_s p_{O_2},$$ (7)

where n is the number of elementary charges transported per molecule; F, Faraday constant; r_0, cathode radius; D_s, diffusion coefficient; S_s, solubility coefficient.

Using the different diffusion coefficients reported in the literature (Table 3), we obtain the highest signals in water or diluted electrolytes such as 0.2 mol dm^{-3} NaCl or KCl. This is a finding of practical significance. Assuming, for instance, that with the sensor a p_{O_2} signal of 2 nA is measured in connective tissue, and plotting this value on the calibration curve determined in 0.2 mol dm^{-3} NaCl solution, a p_{O_2} of about 35 mmHg results ($\hat{=} 4.7$ kPa). However, according to the calibration curve calculated by using the proper diffusion coefficient, 2 nA correspond to 112 mmHg ($\hat{=} 15.0$ kPa) in connective tissue. This means that for this particular case, the absolute p_{O_2} determined was too low by a factor of about 3.2. Since it cannot be excluded that the membrane of a sensor has been damaged during an experiment, mistakes can arise. p_{O_2} values determined in the cortex may be too low by a factor of up to approx. 1.4,

Table 3. Coefficients for O_2 diffusion and solubility of various biological media ($37°C$). The values given in parentheses have not been measured but estimated with the aid of $D_S = P_S/S_S$ and then converted to $37°C$ according to the dependence on temperature [25]. The corrected values of P_S are: in muscle $= 9.73 \times 10^{-12}$ mol O_2 cm^{-1} s^{-1} atm^{-1} $\hat{=}$ 9.59×10^{-14} mol O_2 cm^{-1} s^{-1} kPa^{-1}; in connective tissue $= 7.95 \times 10^{-12}$ mol O_2 cm^{-1} s^{-1} atm^{-1} $\hat{=}$ 7.83×10^{-14} mol O_2 cm^{-1} s^{-1} kPa^{-1}. In both cases S_S was assumed to be 8.93×10^{-7} mol O_2 cm^{-3} atm^{-1} $= 0.02$ cm^3 O_2 cm^{-3} atm^{-1} $\hat{=}$ 8.8×10^{-9} mol O_2 cm^{-3} kPa^{-1}

Medium	Diffusion coefficient $D_S \times 10^5$ $\left[\dfrac{cm^2}{s}\right]$	Solubility coefficient $S_S \times 10^8$ $\left[\dfrac{mol\ O_2}{cm^3\ kPa}\right]$	$\alpha \times 10^2$ $\left[\dfrac{cm^3\ O_2}{cm^3\ atm}\right]$	Ref.
Water	2.5	1.06	2.4	[27]
Water	3.3	1.06	2.4	[26]
Serum	2.54	0.94	2.13	[25]
Brain	2.0	0.99	2.25	[74]
Lung tissue	2.3	0.79	1.8	[26]
Heart muscle	1.95	0.93	2.1	[27]
Erythrocytes	1.15	1.11	2.5	[27]
Muscle	(1.09)	–	–	[41]
Connective tissue	(0.89)	–	–	[41]

in the lungs to 1.5, in heart to 1.7, and in muscle to 2.6. These errors cannot be eliminated by intermediate calibration, and hardly by correction factors. The most effective remedy is coordination of the O_2 diffusion properties of the calibration solution with those of the medium to be measured.

With exactly defined Pt-cathode surfaces we measured the p_{O_2} signal in water and serum (cf. Chap. II.10) under standardized conditions and compared the values with those calculated. The values were in good agreement, and the values for serum were almost identical. The large scattering range for water shown in Fig. 10a (dotted area) was caused by the different O_2 diffusion coefficients given in the literature. The dotted area is limited by the lowest and highest D_S value.

In other experiments the p_{O_2} signal was measured with bare circular Pt-cathodes (diameter 15 μm) in 0.2% agarose as well as KCl and NaCl solutions of different concentration at $37°C$. The result calculated for a 10-μm radius of the cathode shows that the effects of different O_2-diffusion properties of tissue are easily simulated with mixtures of fluids (Fig. 10b).

Another shift of the calibration curve can result when the pH-values of the calibration solution are in the acid range. We observed that the signal increased in the acid milieu and that the N_2 value shifted to higher signals. The N_2 value often increased already by changing pH from about 7.4 to 6.0. When pH further decreased to about 4.0, the N_2 value distinctly increased. As a consequence, when calibrating the POS in an acid medium and using the calibration curve for measurement in neutral or slightly alkaline solutions, the absolute p_{O_2} values become too low. Since physiological NaCl solutions or 0.2 mol dm^{-3} KCl may have acidic pH values under normal calibration conditions, they cannot be used to calibrate needle sensors. It is therefore necessary to adjust the pH of the calibration solution to that of the measuring medium. Such calibration problems are most important with regard to precise polarographic measurements with any kind of electrode systems. They have been discussed by other authors as well [44].

4 Application of p_{O_2} Needle Sensor

Up to now, the microcoaxial needle sensor presented here and some of its modifications have been applied to physiological basic research and used for studies such as

- Estimation of oxygen supply and microcirculation of kidney in vivo and of isolated perfused organs [12, 45].
- Determination of the diffusion coefficients for hydrogen and oxygen in agar-agar-layer and blood as well as in brain and liver homogenates [13, 33].
- Local measurement of p_{O_2} and p_{H_2} in the three bones of the inner ear of guinea-pig and cat under conditions of upper cervical sympathectomy as well as of hemorrhagic hypotension and noise [58, 61].
- p_{O_2} and p_{H_2} clearance in the central nervous system [72].
- Determination of O_2-supply in duck egg to investigate the oxygen partial pressure distribution in albumen, yolk, and embryo in dependence on the breeding period (Lomholt, Baumgärtl, Lübbers, unpubl.).
- Local p_{O_2}-measurement in tissue and the hemolymph of *Tarantula eurypelma* [5, 6].
- Examination of O_2-transport conditions in the pericardium of the crayfish [7].
- Direct determination of diffusion layers in biological systems [42].
- Measurement of oxygen supply and microcirculation in the liver [38, 80].
- Measurement of circulatory disturbance in the hypothalamus of freely moving cats (Betz, Baumgärtl, Reschke, Lübbers, unpubl.).
- Measurement of oxygen consumption in the retina of the crayfish eye (Lues, Baumgärtl, Lübbers, in prep.).
- Measurement of O_2 and H_2 – diffusion in the anterior chamber and vitreous body of the cat eye [17, 18, 64].
- Measurement of p_{O_2} fields and flow-through rates in the carotid body [1–3, 16, 36, 76].
- Estimation of O_2 transport in Walker-tumor in connection with the influence of O_2 supply and p_{O_2} distribution through the cancer inhibitor "ICRF-159" [79].
- Measurement of p_{O_2} profiles for investigation of capillary structure and O_2 regulation in the brain cortex [46, 48, 50].
- Differentiation of flow parameters and O_2 supply to the gray and white matter of the brain [47, 73, 78, 80].

Figure 11 shows an example of p_{O_2}-profiles recorded with needle sensors. The profiles were recorded by continuously advancing the sensor into the rat brain cortex. When the tip of the needle sensor (a) approached the arterial beginning of a capillary, high p_{O_2} values were measured, whilst on other sites p_{O_2} was fairly constant or relatively low. The p_{O_2} measured on the brain surface with a stationary needle sensor (b) was almost constant throughout the experiment. This means that the fluctuations

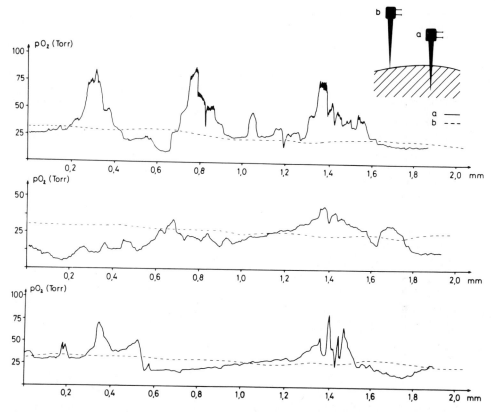

Fig. 11. p_{O_2} profile recorded with a needle sensor vertically inserted into rat cerebral cortex. With continuous speed of 150 $\mu m/min$ the sensor (*a*) is impaled into the right hemisphere of the exposed cortex at three different points. Apart from arterial p_{O_2} values, a rather steady and relatively low p_{O_2} course is seen at other sites. Throughout the experiment the sensor (*b*) registers the p_{O_2} stationarily on the surface of the cerebral cortex showing a more or less steady course

measured with the sensor (a) were not caused by changes in blood perfusion. The experiment shows quite clearly that oxygen pressure is not uniform in vascularized tissue, but that a p_{O_2} field with different p_{O_2} levels exists.

To estimate reliably the oxygen supply to a tissue, it is useful to investigate the p_{O_2} frequency distribution with the aid of a p_{O_2}-histogram [37, 39, 54, 68], since each stationary situation of oxygen supply has a characteristic p_{O_2} distribution. Figure 12 shows a p_{O_2}-histogram recorded from the gray matter of the brain cortex during normoxia. The figure shows the mean value of seven puncture channels (three animals) measured in a tissue layer of 0.2–2.1 mm over steps of 10 μm each and united to classes of 0.65 kPa $\triangleq$ 5 mmHg (abscissa). The ordinate shows the frequency of the classes. More than half of the p_{O_2} values are in the range of 11–30 mmHg ($\triangleq$ 1.5– 4.0 kPa), 15% of values are below 10 mmHg ($\triangleq$ 1.33 kPa). 25% are between 31 and 90 mmHg ($\triangleq$ 4.0–12.0 kPa). To interpret the p_{O_2} histograms exactly, we calculate the mean p_{O_2} value and determine the sites of maximum (module) and median indi-

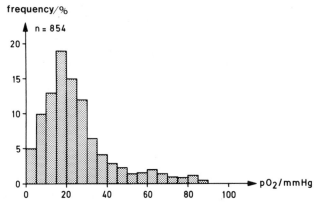

Fig. 12. p_{O_2} histogram of the rat cerebral cortex (depth 0.2–2.1 mm) during normoxia. *Abscissa:* 5 mmHg (0.65 kPa) classes; *ordinate:* percentage frequency (n = 854 out of seven puncture channels, three experiments). Under normal conditions the p_{O_2}-histogram is left-shifted and bell-shaped, characterized by the median (19 mmHg $\triangleq$ 2.5 kPa) and modulus (16–20 mmHg $\triangleq$ 2.1–2.7 kPa). The mean p_{O_2} = 24.4 mmHg $\triangleq$ 3.2 kPa)

cating which values are above or below 50% of the characteristic values. For the histogram shown in Fig. 12 we obtain a mean p_{O_2} of 24.4 mmHg ($\triangleq$ 3.2 kPa), a median of 19 mmHg ($\triangleq$ 2.5 kPa) and a module of 16–20 mmHg ($\triangleq$ 2.1–2.7 kPa).

When plotting the p_{O_2}-differences measured from one 10 μm-step to another in a two-dimensional coordinate system, the p_{O_2} gradient histogram shown in Fig. 13 is obtained. The steepest gradient measured over a length of 10 μm in the rat cortex was 24 mmHg ($\triangleq$ 3.2 kPa). 34% values do not show a pressure difference over is distance, while about 50% have a p_{O_2} decrease of 1–4 mmHg/10 μm ($\triangleq$ 0.13–0.53 kPa). According to results of experiments on oxygen supply to tissue, we conclude that oxygen is mainly transported by diffusion.

As well as a carefully prepared sensor, a stable polarization voltage without leaking currents and an exact experimental setup (suitable electrical shielding, stable and sensitive amplifiers) are necessary for reliable p_{O_2} measurements. In some cases, interpretation of the signal is difficult or even misleading because the signal is superimposed by artifacts. To exclude such mistakes, the polarographic setup has to be checked before each experiment by interrupting the measuring circuit by disconnecting the reference electrode for a short time. Whith an adequate measuring arrangement, the signal must immediately decrease to zero when switched on after a short period of polarization, and then return to initial p_{O_2} values.

The damage to tissue through puncturing with needle sensors is one of the main problems. Apart from possible methodological errors arising with the polarographic technique, the fact has to be considered that every puncture causes changes of the physiological milieu of a living object. So far, only very few systematic studies on histologically visible tissue damage have been performed. Together with E. Seidl, we carried out light microscopic investigations to identify the puncture channels in brain preparations. Sensors with an outer diameter of about 3 μm were found to leave channels in the cortex that were marked by vacuolization of the interstitial substance, pyknosis of ganglia cells, and bleeding when vessels were punctured.

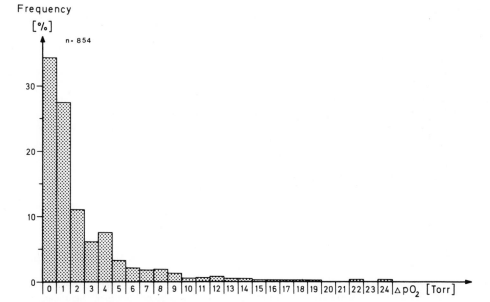

Fig. 13. p_{O_2}-gradients over 10 μm in the gray matter of the rat brain (0.2−2.1 mm). 34% of the values (n = 854) show a steady p_{O_2}. The highest p_{O_2} decrease was 24 mmHg/10 μm $\triangleq$ 3.2 kPa/10 μm and was only found twice. 50% of values range between 1 and 4 mmHg/10 μm $\triangleq$ 0.13− 0.53 kPa/10 μm

In the region of both the sensor tip and lateral shaft, damage to tissue was seen which may influence the experimental results. Damaged tissue was also seen in liver preparations after inserting a simple glass needle sensor (tip diameter about 2 μm) such as used for measurements of ionic activity [66]. In contrast, histological investigation of kidney tissue showed only little damage. In any case, tissue damage is kept at a minimum if the tip diameter of the sensor ranges below 1 μm and if the shaft is very thin. We think that these requirements are satisfactorily fulfilled by the microcoaxial needle sensor developed in our laboratory.

The results of our investigations have shown that micro needle sensors have found broad application to the medico-biological field.

Lately, microelectrodes have also been used for p_{O_2} measurements in marine sediments for ecologico-physiological purposes (Chap. III.2 [35, 62, 63, 71]).

Acknowledgment. The authors wish to thank Mrs. G. Blümel and Mrs. E. Menne for preparing the English translation.

References

1. Acker H (1977) Possible excitation mechanism in the chemoreceptor of the carotid body. Verhandlungsber Ges Lungen- und Atmungsforsch 6:1–13
2. Acker H, Lübbers DW (1976) Oxygen transport capacity of the capillary blood within the carotid body. Pfluegers Arch 366:241–246
3. Acker H, Lübbers DW, Purves MJ (1971) Local oxygen tension field in the glomus caroticum of the cat and its change at changing arterial p_{O_2}. Pfluegers Arch 329:136–155
4. Albanese RA (1973) On microelectrode distortion of tissue oxygen tension. J Theor Biol 38: 143:154
5. Angersbach D (1975) Oxygen pressures in haemolymph and various tissues of the tarantula, *Eurypelma helluo*. J Comp Physiol 98:133-145
6. Angersbach D (1978) Oxygen transport in the blood of the tarantula, *Eurypelma californicum*: p_{O_2} and pH during rest, activity and recovery. J Comp Physiol 123:113–125
7. Angersbach D, Decker H (1978) Oxygen transport in crayfish blood: effect of thermal acclimation and short-term fluctuation related to ventilation and cardiac performance. J Comp Physiol 123:105–112
8. Bartels H, Reinhardt W (1960) Einfache Methode zur Sauerstoffmessung im Blut mit einer kunststoffüberzogenen Platinelektrode. Pfluegers Arch Gesamte Physiol Menschen Tiere 271: 105:114
9. Baumgärtl H (1975) Anwendung der CEAG-SCHIRP-Reinraumtechnik bei der Herstellung von Mikroelektroden zur Aufklärung biologisch wichtiger Reaktionsmechanismen. Staubjournal 17:12–13
10. Baumgärtl H (1978) Nadelelektroden zur Messung der Sauerstoffversorgung, Mikrozirkulation und Ionenaktivität im cellulären Bereich lebender Organismen. Garching Instrumente, München, Analytika
11. Baumgärtl H, Grunewald W, Lübbers DW (1974) Polarographic determination of the oxygen partial pressure field by Pt-microelectrodes using the O_2 field in front of a Pt-macroelectrode as a model. Pfluegers Arch 347:49–61
12. Baumgärtl H, Leichtweiss H-P, Lübbers DW, Weiss Ch, Huland H (1972) The oxygen supply of the dog kidney; measurements of intrarenal p_{O_2}. Microvasc Res 4:247–257
13. Baumgärtl H, Lübbers DW (1973) Platinum needle electrode for polarographic measurement of oxygen and hydrogen. In: Kessler M, Bruley DF, Clark LC Jr, Lübbers DW, Silver IA, Strauss J (eds) Oxygen supply. Urban & Schwarzenberg, München, pp 130–136
14. Baumgärtl H, Lübbers DW (1975) Herstellung von Mikro-p_{O_2}- und p_{H_2}- Elektroden mit der Hochfrequenzkathodenzerstäubungstechnik. Naturwissenschaften 62:572
15. Bicher HI, Knisely MH (1970) Brain tissue reoxygenation time, demonstrated with a new ultramicro oxygen electrode. J Appl Physiol 28:387–390
16. Bingmann D, Schulze H, Caspers H, Acker H, Keller HP, Lübbers DW (1977) Tissue p_{O_2} in the cat carotid body during respiratory arrest after breathing pure oxygen. In: Acker H, Fidone S, Pallot d, Eyzaguirre C, Lübbers DW, Torrance RW (eds) Chemoreception in the carotid body. Springer, Berlin Heidelberg New York, pp 264–270
17. Briggs D (1972) Experimentelle Untersuchungen zum Sauerstofftransport im Glaskörper des Katzenauges. Dissertation, Marburg
18. Briggs D, Rodenhäuser J-H (1973) Distribution and consumption of oxygen in the vitreous body of cats. In: Kessler M, Bruley DF, Clark LC Jr, Lübbers DW, Silver IA, Strauss J (eds) Oxygen supply. Urban & Schwarzenberg, München, pp 265–269
19. Cater DB (1966) The significance of oxygen tension measurements in tissues. In: Payne JP, Hill DW (eds) Oxygen measurements in blood and tissues and their significance. Churchill, London, pp 155–172
20. Cater DB, Silver IA (1961) Microelectrodes and electrodes used in biology. In: Ives DJG, Janz GJ (eds) Reference electrodes. Academic Press, London New York, pp 464–519
21. Davies PW, Brink F (1942) Microelectrodes for measuring local oxygen tension in animal tissues. Rev Sci Instrum 13:524–533

22. Drenckhahn F-O (1956) Über eine Methode zur Messung des Sauerstoffdruckes im Blut mit der Platin-Kathode. Pfluegers Arch Gesamte Physiol Menschen Tiere 262:169–177
23. Erdmann W, Krell W, Metzger H, Nixdorf I (1970) Ein Verfahren zur Herstellung standardisierter Gold-Mikroelektroden für die p_{O_2}-Messung im Gewebe. Pfluegers Arch 319:R 69
24. Fatt I (1964) An ultramicro oxygen electrode. J Appl Physiol 19:326–329
25. Gertz KH, Loeschcke HH (1954) Bestimmung der Diffusionskoeffizienten von H_2, O_2, N_2 und He in Wasser und Blutserum bei konstant gehaltener Konvektion. Z Naturforsch 96:1–9
26. Grote J (1967) Die Sauerstoffdiffusionskonstanten im Lungengewebe und Wasser und ihre Temperaturabhängigkeit. Pfluegers Arch Gesamte Physiol Menschen Tiere 295:245–254
27. Grote J, Thews G (1962) Die Bedingungen für die Sauerstoffversorgung des Herzmuskelgewebes. Pfluegers Arch 276:142–165
28. Grunewald W (1966) Zur Theorie der Ausgleichsvorgänge an Pt-Elektroden und ihre mathematischen Grundlagen. Dissertation, Marburg
29. Grunewald W (1970) Diffusionsfehler und Eigenverbrauch der Pt-Elektrode bei p_{O_2}-Messungen im steady state. Pfluegers Arch 320:24–44
30. Grunewald W (1971) Einstellzeit der Pt-Elektrode bei Messungen nicht-stationärer O_2-Partialdrucke. Pfluegers Arch 322:109–130
31. Grunewald W (1973) Accuracy and errors of the p_{O_2} measurement by means of the platinum electrode and its calibration in vivo. In: Gross JF, Kaufmann R, Wetterer E (eds) Modern techniques in physiological sciences. Academic Press, London New York, p 309
32. Grunewald W (1973) How "local" is p_{O_2} measurement? In: Kessler M, Bruley DF, Clark LC Jr, Lübbers, DW, Silver IA, Strauss J (eds) Oxygen supply. Urban & Schwarzenberg, München, pp 160–163
33. Grunewald W, Baumgärtl H, Reschke W, Lübbers DW (1967) Bestimmung des H_2-Diffusionskoeffizienten mit der palladinierten Pt-Stichelektrode zur Messung der Mikrozirkulation im Gehirn. Pfluegers Arch Gesamte Physiol Menschen Tiere 294:R 40
34. Günther H, Aumüller G, Kunke S, Vaupel P (1974) Die Sauerstoffversorgung der Niere. Verteilung der O_2-Drucke in der Rattenniere unter Normalbedingungen. Res Exp Med 163:251–264
35. Jørgensen BB, Revsbech NP, Blackburn TH, Cohen J (1979) Diurnal cycle of oxygen and sulfide microgradients and microbial photosynthesis in a cyanobacterial mot sediment. Appl Environ Microbiol 38:46–58
36. Keller HP, Lübbers DW (1973) Local blood flow measurement in the carotid body of the cat by means of hydrogen clearance. In: Kessler M, Bruley DF, Clark LC Jr, Lübbers DW, Silver IA, Strauss J (eds) Oxygen supply. Urban & Schwarzenberg, München, pp 233–235
37. Kessler M, Bruley DF, Clark LC Jr, Lübbers DW, Silver IA, Strauss J (eds) (1973) Oxygen supply. Urban & Schwarzenberg, München
38. Kessler M, Thermann M, Lang H, Hartel W, Schneider H (1970) O_2-Versorgung lebenswichtiger Organe im Schock unter besonderer Berücksichtigung der Leber. In: Zimmermann W, Staib J (eds) Schock, Stoffwechselveränderungen und Therapie. Schattauer, Stuttgart New York, pp 117–131
39. Kessler M, Höper J, Krumme AB (1976) Monitoring of tissue perfusion and cellular function. Anaesthesiology 45:184–197
40. Kreuzer F, Kimmich H-P (1976) Recent developments in oxygen polarography as applied to physiology. In: Degen H, Balslev I, Brook R (eds) Measurement of oxygen. Elsevier Scientific Publ Comp, Amsterdam Oxford New York, pp 123–158
41. Krogh A (1919) The rate of diffusion of gases through animal tissues with some remarks on the coefficient of invasion. J Physiol 52:391–408
42. Kuhlmann G (1979) Die direkte Bestimmung von Diffusionsschichten an biologischen Systemen mit Mikroelektroden. Diplomarbeit, Bremen
43. Kunze K (1969) Das Sauerstoffdruckfeld im normalen und pathologisch veränderten Muskel. Schriftenreihe Neurologie, Vol III. Springer, Berlin Heidelberg New York
44. Lee YH, Tsao GT (1979) Dissolved oxygen electrodes. In: Ghose TK, Fletcher A, Blakebrough N (eds) Advances in biochemical engineering, vol 13. Springer, Berlin Heidelberg New York, pp 35–86

45. Leichtweiss H-P, Lübbers DW, Weiss Ch, Baumgärtl H, Reschke W (1969) The oxygen supply of the rat kidney: measurements of intrarenal p_{O_2}. Pfluegers Arch 309:328–349

46. Leniger-Follert E (1976) Die Sauerstoffversorgung und die Anpassung der Mikrozirkulation an den Sauerstoffbedarf des Gehirncortex. Habilitationsschrift, Ruhr-Univ Bochum

47. Leniger-Follert E, Lübbers DW (1979) Significance of local tissue p_{O_2} and of extracellular cations of functional and reactive hyperemia of microcirculation in the brain. In: Zülch KJ, Kaufmann W, Hossmann K-A, Hossmann V (eds) Brain and heart infarct, vol II. Springer, Berlin Heidelberg New York, pp 193–201

48. Leniger-Follert E, Wrabetz W, Lübbers DW (1976) Local tissue p_{O_2} and microflow of the brain cortex under varying arterial oxygen pressure. In: Grote J, Reneau D, Thews G (eds) Oxygen transport to tissue, vol II. Plenum Publ Corp, Oxford, pp 361–367

49. Lübbers DW (1966) Methods of measuring oxygen tensions of blood and organ surfaces. In: Payne JP, Hill DW (eds) Oxygen measurements in blood and tissues and their significance. Churchill, London, pp 103–127

50. Lübbers DW (1967) Kritische Sauerstoffversorgung und Mikrozirkulation. In: Wendt CG (ed) Marburger Jahrbuch 1966/67. Elwerth, Marburg, pp 305–319

51. Lübbers DW (1968) The oxygen pressure field of the brain and its significance for the normal and critical oxygen supply of the brain. In: Lübbers DW, Luft UC, Thews G, Witzleb E (eds) Oxygen transport in blood and tissue. Thieme, Stuttgart, pp 124–139

52. Lübbers DW (1969) The meaning of the tissue oxygen distribution curve and its measurement by means of Pt electrodes. In: Kreuzer F (ed) Oxygen pressure recording in gases, fluids, and tissues. Prog Respir Res 3:112–123

53. Lübbers DW (1973) Local tissue p_{O_2}; its measurement and meaning. In: Kessler M, Bruley DF, Clark LC Jr, Lübbers DW, Silver IA, Strauss J (eds) Oxygen supply. Urban & Schwarzenberg, München, pp 151–155

54. Lübbers DW (1977) Bedeutung des lokalen Gewebesauerstoffdrucks und des p_{O_2}-Histogramms für die Beurteilung der Sauerstoffversorgung eines Organs. Prakt Anaesthesiol 12:184–193

55. Lübbers DW (1977) Quantitative measurement and description of oxygen supply to the tissue. In: Jöbsis FF (ed) Oxygen and physiological function. Professional Information Library, Dallas, pp 62–71

56. Lübbers DW, Baumgärtl H (1967) Herstellungstechnik von palladinierten Pt-Stichelelektroden (1–5 μ Außendurchmesser) zur polarographischen Messung des Wasserstoffdruckes für die Bestimmung der Mikrozirkulation. Pfluegers Arch 294:R 39

57. Lübbers DW, Baumgärtl H, Fabel H, Huch A, Kessler M, Kunze K, Riemann H, Seiler D, Schuchhardt S (1969) Principle of construction and application of various platinum electrodes. In: Kreuzer F (ed) Oxygen pressure recording in gases, fluids, and tissues. Prog Respir Res 3: 136–146

58. Maass B (1977) Tierexperimentelle Untersuchungen des sympathischen Einflusses auf die Innenohrfunktion. Habilationsschrift, Düsseldorf

59. Maass B, Baumgärtl H, Lübbers DW (1976) Lokale p_{O_2}- und p_{H_2}-Messungen mit Nadelelektroden zum Studium der Sauerstoffversorgung und Mikrozirkulation des Innenohres. Arch Oto-Rhino-Laryngol 214:109–124

60. Maass B, Baumgärtl H, Lübbers DW (1978) Lokale p_{O_2}- und p_{H_2}-Messungen mit Mikrokoaxialnadelelektroden an der Basalwindung der Katzencochlea nach akuter oberer zervikaler Sympathektomie. Arch Oto-Rhino-Laryngol 221:269–284

61. Maass B, Baumgärtl H, Lübbers DW (1979) Wirkung einer Sympathektomie auf den Sauerstoffpartialdruck (p_{O_2}) in der Cochlea unter hämorrhagischer Hypotension. Laryngol Rhinol 58: 665–670

62. Revsbech NP, Jørgensen BB, Blackburn TH (1980) Oxygen in the seabottom measured with a microelectrode. Science 207:1355–1356

63. Revsbech NP, Sørensen J, Blackburn TH, Lomholt JP (1980) Distribution of oxygen in marine sediments measured with micro-electrodes. Limnol Oceanogr 25:403–411

64. Rodenhäuser J-H, Baumgärtl H, Lübbers DW, Briggs D (1971) Behaviour of the oxygen partial pressure in the vitreous body under various oxygen conditions. In: Proc XXIth Int Congr Ophthalmol. Excerpta Medica Int Congr Ser, No 222, Amsterdam, pp 1624–1628

65. Saito Yukio (1968) A theoretical study on the diffusion current at the stationary electrodes of circular and narrow bond types. Rev Polarogr 15:177–187
66. Schäfer D, Höper J (1976) The influence of glass needle electrodes on rat liver cells and tissue. 6th Eur Congr Electron Microsc, Jerusalem, pp 304–306
67. Schneiderman G, Goldstick TK (1976) Oxygen fields induced by recessed and needle oxygen microelectrodes in homogeneous media. In: Grote J, Reneau D, Thews G (eds) Oxygen transport to tissue, vol II. Plenum Publ Corp, Oxford
68. Schuchhardt S (1971) p_{O_2}-Messung im Myocard des schlagenden Herzens. Pfluegers Arch 322: 83–94
69. Silver IA (1965) Some observations on the cerebral cortex with a ultra-micro, membrane-covered, oxygen electrode. Med Electron Biol Eng 3:377–387
70. Silver IA (1966) The measurement of oxygen tension in tissue. In: Payne JP, Hill DW (eds) Oxygen measurements in blood and tissues and their significance. Churchill, London, pp 135–153
71. Sørensen J, Jørgensen BB, Revsbech NP (1979) A comparison of oxygen, nitrate, and sulfate respiration in coastal marine sediments. Microbiol Ecol 5:105–115
72. Speckmann EJ, Caspers H (1970) Messung des Sauerstoffdruckes mit Platinmikroelektroden im Zentralnervensystem. Pfluegers Arch 318:78–84
73. Stossek K, Lübbers DW (1970) Determination of microflow of the cerebral cortex by means of electrochemically generated hydrogen. In: Russel RW Ross (ed) Brain and blood flow. Pitman Medical and Scientific Publ Co Ltd, London, pp 80–84
74. Thews G (1960) Die Sauerstoffdiffusion im Gehirn. Pfluegers Arch Gesamte Physiol Menschen Tiere 271:197–226
75. Tsacopoulos M, Lehmenkühler A (1977) A double-barrelled Pt-microelectrode for simultaneous measurement of p_{O_2} and bioelectrical activity in excitable tissues. Experientia 33:1337–1338
76. Weigelt H (1975) Der lokale Sauerstoffdruck im *Glomus caroticum* des Kaninchens und seine Bedeutung für die Chemorezeption. Dissertation, Bochum
77. Whalen WJ, Riley J, Nair P (1967) A microelectrode for measuring intracellular p_{O_2}. J Appl Physiol 23:798–801
78. Wrabetz W, Leniger-Follert E, Baumgärtl H, Seidl E, Lübbers DW (1975) Local tissue p_{O_2} in the white matter of cat brain and its regulation. In: Leniger-Follert E, Lübbers DW (eds) Regulation of microcirculation. Arzneimittelforschung (Drug Res) 25:1675
79. Ziegler H (1974) Beeinflussung des lokalen Sauerstoffpartialdruckes im WALKER-Tumor der Ratte durch den Krebshemmstoff ICRF 159. Dissertation, Münster
80. Zorn H (1972) Der Sauerstoffpartialdruck im Hirngewebe und in der Leber bei subtoxischen Kohlenmonoxydkonzentrationen. Staub-Reinhalt Luft 32:151–155

Chapter I.5   Electrolytes

R. Bucher[1]

1 Introduction

In order to provide a properly functioning polarographic oxygen sensor (POS), the electrolyte must be compatible with the oxidation and reduction mechanisms at the electrodes. Furthermore, it should provide a conductive path for the transport of ionic species between the electrodes. In the design of a POS, the choice of the electrolyte is of the utmost importance.

Generally speaking, there is a specific composition of the electrolyte in an operating sensor at which optimum performance will result. A marked change of the electrolyte composition is undesirable. On the other hand, it is a property of electrochemical cells that the electrolyte changes as a result of the electrode reactions. Depending on the particular electrode and the reactions involved, chemical species may become altered or may accumulate or be depleted in the electrolyte. This is especially pronounced in POS.

It follows from the above that the principal concern is the problem of changes in the composition of the electrolyte. These changes depend not only on electrochemical reactions, but also on geometrical factors, such as the thickness of the electrolyte film, the size of the electrolyte reservoir, and the nature of the electrolyte path between cathode and anode, etc.

The most favored anode material of POS is silver, caoted with silver chloride or silver oxide. These two couples call for a chloride-containing electrolyte in the former case, for an alkaline one or a buffer solution in the latter. The following reactions occur at the anode:

$$Ag + Cl^- \quad \rightarrow \ AgCl + e^- \qquad E_o = -0.222 \text{ V.} \tag{1}$$

$$2\,Ag + 2\,OH^- \quad \rightarrow \ Ag_2O + H_2O + 2\,e^- \ \ E_o = -0.35 \text{ V.} \tag{2}$$

The potential of the anode system depends on the activity of the chloride ions and on the pH value respectively. During measurement, the activity of the ions must therefore be kept constant within certain limits.

1 Dr. W. INGOLD LTD., Industriezone Nord, CH–8902 Urdorf/ZH, Switzerland

Polarographic Oxygen Sensors (ed. by Gnaiger/Forstner)
©Springer-Verlag Berlin Heidelberg 1983

2 Effect of Impurities

The oxygen reaction at the cathode is so irreversible or, in other words, the exchange current density so low (for platinum 10^{-10} to 10^{-9} A/cm^2) that even minute traces of reducible impurities in the electrolyte may cause a depolarization [5]. The impurities may either pass from the measuring solution into the electrolyte through the membrane or they are already present in it. In principle they may act in either of two different manners:

- The impurity leaves the cathode surface unaffected but causes an increase of the current measured, or
- it deactivates the cathode by forming a deposit on its surface.

In the former case the disturbance is normally of short-term duration as the impurities are reduced during the early stage of operation.

Solid deposits on the working electrode surface, which inhibit its catalytic activity, are undoubtedly deleterious to the operation of the sensor and should be avoided. With an ideal electrode all electrochemical problems are eliminated or well defined. Nevertheless, some electrode contamination reactions may occur at rates so low that the sensor will operate satisfactorily over a reasonable length of time, despite their presence.

2.1 Deactivation of the Cathode by Internal Impurities

Oxidation of the anode material causes reaction products which are for several reasons undesirable, but cannot be avoided. On a silver anode oxidized silver forms a precipitation with the chloride ions present in the solution. However, some silver remains dissolved in the electrolyte and diffuses to the cathode where it will be reduced.

Even tiny concentrations of silver ions ($\approx 10^{-8}$ mol dm^{-3}) cause deactivation of the platinum cathode [13]. In particular electrolytes of POS which are in equilibrium with silver chloride have a critical silver ion concentration. It is therefore not surprising that great efforts have been made to reduce the poisoning of the cathode surface by silver ions.

Several attempts have been made to slow down this process either by decreasing the silver concentration in the electrolyte, by extending the transport path of the silver ions to the cathode [12], by using a larger cathode surface, or by reducing its speed of diffusion.

The application of a gel or a paste instead of a liquid low-viscosity electrolyte is an example of a measure to reduce silver flow [8]. Simultaneously the resistance to pressure fluctuations is increased with a gelled electrolyte and results in a more stable signal level. On the other hand some serious disadvantages arise as, e.g., a considerably prolonged polarization time, the danger of drying out, and an enclosure of air bubbles. Gel electrolytes are therefore hardly applied today.

The process of deactivation may also be partly suppressed with the aid of anions which form a more insoluble precipitation with silver ions than chloride does.

2.2 Other Impurities Affecting the Cathode Reactions

CO_2 is present in biological systems. To ensure an accurate oxygen determination, an interference with CO_2 should be avoided. However, it has often been observed that the POS becomes less sensitive to oxygen upon even short exposure to high CO_2 partial pressures.

Radhakrishna and Roggenkamp [9] have observed the following effect: When the POS was first exposed to a mixture of high CO_2 and low O_2 content and subsequently to an ambient air sample, the output exceeded the 21% level, then reversed and fell below the value, then approached the theoretical value asymptotically. The recovery period was dependent on the amount of CO_2 and the duration of exposure to CO_2.

At standard polarization voltages CO_2 will not be reduced, but when the gas enters carbonic acid is formed, which results in a pH change in the electrolyte adjacent to the cathode. The time response characteristics of the electrode will be correspondingly impaired until the pH value can attain its normal value by diffusion of the electrolyte from the reservoir. The poisoning of the electrolyte by CO_2 is therefore an example of a reversible process.

The problem of CO_2 interference was solved by introducing a spacer made of a porous material between the cathode and the membrane or by a porous cathode, allowing a faster interchange of electrolyte film with the reservoir [9]. Nowadays spacers are often used but mainly for other reasons which will be discussed below.

Other gases, such as SO_2, which are commonly encountered in the environment, will not interfere with oxygen measurements. This is clearly shown by comparing the standard potential for the electrode reaction involving SO_2 with that of oxygen:

$$H_2SO_3 + 4\,H^+ + 4\,e^- \rightarrow S + 3\,H_2O \qquad E_o = +0.4\ V. \tag{3}$$

$$O_2 + 4\,H^+ + 4\,e^- \rightarrow 2\,H_2O \qquad E_o = +1.229\ V. \tag{4}$$

In principle, SO_2 could act as an electrode poison in a manner similar to CO_2 but the SO_2 permeability of membranes compared to O_2 is smaller by some orders of magnitude.

2.3 Impurities Poisoning the Anode

Serious poisons for the POS are H_2S and thio-organic compounds. While the cathode remains unaffected, a silver anode reacts readily with these species, resulting in a considerable potential shift of several hundreds of millivolts (Chap. I.6). Schmid and Mancy [11] have recommended a cadmium nitrate electrolyte besides the standard one to eliminate H_2S. The former was sandwiched between two membranes and H_2S was removed by precipitation. From time to time CdS has to be replaced by a fresh electrolyte. Oxidizing agents show an effect similar to that of $Cd(NO_3)_2$.

A silver-silver chloride anode also undergoes a potential shift as the pH value is raised by the reduction of oxygen. This reduction leads to an excess of OH^- ions which can be removed by precipitation at the silver anode, stabilizing the pH value of the electrolyte. In a chloride-containing electrolyte the chloride concentration will drop as it is consumed by the anode reaction and will gradually be replaced by OH^-, generated by the oxygen reaction. The reference couple will therefore change from Ag/AgCl to Ag/Ag$_2$O, resulting in a shifted potential (~ 120 mV). An electrode with a narrow polarization plateau probably will then suffer from nonlinearity, but in general the accuracy of the measurement will not be affected.

3 Loss of Electrolyte

The loss of electrolyte solvent by diffusion through the membrane is probably one of the most common wear-out modes for POS. Measurement in gases or storage of the sensor in the atmosphere are reasons for such a solvent depletion. The wear-out can be slowed down by using a solvent of low vapor pressure, a small membrane surface, a small permeability of the solvent in the membrane, a high viscosity of the solvent, a small difference of the partial pressures of the solvent between environment and electrolyte, or a large electrolyte reservoir [6].

Since the electrolyte must provide a good electrical conduction path between the anode and the cathode, the choice of a suitable solvent is mainly restricted to water. Its vapor pressure can be reduced by a high salt content. Hitchman [4] has studied the solvent evaporation of electrolytes with and without deliquescent salts. KH$_2$PO$_4$ proved to be excellently suited.

If the solvent lost from the electrolyte layer is not replaced at a given minimum rate the electrolyte will gradually dry out, resulting in an increased ohmic resistance of the electrical conduction path. The electrical current may then be so low that a diffusion-controlled oxygen reduction is no more guaranteed.

To prevent this, the solvent transfer from the electrolyte reservoir may be accomplished by a spacer which secures a minimum electrolyte film thickness. Another example is the use of a porous solid component between the periphery of the electrolyte layer and the electrolyte reservoir to transport solvent by capillary pumping action [10].

4 Electrolyte Layer, Sensitivity and Response

The signal level of a POS is determined mainly by the impedance of the membrane and the cathode area. The electrolyte layer above the cathode plays a minor role and its contribution constitutes usually not more than a few percent of the total current (Chap. I.1). However, the electrolyte layer between cathode and reservoir influences the time rate of response characteristics and the sensitivity.

Lucero [6] has calculated the detector time response characteristics by means of a network of impedances and capacities. He concluded that an increase of the electrolyte impedance between cathode and reservoir, by way of example a decrease of the thickness of the electrolyte film or a longer electrolyte path, would improve the sensitivity and the time rate of response characteristics. This mode of action lowers the background current which arises from a number of sources. One possibility is the reduction of oxygen diffusion from the electrolyte reservoir to the cathode. However, there is a practical limit to this impedance. It is established when the current is so low that oxygen reduction is no longer controlled by diffusion.

In addition, an improvement of the time response characteristics can also be obtained by minimizing the electrolyte reservoir, but this has a serious drawback: the wear-out time of the electrode is drastically shortened because a large volume is needed to compensate the solvent loss by diffusion through the membrane and to dilute species from the anode reactions which may, in the case of a silver anode, deactivate the cathode. To keep the thickness of the electrolyte film within desired limits, either a spacer may be applied or the surface of the inner body be roughened.

The problem of maintaining a well-defined geometry is even more pronounced in miniaturized cells which are often desirable to control biological systems. The use of a porous material, such as ceramic, soaked with electrolyte, helps to overcome this problem [3].

A very unconventional electrolyte was developed [7] where, instead of a liquid electrolyte, a solid ion exchanger membrane is used. The ion exchange material (e.g., quaternary ammonium compounds) is homogeneously incorporated in a polymer matrix. The hydroxide produced at a porous cathode by oxygen reduction is conveyed, under the influence of an electrical field, across the membrane to the anode where it reacts with oxidized silver.

5 pH Value of the Electrolyte

The chemistry of the cathodic dissolution of oxygen is extremely complicated where peroxide is an important intermediate. The overall cathodic process

$$O_2 + 4\,H^+ + 4\,e^- \quad \rightarrow 2\,H_2O \tag{5}$$

may be separated into two processes which in turn are only overall reactions:

$$O_2 \quad\quad + 2\,H^+ \ + 2\,e^- \ \rightarrow \ H_2O_2. \tag{6}$$

$$H_2O_2 \quad + 2\,H^+ \ + 2\,e^- \ \rightarrow \ 2\,H_2O. \tag{7}$$

Despite the general great interest in research into the mechanisms of the oxygen reaction, it has not yet been possible to throw light upon this situation [2]. It is therefore not worthwhile to discuss here the numerous propositions for reaction mechanisms.

It is remarkable that, in most publications on reaction mechanisms, the proton is involved, so that the pH value will partly determine which reaction path will be favored. Depending on the relative rates of these processes four or fewer electrons will be required for the cathodic reduction of an oxygen molecule. It is assumed that the four-electron reduction dominates at elevated pH values, causing a higher current [5]. On the other hand, with an increasing pH value, the solubility of oxygen becomes smaller, resulting in a decreasing current.

Whether the salting-out effect or an alteration of the reaction mechanisms is mainly responsible for a change in the read-out if the alkalinity of the electrolyte is raised was discussed by Hitchman [4]. Experiments show that the current steadily drops with increased alkalinity of the electrolyte. If it is true that the four-electron reduction of oxygen is dominant in alkaline solutions, it is tempting to attribute the dominating effect to the changing diffusion conditions in the electrolyte at the cathode surface.

It follows from the above that the pH value plays an important role in achieving a stable read-out. A steady state presupposes a constant gradient of the hydroxide ions near the cathode. There the reduction of oxygen will always occur in an alkaline medium, apart from an initial short period when the detector is first used. In a rather acidic or in a buffered electrolyte, it will take considerable time for the whole electrolyte volume to follow suit in achieving a stable read-out.

Hitchman [4] has measured the switching-on transients for POS with electrolytes of different pH values (pH 7; 13.7; 14.0) and recognized a considerable effect on the time rate of response characteristics. Alkaline electrolytes showed a shorter polarization time. On the other hand, our own experiments with electrolytes of different pH values (10 to 13) yielded no significant effect. The rather contradictory results demonstrate that further parameters must be considered.

While an elevated pH value may shorten the polarization time it also has a drawback: Bergman [1] relates the interference of CO_2 to strongly alkaline electrolytes. He recommended a saturated solution of $KHCO_3$ to lower the effect. However, alkaline electrolytes are in general accepted for POS.

6 Final Observations

The interrelation between geometrical factors and the electrochemistry of the electrolyte makes every POS a rather complex system and no general formula exists for its optimization. An improvement of one property of the POS often affects one of its other properties detrimentally. This means that the development of a POS must always be adapted to the specific measuring problem.

References

1. Bergman I (1970) Improvements in or relating to membrane electrodes and cells. US Patent 1 200 595
2. Erdey-Gruz T (1975) Kinetik der Elektrodenprozesse. Akadémiai Kiado, p 273
3. Friese P, Rösel K-H, Schneiderreit R, Wessler G-R (1975) Meßsonde zur schnellen quantitativen electrochemischen Bestimmung von Gasen, insbesondere von Sauerstoff in flüssigen und/oder gasförmigen Medien. DDR, Patentschrift 114462
4. Hitchman ML (1978) Measurement of dissolved oxygen. John Wiley & Sons and Orbisphere Laboratories, New York, Maine, p 84, 92, 271
5. Hoare JP (1968) The electrochemistry of oxygen. Interscience Publ, New York, p 148
6. Lucero DN (1969) Design of membrane-covered polarographic gas detectors. Anal Chem 41:613
7. Niedrach LW, Stoddard WH (1973) Sensor with ion exchange resin electrolyte. US Patent 3719575

8. Nösel H (1978) Technologie und Methodik zur Messung des Gelöst-Sauerstoffs mit Membran-Elektroden. Selbstverlag Wissenschaftlich-Technische Werkstätten GmbH, Weilheim, p 26
9. Radhakrishna MN, Roggenkamp RL (1978) CO_2 interference free O_2 electrode. US Patent 4078981
10. Scheidegger AE (1960) The physics of flow through porous media, ch 3. Univ Press, Toronto
11. Schmid M, Mancy KH (1969) The electrochemical determination of dissolved oxygen in water in the presence of hydrogen sulfide. Chimia 23:398
12. Scott TF, Brushwyler GR (1976) Dissolved oxygen cell. US Patent 3997419
13. Tindall GW, Cadle SH, Bruckenstein S (1969) Inhibition of the reduction of oxygen at a platinum electrode by the deposition of a monolayer of copper at underpotentials. J Am Chem Soc 91:2119

Chapter I.6   The Action of Hydrogen Sulfide on Polarographic Oxygen Sensors

J.M. Hale[1]

1 Introduction

It is often necessary to measure oxygen concentrations in water containing dissolved hydrogen sulfide (Chap. III.1). Conventional polarographic oxygen sensors (POS) fail within a very short time under such conditions, for a variety of reasons to be described in this article. The principle of the modified sensor made by Orbisphere which may be exposed to high concentrations of H_2S is also described.

2 The Chemistry and Electrochemistry of Solutions of Hydrogen Sulfide

In aqueous solutions, hydrogen sulfide enters into the following equilibria:

$$H_2S = H^+ + HS^- \qquad\qquad K = 1.1 \times 10^{-7}$$
$$HS^- = H^+ + S^{2-} \qquad\qquad K = 10^{-14}.$$

Hence the predominant species present in solutions of pH less than 7 is undissociated H_2S, in solutions of pH between 7 and 14 it is HS^-, and in solutions of pH greater than 14 it is S^{2-}.

   All of these species can be oxidized, in the first place to elemental sulfur, at an "indifferent" metal electrode (that is, an electrode which acts only as a reservoir of electrons, as distinct from an electrode which itself participates in the electrochemical reaction) viz:

$$H_2S = S + 2\,H^+ + 2\,e^- \qquad\qquad E_o = +0.141 - 0.059 \text{ pH}$$
$$HS^- = S + H^+ \;\; + 2\,e^- \qquad\qquad E_o = -0.065 - 0.029 \text{ pH}$$
$$S^{2-} = S + 2\,e^- \qquad\qquad\qquad E_o = -0.48 \text{ V}.$$

Further oxidations occur at more positive potentials to oxyanions, but these reactions are irrelevant in the context of POS.

---

1 Orbisphere Laboratories, 3, Chemin de Mancy, CH–1222 Vesenaz, Geneva, Switzerland

Polarographic Oxygen Sensors (ed. by Gnaiger/Forstner)
©Springer-Verlag Berlin Heidelberg 1983

At nonindifferent electrodes (Chap. I.5), metal sulfides are formed at lower, that is more negative, potentials than the oxidation potential for sulfide at an indifferent electrode:

$$2 \text{ Ag} + \text{S}^{2-} = \text{Ag}_2\text{S} + 2 \text{ e}^- \qquad E_o = -0.69 \text{ V}$$
$$\text{Zn} + \text{S}^{2-} = \text{ZnS} + 2 \text{ e}^- \qquad E_o = -1.44 \text{ V}$$
$$\text{Cd} + \text{S}^{2-} = \text{CdS} + 2 \text{ e}^- \qquad E_o = -1.21 \text{ V}$$
$$\text{Pb} + \text{S}^{2-} = \text{PbS} + 2 \text{ e}^- \qquad E_o = -0.98 \text{ V}$$
$$2 \text{ Tl} + \text{S}^{2-} = \text{Tl}_2\text{S} + 2 \text{ e}^- \qquad E_o = -0.96 \text{ V}$$
$$\text{Pt} + \text{S}^{2-} = \text{PtS} + 2 \text{ e}^- \qquad E_o = -0.95 \text{ V}$$

These potentials and equilibrium constants are taken from Latimer (1952).

3 Interfering Effects of H_2S Upon the Operation of POS

From the previous section it is clear that the details of the interaction between H_2S and an electrochemical cell depend upon the specific choices of cathode and anode metals and of electrolyte composition.

Consider, as an example, the behavior of a POS constructed with a silver cathode, a lead anode, and containing an electrolyte of potassium hydroxide solution. The potential of the $Pb/HPbO_2$ electrode is -0.54 V, and such a sensor is normally operated in the "galvanic" mode, that is, the applied voltage is zero. Then, assuming no "ohmic" loss of potential in the electrolyte, the cathode potential is also at -0.54 V versus a normal hydrogen electrode.

Hydrogen sulfide dissolved in the alkaline electrolyte solution is present as HS^- or S^{2-}, and is not oxidizable to elemental sulfur at either the cathode or the anode of the cell, because the potentials of both of these electrodes are too low in this example. However, it can enter into electrochemical reactions at both of these electrodes with participation of the metals. This happens because the potentials of both the silver and the lead electrodes are more anodic than the corresponding thermodynamic potentials for the formation of silver and lead sulfides.

The effect of the hydrogen sulfide upon the current output from the sensor also depends upon several factors. If the anode is initially unaffected, because it is remote from the membrane through which the H_2S enters the cell, then the negative current from the oxidation reaction occurring at the cathode surface subtracts from any oxygen reduction current which may be generated, and so produces an anomalously low or even negative signal. Also, the formation of an insoluble sulfide film on the cathode surface interferes directly with the oxygen reduction reaction, so that the sensor produces erroneously low readings even after the H_2S has been eliminated from the system.

When the anode is also accessible to the H_2S, the situation becomes more complicated, because the formation of lead sulfide as an anode product is thermodynamically favored over the formation of the plumbite ion and the anode potential may therefore shift to -0.98 V from its initial value of -0.54 V. This means that the cathode potential, which shifts concomitantly, might become sufficiently cathodic to enable the reduction of water to hydrogen. Erroneously *high* oxygen signals are observed in this case. The anode potential shifts if the Pb/PbS reaction can supply all of the current demanded by the cathode reaction. If it cannot, a mixture of PbS and $HPbO_2^-$ products are formed at the anode but the potential is then that of the $Pb/HPbO_2^-$ couple.

These effects of H_2S, which were described in connection with the example of a silver cathode and lead anode, are also found with other choices of electrodes, although silver is the most severely tarnished of the metals commonly chosen as cathodes in POS.

4 An H_2S-Resistant Sensor

Orbisphere Laboratories manufactures a sensor which is totally insensitive to H_2S, and which therefore is ideal for monitoring oxygen in those biologically interesting situations where oxygen concentrations are low and H_2S concentrations high. It has a gold cathode, which does not tarnish in the presence of S^{2-}. The electrolyte is composed of alkaline sodium sulfide solution, and the anode is of silver; hence the anode couple is Ag/Ag_2S, having a potential of -0.69 V(NHE). The cathode is held at a potential of -0.1 V relative to this anode. The entry of hydrogen sulfide into the sensor does not change the composition of the electrolyte significantly, hence no changes of potential occur at the electrodes. The sensor has been shown to operate quite normally even in H_2S-saturated water.

Reference

Latimer WJ (1952) The oxidation states of the elements and their potentials in aqueous solutions, 2nd edn. Prentice Hall Inc, Englewood Cliffs, NJ

Chapter I.7   A Double-Membrane Sterilizable Oxygen Sensor

H. Bühler[1]

1 Introduction

Oxygen plays an important role in the growth and metabolism of microorganisms. For many organisms oxygen is essential to life, others can grow only in its complete absence and intermediate types can grow with or without it. The decisive factor in an aerobic process is the rate of oxygen supply and dissolution. A proper understanding of oxygen transfer phenomena thus calls for the measurements of the oxygen dissolved in the fermentation solution.

The oxygen content in a fermentation broth can be determined using standard chemical methods as well as polarographic cells. But these methods do not permit measurement in situ and are tedious and time-consuming. The electrochemical method using the Clark polarographic oxygen sensor (POS) is therefore very promising for the control of fermentation processes.

The basic operation of a POS is relatively simple. The active electrode, the cathode at which the oxygen is reduced, is separated from the test medium by a thin layer of electrolyte and a gas-permeable membrane. The amount of current produced at a constant temperature is proportional to the partial pressure of oxygen in the measuring solution. This condition is largely met by a suitable electrode design and a proper polarization voltage between anode and cathode.

Fermentation sensors must withstand the conditions of steam sterilization, i.e., temperatures between 120°C and 135°C and the corresponding water vapor pressures. Therefore sturdy POS capable of operating under these adverse conditions was designed.

2 The Concept of the Double Membrane

The detailed theory of POS has been described by many authors (Hitchman 1978, Krebs and Haddad 1972; Chaps. I.1, I.2). Some characteristics of oxygen sensors must,

---

1 Dr. W. INGOLD LTD., Industriezone Nord, CH–8902 Urdorf/ZH, Switzerland

Polarographic Oxygen Sensors (ed. by Gnaiger/Forstner)
©Springer-Verlag Berlin Heidelberg 1983

however, be borne in mind if the design of the sterilizable POS described here is to be understood.

The current, I_1 of a POS depends on various parameters,

$$I_1 = K \times \frac{A \times D_m \times S_m \times p_{O_2}}{z_m}, \tag{1}$$

in which K is a constant; A is the cathode surface; D_m, S_m, and z_m are the diffusion coefficient, the solubility coefficient, and the thickness of the membrane respectively; and p_{O_2} is the oxygen partial pressure.

The response time is proportional to z_m^2/D_m, and thin, highly permeable membranes are of advantage. On the other hand, the flow dependence of the electrode current should be small, which is the case with low electrode currents. POS with a short response time and low flow dependence therefore have a thin, medium-permeable membrane and a small cathode. The stability of the electrode current, however, grows with increasing cathode surface. These facts reveal that no ideal POS can be developed.

All common POS for measurements in water have a relatively large cathode and a thin membrane in order to achieve good long-time stability and fast response. The flow dependence of the current is large and measurement must be effected in stirred solutions. This compromise is not applicable to sterilizable POS. Mainly the thin membrane (z_m between 10 and 25 μm) is problematic because it will not withstand large pressure differences. In addition, the sterilization temperature of at least 120°C causes a pronounced irreversible change in the thickness of the membrane. Furthermore, the tension of the membrane against the cathode will decrease, causing higher zero currents and inferior linearities.

A relatively thick homogeneous membrane of sufficient mechanical strength is not applicable due to intolerable response times. A patented double-layer membrane is by far best suited for sterilizable POS. The inner current-determining membrane is a 25 μm Teflon film. The outer membrane, approximately 150 μm thick, is a highly permeable silicone membrane reinforced by a thin stainless steel mesh.

Equation (2) gives the current for a POS with a double membrane:

$$I_1 = K \times \frac{A \times p_{O_2}}{z_m/P_m \text{ (Teflon)} + z_m/P_m \text{ (silicone)}}. \tag{2}$$

z_m/P_m may be called the mass transfer impedance for the oxygen molecule where P_m is the permeability coefficient [mol m^{-1} s^{-1} kPa^{-1}] (Chap. I.3).

The permeability of silicone is roughly 90 times higher than Teflon. Taking the above-mentioned thickness of the double membrane, the two mass transfer impedances can be easily compared

$$z_m/P_m \text{ (Teflon)} = 15 \times z_m/P_m \text{ (silicone)}. \tag{3}$$

About 94% of the total mass transfer impedance is thus located within the thin Teflon membrane. In other words, the current is only slightly reduced by the thick silicone membrane.

The good behavior of a POS with a 250 μm platinum cathode and this double membrane has been proved using an autoclave. The sterilization tests (121°C/20 min) are summarized in Table 1.

Table 1. Mean values of 2 POS before and after 10 and 16 sterilizations

	Sterilizations		
	0	10	16
Current in air (20°C)	39.3×10^{-9} A	48.6×10^{-9} A	52.7×10^{-9} A
Current in nitrogen	0.03×10^{-9} A	0.08×10^{-9} A	0.1×10^{-9} A
Response time (98%) for nitrogen-air	51 s	51 s	49 s

These POS normally indicate higher currents after sterilization, which is probably attributable to a slightly reduced membrane thickness. The change in the electrode output is highest during the first two sterilizations. The stable currents in nitrogen and the unchanged response times strongly suggest that the important thickness of the captive electrolyte film between cathode and Teflon membrane does not change significantly.

Another important property of a POS with such a double membrane is its greatly reduced flow sensitivity of the current. The phenomenon of flow sensitivity is related to the ease with which the sample can provide the oxygen required by the electrode. Its magnitude is easily determined by measuring the sensor current in agitated and stagnant water.

The degree of flow dependence increases directly with the cathode area, membrane permeability and, inversely, with membrane thickness. These factors, which tend to produce a high output current and a short response time, also tend to increase flow sensitivity. The outer silicone layer of the double membrane acts as an oxygen reservoir or stagnant layer with high oxygen-dissolving power. The thick silicone membrane therefore improves the supply of oxygen reduced by the cathode in stagnant water. This fact is clearly demonstrated by the following experiment:

A POS with a 250 μm platinum cathode has been evaluated first with the 25 μm Teflon membrane and then with the double membrane (25 μm Teflon + 150 μm silicone):

Other POS of low flow sensitivity generally have a micro- or multi-cathode. These sensors are difficult to manufacture and show unsatisfactory long-term stability. But due to their fast response they are normally used in blood-gas analyzers (Chap. II.10). The problem of stability is eliminated by frequent recalibration. Sensors with the double membrane have a stable electrode output because of the rigid membrane and a medium size cathode.

Table 2. Parameters of a POS with a single membrane and one with a double membrane in air-saturated water at 25°C

	Single membrane	Double membrane
Current in agitated water	54.6×10^{-9} A	50.3×10^{-9} A
Current in stagnant water	42.3×10^{-9} A	48.9×10^{-9} A
Flow dependence	22.5%	2.8%
Response time (95%) for nitrogen-air	20 s	50 s

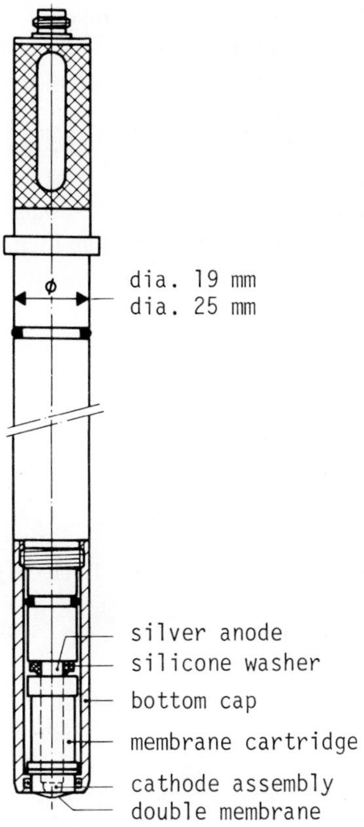

Fig. 1. Sterilizable oxygen sensor

dia. 19 mm
dia. 25 mm

silver anode
silicone washer
bottom cap
membrane cartridge
cathode assembly
double membrane

Besides fermentation processes there are other applications of the double membrane sensor. Special importance attaches to measurements where stirring is undesirable:

- Oxygen measurements in sediments (Chap. III.2);
- Oxygen profiles in lakes where sharp changes of the oxygen concentration occur (Chap. III.1);
- Respiratory studies on living plants and animals (Part II).

3 Design of the Sterilizable POS

The electrodes (Fig. 1) are contained within a sturdy stainless steel housing. The cathode assembly, which is made of glass, is reinforced by the tubular silver anode. The disposable membrane cartridge with the prefixed double membrane is partly covered by a silicone rubber sleeve which allows expansion of the electrolyte at high temperatures. The stainless steel bottom cap presses the cartridge against the silicone washer and provides the correct pressure of the double membrane against the cathode. Silicone

O-rings are used as seals in critical locations. The sensor contains a built-in temperature compensating element. It responds rather slowly to temperature changes (several minutes). This is not a serious disadvantage; most fermentations are carried out at a constant temperature.

The large anode, together with a small electrode output, guarantees an almost unlimited life time of the sensor. The membrane cartridges have to be replaced after several sterilizations. Despite the reinforcing stainless steel mesh the tension of the double membrane will slightly decrease because of the large differences of pressure and temperature during steam sterilization. Nevertheless some users get 15 or more sterilizations with the same cartridge.

4 Sensor Specifications

Cathode-diameter:	Pt; 0.25 mm
Anode-area:	Ag/AgCl; 5.5 cm^2
Electrolyte:	0.05 mol dm^{-3} KCl; 0.3 mol dm^{-3} KNO$_3$; pH = 10 (KOH)
Polarization voltage:	0.67 V
Membrane:	Teflon-silicone double membrane
Response time (95%):	45–60 s
Long-time stability:	better than 2% per week at constant temperature and pressure
Flow sensitivity:	2%–5%
Current in air:	40–80 × 10^{-9} A (20°C)
Current in nitrogen:	less than 1% of the air current
Temperature range:	0° to 130°C
Pressure range:	0 to 600 kPa

References

Hitchman ML (1978) Measurement of dissolved oxygen. John Wiley & Sons, New York, p 255

Krebs WM, Haddad IA (1972) The oxygen electrode in fermentation systems. Ind Microbiol 13: 113–127

Chapter I.8 Construction of a Polarographic Oxygen Sensor in the Laboratory

T.J. Mickel, L.B. Quetin, and J.J. Childress[1]

1 Introduction

One important advantage of constructing polarographic oxygen sensors (POS) in the laboratory is the potential of custom-designing a sensor to suit the specific requirements of an experimental design or apparatus. The final dimensions of the POS, its approximate output, and ability to withstand submergence and high hydrostatic pressure can be controlled. In addition the cost is reduced, an important consideration if many POS of different design are required.

The POS we describe here consists of a platinum cathode and a silver-silver chloride anode. Its body is cast of epoxy resin and the tip shaped to accept an oxygen-permeable membrane that encloses the anode and cathode in an electrolyte solution.

2 Design of the POS

Two important criteria must be considered when designing a POS for a specific application: (1) the final dimensions and construction of the sensor body, and (2) the diameter of the cathode. If the membrane O-ring of the POS forms the seal between the sensor and the sensor port in a respirometer, the dimensions of both must be precisely matched. The diameter of the cathode affects the output of the POS as well as the magnitude of the stirring effect (Chaps. I.1, I.4). We construct one version of POS with a 0.5 mm (24 ga.) diameter cathode (macro-cathode) that has an output of approximately 2×10^{-6} A at a p_{O_2} of 21.3 kPa if covered with a 0.025 mm thick polypropylene membrane. A 0.0152 mm micro-cathode covered with the same membrane yields an output of about 2×10^{-9} A. While POS with macro-cathodes yield signals large enough to be recorded without preamplification (Chap. I.10), a picoammeter is needed to amplify the output of POS with micro-cathodes. However, the advantage of the micro-cathode POS is the small stirring effect which may be an important consideration for experimental design. For example in respirometers many aquatic invertebrates cannot tolerate the vigorous stirring needed for macro-cathode POS (Part II).

---

1 Marine Science Institute, University of California at Santa Barbara, Santa Barbara, CA 93106, USA

Polarographic Oxygen Sensors (ed. by Gnaiger/Forstner)
© Springer-Verlag Berlin Heidelberg 1983

3 Construction of the POS

We will describe the simplest method we have found to construct both macro- and micro-cathode POS. Construction can be divided into the four following steps: (1) construction of the cathode by sealing a platinum wire in glass; (2) connection of the cable, cathode, and silver wire anode; (3) casting the cathode-anode assembly in epoxy resin; and (4) shaping the sensor to the desired form.

3.1 The Cathode

The cathode is 99.99% pure platinum wire (0.5 mm, 24 ga.) sealed in 3 mm outer diameter flint glass tubing. Flint glass tubing has a melting point low enough to allow it to be easily closed in a bunsen burner flame. We have had problems in the past using harder glass tubing with higher melting points. It cracks upon cooling when a large diameter cathode is used and it is difficult to seal the fine diameter wires in the tubing without melting the platinum.

The cathode is constructed in the following way. One end of a glass tube is heated in a flame until the end begins to seal. Before it closes, a 1.0- to 1.5-cm long piece of platinum wire is inserted into the tube and the tube closed by slowly turning it in the flame. We seal the glass around 75% of the length of the platinum wire. The glass tubing is then cut several cm above the free end of the platinum wire (Fig. 1a). The sealed tip is ground with successively finer grades of emery cloth and finally polished with crocus paper to produce a smooth rounded tip.

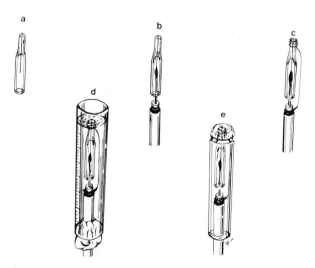

Fig. 1a–e. Construction of a POS: **a** cathode, **b** cathode soldered to co-axial cable, **c** cathode-anode assembly, **d** casting the oxygen sensor, **e** completed POS

3.2 Cable and Anode Connection

The cathode is attached to co-axial cable with a single center wire and surrounding wire shield in either of two diameters, 2.54 mm or 4.95 mm (RG 58 c/U Type 6058C1A and RG 174/U Type 61741A, Dearborn Wire and Cable Co, Chicago, ILL). The external insulation, shield wire, and insulation of the center wire of the cable are removed to expose a length of the center conductor long enough to fit into the open end of the glass cathode tube and contact the platinum wire. A small piece of electrical solder is placed in the end of the tube followed by the tinned center conductor of the cable. The tube is briefly heated in a flame until the platinum cathode and center conductor are soldered together (Fig. 1b). The anode (99.99% pure silver, 1.25 mm diameter, 16 ga.) is now attached to the shield wire with solder. The silver wire is then wound around the tip of the cathode assembly, securing it in place (Fig. 1c).

3.3 Casting the Sensor

The cathode-anode assembly is cast in epoxy resin in a mold determining the approximate size and shape of the POS. To ensure a good bond the cathode-anode assembly is cleaned thoroughly with organic solvents to remove any grease. The epoxy body of the sensor covers and insulates all the exposed wire from the last 2 cm of the co-axial cable to the tip of the cathode-anode assembly. The seal between the co-axial cable and epoxy body is especially important if the POS is to be immersed at high hydrostatic pressure.

Various kinds of molds can be devised for casting POS. One of the simplest is made from a disposable plastic syringe body (2.5 or 5.0 cm^3). The end of the syringe is cut off, leaving a plastic tube as the mold. The tube is placed over the cathode-anode assembly and then sealed to the co-axial cable with silicone aquarium cement. The cathode-anode assembly thus is enclosed and centered in the tube with the tip of the assembly near the open end of the mold. After the silicone aquarium cement has dried, epoxy resin is mixed, degassed under a vacuum of 7 kPa for 20 min, and then poured into the tube to a level just below the tip of the cathode-anode assembly and allowed to harden (Fig. 1d). We use either Hysol epoxy resin and hardener (R9-2039, HD 3561; K.R. Anderson Co. Inc., Mountain View, CA, USA) or TRA-CON epoxy resin (Type 2113; K.R. Anderson Co. Inc.).

3.4 Shaping the Sensor

Once the resin has hardened, the oxygen sensor can be removed from the mold by peeling off the silicone aquarium cement and gently squeezing the plastic syringe barrel to break the seal between the resin and the plastic barrel. The sensor should then slide out of the mold easily. The tip must be rounded and polished with successively finer grades of emery cloth to form the end of the sensor and assure that the anode and cathode are exposed. The next step is to carve an O-ring groove near the tip of the sensor using either a lathe or file and sandpaper. Care must be taken not to put too much stress on the sensor and separate the glass-epoxy bond. When this step is completed (Fig. 1e) the oxygen sensor is ready to be tested.

4 Using the Self-Constructed Oxygen Sensor

4.1 Membranes and Electrolyte Solutions

The sensor is held in a vertical position with the membrane poised next to the barrel, a few drops of electrolyte solution are placed on the tip, and the membrane is rapidly but smoothly stretched over the tip of the sensor and secured with an O-ring. Excess membrane may be trimmed off from around the O-ring. It is important that the membrane be tight and smooth with no gas bubbles trapped beneath it.

The materials used most frequently for POS membranes are polyethylene and Teflon. We usually use polypropylene because we find it easier to obtain. The characteristics of these materials are discussed in Chap. I.1.

The electrolyte solution is potassium chloride dissolved in water. A half-saturated solution is used with macro-cathode sensors and a 3% w/v solution works well with micro-cathode sensors (Chap. I.5).

4.2 Troubleshooting the Self-Constructed POS

Several POS are usually constructed simultaneously. Some may operate poorly, or not at all, due to inconsistencies in workmanship or materials. The most obvious problem that could occur when calibrating or testing a newly constructed POS is that it will not respond to changes in oxygen tension or has no output at all. The electrical connections between the POS and the recording device should first be checked. The polarizing voltage power supply should be checked and one should make sure that the cathode is polarized negatively with respect to the anode. A problem that sometimes occurs in a self-constructed POS is that the soldered electrical connection between the co-axial cable and the cathode is poor. This can be tested by measuring the resistance across the leads from the sensor. If the resistance is infinite, then the solder joint is bad and the POS must be discarded.

Another symptom of trouble with a POS is that the output will not approach zero when the sensor is placed in an oxygen-free medium (99.99% N_2 saturated) or the output will drift or slowly increase during zero calibration. The epoxy resin of the sensor body sometimes absorbs water if immersed for long periods of time. This leakage reduces the insulating quality of the epoxy resin and the resulting current causes high zero currents. A remedy for this type of drift is to dry the POS at 60°C for at least 24 h. If the zero current slowly increases it may also indicate a crack in the insulating glass. These cracks are sometimes produced during construction of the cathode assembly. In this case the POS should be discarded. Drift may also occur when calibrating the POS at constant p_{O_2}. This does not necessarily mean that the sensor is defective. If it "zeros" well, has a reasonable output, and responds to changing oxygen tensions, the problem may be due to the membrane (Chap. I.1). Proper application of the mem-

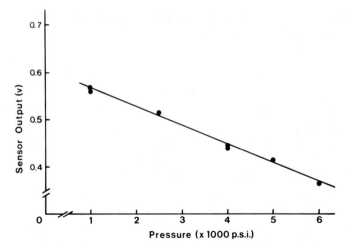

Fig. 2. The output response of a macro-cathode POS to hydrostatic pressure

brane so that it is tight and has no wrinkles requires practice. Also if the tip was not properly prepared during construction it may have sharp areas that will puncture the membrane as it is applied. The tip should be checked and a new membrane applied. Another symptom of poor membrane application is an electrically "noisy" output from the POS. In our experience membrane application has been the major cause of a POS operating poorly. The membrane should always be changed as the first step in trying to improve the operation of a POS.

The output of our POS decreases under increased hydrostatic pressures (Fig. 2). We suspect that it may be due to a decrease in membrane permeability with pressure. Oxygen solubility changes little with pressure and the p_{O_2} increases with pressure which would be expected to increase POS output (App. A). However, unless the POS drifts at high pressure the lower output poses little problem and merely requires that a calibration curve be constructed whenever a new membrane is applied. POS drifts at high pressure may be due to improper membrane application or the absorption of water by the POS body which is accelerated by high hydrostatic pressure.

5 Conclusion

With every manufacturing process products have to be tested and, depending on the complexity of the process, a certain percentage will not meet specifications and will have to be rejected. Commercially obtained POS have gone through this screening procedure, but self-constructed POS must be screened in the laboratory. We are successful with at least 85% of our POS, an acceptable rate of discard when balanced by the advantages of being able to design a sensor to specific requirements at very low cost.

Chapter I.9   A Polarographic Oxygen Sensor Designed for Sewage Work and Field Application

W. Grabner[1]

Numerous designs of polarographic oxygen sensors (POS) have been described in the literature [1] and it may at first sight seem superfluous to add to this extensive list. A survey of the sensors available from various manufacturers reveals no striking differences in design. Most sensors have a gold or platinum cathode and an Ag/AgCl anode connected by an electrolyte. The choice of materials for the electrodes and the electrolyte is quite independent of sensor geometry; optimizing these parameters is not the aim of this article. I shall restrict myself, as a designer, to the geometry.

Let us consider the most important attributes expected of a POS designed for sewage work and field applications: (a) long term stability; (b) insensitivity to poisoning; (c) rapidity of response; (d) effective temperature compensation; (e) easy handling; (f) minimum stirring requirements to avoid mixing effects; and (g) insensitivity to particulate matter in the sample.

Points (a) to (c) are determined by the choice of the electrode material, membrane and electrolyte (Chaps. I.1, I.2, I.7).

d) Temperature transducers are necessary for compensating changes of membrane permeability and oxygen solubility with temperature (Chap. I.10). Ideally the temperature variation of the membrane should be completely compensated by the output of the temperature sensor. This is not easy to achieve, since the temperature of the membrane is affected both by the surrounding medium and the temperature of the sensor body. If a POS is exposed to a temperature change, a temperature gradient is built up across the membrane. Complete temperature equilibrium is only achieved when the cathode has reached the temperature of the environment. The most accurate temperature compensation would therefore seem to be provided by measuring the temperature of either the cathode or the surrounding electrolyte.

e) Filling a sensor with electrolyte and applying a new membrane is a considerable problem for the untrained. An even tension is important for the operation of the sensor (Chap. I.1) and every designer has introduced modifications aimed at overcoming problems connected with handling.

f) Due to the low diffusion rate of oxygen in water the concentration around the cathode decreases as oxygen is consumed. Reducing the size of the cathode diminishes this effect but increases the problem of signal conditioning as output current dimin-

---

1 Technisches Büro, Nußdorferstraße 4/11, A–1090 Wien, Austria

Polarographic Oxygen Sensors (ed. by Gnaiger/Forstner)
© Springer-Verlag Berlin Heidelberg 1983

ishes correspondingly. Attempts to reduce oxygen depletion by pulsing the polarizing voltage did not give satisfying results an account of transient effects which are not, as yet, well understood. The use of a ring-shaped cathode increases the current per unit cathode area by minimizing diffusion effects [1]. Of course the best way to overcome the above problems is to produce a vigorous water flow around the membrane by means of a stirrer. Moving the POS in the water by hand in order to produce adequate stirring hardly gives reproducible results and should be avoided. In many cases mixing of water should be minimized, for instance in the analysis of oxygen microstratification (Chap. III.1). Stirring destroys the structure of the layers and may lead to values averaged over a thickness of up to 50 cm.

As far as hydrodynamics is concerned a flat, disk-shaped sensor face with a flow normal to the cathode area is the worst choice, since the velocity of the medium is lowest near the cathode where it should be maximal.

g) Another source of difficulty in the use of POS in sewage and natural waters is larger particulate matter. Mud and small particles do not impair the function of a sensor as long as the diffusion rate of oxygen is high enough in the layers above the membrane. Leaf debris, pieces of grass, or other large particulate matter severely affects the signal if parts of the cathode are covered. The use of a large cathode greatly reduces the probability that a significant portion of it will be covered.

A consideration of all the points mentioned above, especially those concerning hydrodynamics, suggests a design quite different from the standard POS. Priorities are hard to establish, but it appears that Tödt [2–4] was the first to use the polarographic method for the determination of oxygen in natural waters. Around 1934 he developed the so-called Sauerstofflot. It consisted of a hollow cylinder, with an inner cross-section tapering toward the middle, similar to a venturi tube. The two electrodes were situated on the inside of this cylinder at the narrowest point. By raising and lowering this cylinder, water entered the wide opening and was accelerated in the construction, where the electrodes were situated. The Sauerstofflot had bare electrodes and it would have been very difficult indeed to cover these electrodes with a membrane. This was, nevertheless, the basic idea I adopted in designing my POS. A tapering cross-section was produced by mounting a cone within a hollow cylinder. A stirrer forces water through this construction, maximum flow occurring near the base of the cone where the cross-section is smallest. A ring-shaped cathode is placed just ahead of this constriction in the region of maximum flow. Figure 1 shows the sensor assembly together with the stirrer. The ring-shaped cathode is covered by a cone-shaped membrane made of polythene or PTFE. The system is sealed with an O-ring. For measurements at greater depth, the use of a second membrane ensures that the pressure inside and outside the POS is practically the same. The POS can be filled with electrolyte through two holes which become accessible removing the filling cap. Due to the highly effective pressure compensation, completely air-free filling is not necessary. The disadvantages of this more complicated design have to be borne by the manufacturer: a cone-shaped membrane has to be produced by welding polyethylene foils or sintering PTFE films. On the other hand, this unusual shape of membrane is an advantage for the user, since the membrane exactly fits the sensor and no stretching is necessary. The pressure of the membrane on the cathode is determined mainly by the inherent tension of the membrane during molding, and this is largely independent of handling during applica-

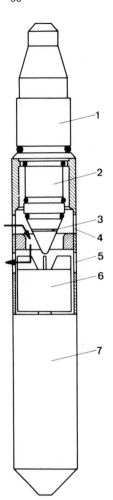

Fig. 1. Conical POS and stirring chamber. *1* filling cap, *2* membrane for pressure compensation, *3* ring cathode, *4* water inlet, *5* water outlet, *6* centrifugal pump, *7* motor with magnetic clutch

tion of the membrane. The anode and the electrolyte reservoir are contained within the body of the sensor, small channels leading to the surface of the cone. In this way the membrane is supported over its entire surface. The surface of the membrane can be cleaned with tissue paper without damage. This design combines the high mechanical stability of a fully supported membrane with the large electrolyte reservoir necessary for long-term use.

The stirrer is a small impeller driven by a miniature electric motor via a magnetic coupling. It functions as a centrifugal pump: water is sucked in horizontally through holes in the side of the stirrer housing and is discharged again horizontally through similar holes further down. Thus, disturbances to the water layers around the POS are kept to an absolute minimum, since the overall pumping rate is very low. The stirrer is attached to the POS by a threaded fitting. Water-tight electric couplings connect the POS and the stirrer motor to a cylindrical housing that contains the electronics for signal conditioning and buffering the sensor output. Any influence of cable length on the measurement is thus avoided.

This new design of a POS with a ring electrode situated on a cone-shaped sensor head offers special advantages for application in sewage works and in the field.

References

1. Fatt I (1976) Polarographic oxygen sensor. CRC Press, Cleveland, 278 pp
2. Tödt F (1928) Korrosion and Reststrom I–III. Z Elektrochemie
3. Tödt F (1933) Verfahren zum fortlaufenden Messen und Anzeigen von in strömendem Wasser gelösten Sauerstoff. DRP. 66 30 80
4. Tödt F (1958) Elektrochemische Sauerstoffmessung. Walter de Gruyter, Berlin

Chapter I.10 Electronic Circuits for Polarographic Oxygen Sensors

H. Forstner[1]

1 Introduction

A polarographic oxygen sensor (POS) is a two-terminal chemoelectric transducer; a suitable polarizing voltage across the sensor leads produces an output current which, ideally, is proportional to the partial pressure of oxygen in the sample. In membrane-covered sensors this signal is quite small and current is first converted into a voltage and then amplified to drive either a meter or a strip chart recorder or for input to an analog-to-digital converter. Membrane-covered POS show a large dependence on temperature, the signal increasing by 1%–6% for a rise of 1°C [5]. Consequently measurements have to be made under thermostatic conditions or there has to be some method of compensating the temperature effect. If a readout in units of concentration is desired, a correction for the variation of oxygen solubility as a function of temperature (App. A) must also be made. A block diagram for a circuit for a temperature-compensated oxygen detector is shown in Fig. 1.

Oxygen meters with matching sensors, incorporating all necessary functions, are available from many manufacturers at reasonable cost. In general, self-construction is a waste of both time and money if a standard type of instrument is needed. Still there are applications in the laboratory where it is advantageous or even necessary to design an electronic circuit for a POS. Working in a well-thermostated environment tempera-

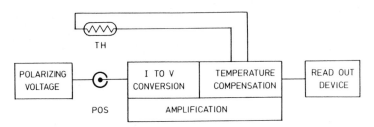

Fig. 1. Block diagram of a circuit for a temperature-compensated oxygen meter: *TH* temperature sensor

1 Institut für Zoologie, Abteilung Zoophysiologie, Universität Innsbruck, Peter-Mayr-Str. 1A, A–6020 Innsbruck, Austria

Polarographic Oxygen Sensors (ed. by Gnaiger/Forstner)

ture compensation is not needed. If in addition a sensitive strip-chart recorder is used, a very simple circuit will perform satisfactorily. A similar case is input to data-loggers or microprocessor based instruments (Chap. I.11). Sensor output and temperature are measured as separate variables, the necessary corrections for temperature compensation can be done numerically by software and the analog part of the circuitry can be kept straightforward. There are applications where special function circuits are needed which are not available commercially; for instance twin-sensor systems where a difference signal must be formed or systems with nonstandard grounding requirements.

In this article the basic functions of electronic circuits for operating a POS will be considered, examples of simple circuits will be given and the selection of suitable components will be discussed. Understanding the function of oxygen meters should also be helpful for troubleshooting and for evaluating the performance of existing instruments.

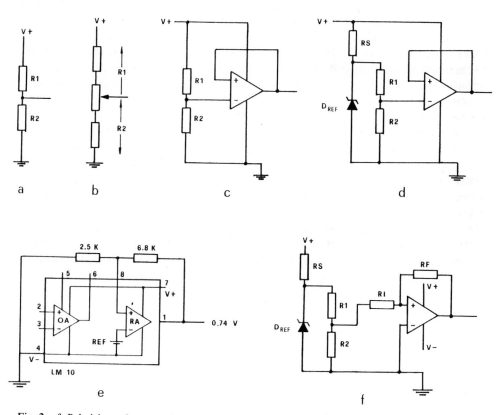

Fig. 2a–f. Polarizing voltage supply: a fixed voltage divider; b precision adjustable divider; c buffered reference; d buffered supply with reference diode; e LM 10 op amp and voltage reference; f generating a negative polarizing voltage from a positive reference

2 The Polarizing Voltage

In order for a POS to function as an oxygen-sensing device, a polarizing voltage of
0.7–0.8 V must be applied between the anode and cathode. In this range the sensor is
operating in the plateau region of the current-to-voltage curve for oxygen reduction
and steady-state output is fairly insensitive to small changes of the polarizing voltage
of the order of a few tens of millivolts. Fast transients will feed through and, therefore, short-term stability must be assured. Equipment operated from the mains usually
has a well-regulated power supply. With battery operation a stable primary cell, for
instance a silver-mercury battery, can be used. The required voltage can then simply be
derived from a resistive divider (Fig. 2a,b). According to Kirchhoff's law, the current
drawn by the electrode will influence the voltage at the node of the divider. To minimize this effect, transversal current through the resistors must be at least 100 times
the maximum electrode current. As this current amounts to a few microamperes at
most, a total resistance of 1–5 kΩ is usually adequate. If an operational amplifier buffer is added (Fig. 2c), a high resistance divider can be employed and power consumption is reduced. If no stable power supply is available, the constant voltage can be
supplied by a zener diode or a band-gap reference diode (Fig. 2d). Although based on
different physical effects, both devices are similar in their function. Band-gap references can be operated at much lower voltage and current levels and are available with
break-down voltages down to 1.2 V (INTERSIL ICL 8069, NATIONAL SEMICON-
DUCTOR LM 385). This is still more than the required 0.7–0.8 V and again a voltage
divider and a buffer amplifier are needed. The LM 10 device (NATIONAL SEMICON-
DUCTOR) combines a buffered reference and an operational amplifier in one package.
The built-in reference produces 200 mV, the required polarizing voltage is adjusted by
connecting appropriate feedback resistors on the buffer amplifier (Fig. 2e). The second amplifier in the package is available for other purposes in the system.

Normally, a POS is operated with the cathode at or near system ground and the
anode is connected to the positive reference voltage. With some types of POS, when
two or more sensors are operated in a common electrical environment, this can lead to
interference. The problem can be avoided by having the anode at ground potential,
but a negative polarizing voltage is then required. This can be done with a modification of the circuit in Fig. 2d by using an inverting amplifier configuration (Fig. 2f).

3 A Simple Oxygen Meter

With the exception of temperature compensation, all functions necessary for an oxygen meter are incorporated in the circuit shown in Fig. 3 [9]. The polarizing voltage is
supplied by the battery and voltage divider, current to voltage conversion is done by
the load resistor in series with the sensor and the voltage developed across this resistor
is further amplified in the recorder. The total system is not so simple as it may appear.
Actually the recorder is more complex than most oxygen meters, but if graphic output is required in any case it is only the additional circuitry which counts. For a num-

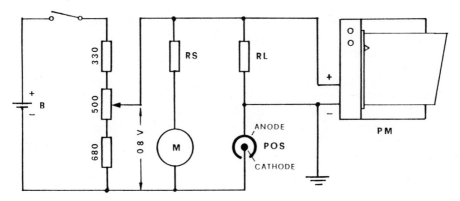

Fig. 3. A simple oxygen meter for output to a potentiometric recorder: B battery 1.35–1.5 V (silver-mercury cell is preferred, RM 502 R or ZM 9 C), the LM 10 op amp and voltage reference can be substituted for the divider (see Fig. 2e); M moving coil meter 100 μA; RS meter series resistance; RL sensor load resistor; PM potentiometric recorder

ber of reasons the load resistor must not be made too large and a value of 40 kΩ should not be exceeded. With a large source resistance the recorders response will become sluggish, dead-band error will increase and the circuit will become more sensitive to electric interference. The voltage across the resistor should also be restricted to less than 25 mV, because this voltage is subtracted from the polarizing voltage as seen by the POS. If the output signal varies, the polarizing voltage is affected correspondingly. For a sensor operated in the plateau region, this effect is not noticeable if a few millivolts are not exceeded.

4 Amplifier Circuits

The operational amplifier has become the universal building block in the design of analog circuits and the basic principles and various aspects of performance have been treated exhaustively [3, 4, 7, 8]. The application of operational amplifiers to POS systems has also been described [6] and here a short survey of commonly used circuits will be given.

The current amplifier (current-to-voltage converter) is the configuration most commonly used as input stage (Fig. 4a). The output voltage, U_o, is related to the input current I_1 by the expression:

$$U_o = -I_1 \times R_f, \tag{1}$$

where I_1 is the current from the sensor and R_f is the feedback resistance. Current-to-voltage conversion and amplification are achieved simultaneously. For a sensor current of 10 nA and a feedback resistor of 20 MΩ, the output voltage will be 200 mV, sufficient to drive a microprocessor-compatible analog-to-digital converter (ICL 7109 INTERSIL). As the anode of the sensor is connected to a positive voltage, current flows to the circuit and according to Eq. (1) a negative output voltage results. If a positive voltage is needed a second inverting amplifier must be added (Fig. 4c).

A method for achieving a high current gain without the use of a large feedback resistor is shown in Fig. 4b. The feedback resistor is connected to a voltage divider at the output (fractional feedback) and the gain is

$$U_o = -I_1 \times \left(R_f \times \frac{R_1 + R_2}{R_2} + R_1 \right) . \tag{2}$$

If R_f is much larger than R_1 and R_2 this reduces to

$$U_o = -I_1 \times R_f \times \frac{R_1 + R_2}{R_2} . \tag{3}$$

With $R_f = 2\ M\Omega$, $R_1 = 9.9\ k\Omega$ and $R_2 = 100\ \Omega$ the circuit of Fig. 4b is equivalent Fig. 4a with $R_f = 20\ M\Omega$. However, there is also a disadvantage: it has been shown that noise voltage at the output increases by the same factor by which the size of the feedback resistor is reduced [6].

A noninverting operational amplifier configuration is shown in Fig. 4d. The voltage follower with gain is the fractional feedback version of the unity-gain voltage follower of Fig. 4e. Input signal is a voltage and the gain is

$$U_o = U_i \times \frac{R_1 + R_2}{R_2} . \tag{4}$$

The sensor current flowing through the series resistance R_s produces the input voltage U_i and the total gain is then

$$U_o = I_1 \times R_s \times \frac{R_1 + R_2}{R_2} . \tag{5}$$

This circuit is an op amp version of the circuit shown in Fig. 2, with the amplifier in place of the potentiometric recorder, with all the restrictions applying which have been mentioned there. There is an additional disadvantage: with a single-ended amplifier neither anode nor cathode can be connected to ground and the circuit is even more susceptible to noise. The amplification scheme is useful for a single-stage design in portable equipment and with sensors delivering more than 10 μM of output current. Only moderate gains will then be needed and the noise problem is less severe.

5 Temperature Compensation Networks

The sensitivity of a POS increases with temperature and it has been shown that there is a reasonable linear relationship between the logarithm of output current and the inverse of absolute temperature [5]. The dependence on temperature is caused by two effects: at constant p_{O_2} the current varies as a function of membrane permeability; at constant

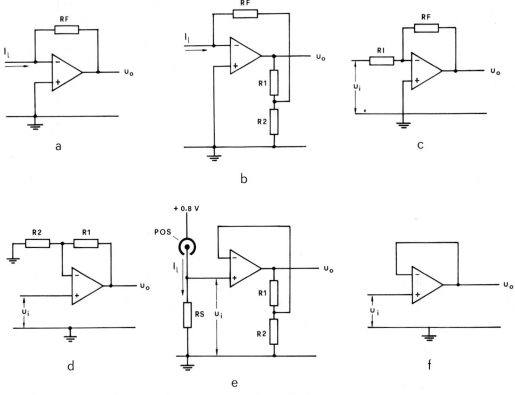

Fig. 4a–f. Basic op amp circuits; **a** current amplifier input stage; **b** current amplifier input stage with fractional feedback; **c** inverting amplifier; **d** voltage follower with gain; **e** voltage follower with POS and series resistance, feedback configuration is identical to **d**, but has been redrawn to show fractional feedback more clearly; **f** basic voltage follower (unity gain noninverting amplifier, buffer amplifier)

concentration p_{O_2} increases since solubility decreases with temperature (Fig. 5). If temperature-dependent elements are incorporated in the amplifier circuits of an oxygen meter (Fig. 1) and if the gain is varied in an appropriate fashion, the sensitivity of a POS can be held constant. Negative-temperature-coefficient resistors (NTC-resistors, thermistors) exhibit the required behavior. Their resistance varies according to

$$R_T = R_o \times \exp(\beta\, T^{-1}), \qquad (6)$$

where R_o is the resistance at some reference temperature, usually taken as 25°C and β is a material constant of the thermistor. The slope factor β can also be expressed more conveniently for the present purpose as temperature coefficient of resistance, α_T, and is commonly given in % °C^{-1} ($\Omega\Omega^{-1}$ °C^{-1}). For the thermistors normally incorporated in a POS, α_T ($T = 25$°C) ranges from -3.3% to -4.5% °C^{-1} and is always higher for larger resistance values. As can be seen from the temperature dependence curves in Fig. 5, a single thermistor can compensate for only one of the temperature effects. Two or more thermistors are needed to achieve full compensation.

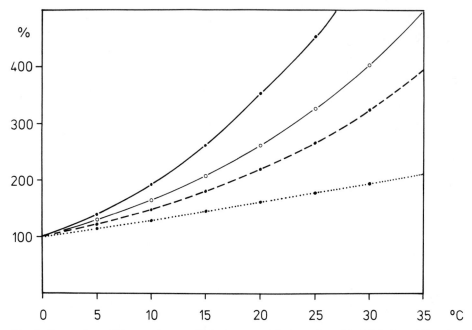

Fig. 5. Comparison of temperature coefficients, percent increase from $0°-35°C$; *dotted line:* increase of p_{O_2} for constant concentration; *dashed line:* increase in sensor current at constant p_{O_2}, temperature coefficient of membrane permeability is 4%; *solid line open circles:* conductance increase of a thermistor (YSI part 44030); *solid line full circles:* combined effect of variation of membrane permeability and solubility of oxygen on sensor current

Achieving adequate compensation over the full environmental temperature range requires careful choice and matching of components. For good thermal tracking, the temperature-sensing elements should be placed as close as possible to the membrane and the working electrode. In order to reduce electrical interference and problems with leakage currents, resistances of less than 10 kΩ should be selected. Whether a one- or two-element compensation circuit is used, an additional thermistor should be provided for temperature measurement. Even when the electronics is designed specifically, it is advisable to use a commercially available sensor because the selection of suitable components has already been made by the manufacturer.

In principle, any of the basic amplifier circuits discussed in the preceding chapter can be modified for temperature compensation by substituting a NTC-resistor for an element which appears only in the numerator of the gain equation [Eqs. (1−5)]. For reasons already mentioned, replacing a high resistance element should be avoided and the basic current amplifier (Fig. 4a) is therefore not used.

Unless the membrane temperature coefficient is exceptionally high, a single thermistor is sufficient to compensate for the variation of membrane permeability with temperature. In the circuit shown in Fig. 4e this can be done by adding a thermistor in series with RS. RS must then be adjusted to tailor the slope of the compensation curve to align with the temperature dependence curve of the sensor. The voltage produced across the total series resistance is then

$$U_o = I_1 \times (RT + RS), \tag{7}$$

where RT is the thermistor resistance. The required value of RS is determined by measuring the sensor current at constant p_{O_2}, and at two temperatures, T_u and T_1, near the upper and lower end of the range in which measurements are performed. If I_u and I_1 is the sensor current at temperatures T_u and T_1 and RT_u and RT_1 is the respective thermistor resistance then

$$U_u = I_u \times (RT_u + RS) \tag{8}$$

and

$$U_1 = I_1 \times (RT_1 + RS). \tag{9}$$

Since output should remain constant

$$I_u \times (RT_u + RS) = I_1 \times (RT_1 + RS) \tag{10}$$

and solving for RS

$$RS = \frac{I_1 \times RT_1 - I_u \times RT_u}{I_u - I_1}. \tag{11}$$

It can be seen from Eq. (11) that for RS to be positive, the temperature coefficient of the thermistor must be larger than the temperature coefficient of membrane permeability. The series resistance compensation with a single thermistor is well suited for data acquisition systems when the signal from more than one POS is multiplexed into a common amplifier and analog-to-digital converter. The magnitude of the POS output signal is kept approximately constant over a wide temperature range and only a thermistor and a resistor is required for each sensor.

Another circuit widely used with a single thermistor is the current amplifier with fractional feedback. The compensating element is substituted for resistor $R1$ in Fig. 4b. The temperature-dependent gain is

$$U_o = -I_1 \times RF \times \frac{RT + R2}{R2}. \tag{12}$$

Here $R2$ is the slope-matching resistor and proceesing as in the previous example we obtain

$$R2 = \frac{I_1 \times RT_1 - I_u \times RT_u}{I_u - I_1}, \tag{13}$$

which is the same as Eq. (11). Since $R2$ appears in the denominator of the gain equation and will be much smaller than RT, the gain factor of the divider will be very large, emphasizing the noise-enhancing features of this circuit.

Two-thermistor compensation networks are mostly realized around voltage-followers with fractional feedback. One example based on the circuit of Fig. 4e is shown in Fig. 6a. In addition to the thermistor in the series resistance a second temperature dependent resistance has been substituted for $R1$. The gain is now

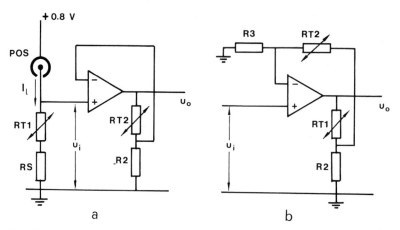

Fig. 6a,b. Two-thermistor compensation networks; a series resistance and noninverting amplifier; b noninverting amplifier with cascaded fractional feedback

$$U_o = I_l \times (RT1 + RS) \times \frac{RT2 + R2}{R2} \tag{14}$$

and three equations are needed to solve for RS and $R2$. The detailed calculations are beyond the scope of this article and only the general procedure will be outlined.

With two thermistors also the change in solubility can be accounted for, and full compensation of POS output can be achieved. Substituting $f(T)$ for the right hand side of Eq. (14)

$$C_1^* = f(T_1); \quad \frac{C_m^*}{C_1^*} = f(T); \quad \frac{C_u^*}{C_1^*} = f(T_u). \tag{15}$$

C_1^*, C_m^* and C_u^* are the unit standard concentrations (App. A) of oxygen at three temperatures at the low end, in the middle and at the upper end of the temperature range of interest. Inserting known values for POS output current and the resistance of the two thermistors into Eqs. (11–15), RS and $R2$ can be calculated; a tedious but straightforward procedure. For given thermistors there is only one solution for RS and $R2$. As will be shown later, the overall gain can be adjusted independently without affecting the compensation.

A two-thermistor compensation network which can be totally separated from the input stage can be realized with the voltage-follower with cascaded fractional feedback from Fig. 6b. The basic current amplifier can then be used as an input stage and this is the preferred design for an oxygen meter. Separating the temperature compensation from the input stage, values for the network components can be much more freely selected and compensation to within $\pm 2\%$ over an operating range $0°–40°C$ can be achieved [5]. The temperature dependent gain is

$$U_o = U_i \times \frac{RT1 \times (R2 + RT2 + R3) + RT1 \times (RT2 + R3)}{R2 \times R3}. \tag{16}$$

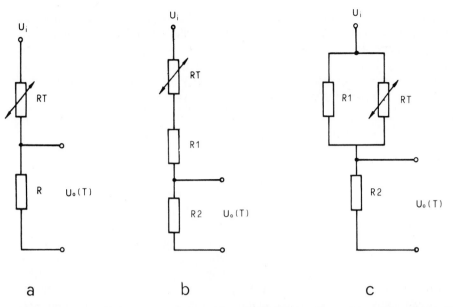

Fig. 7a–c. Thermistor voltage dividers; **a** basic divider; **b** thermistor with series resistor; **c** thermistor with parallel resistor

To determine $R2$ and $R3$ the procedure is the same as in the last example [Eq. (15)] but the calculation is even more cumbersome because both unknown resistances appear in the denominator.

With the simple voltage divider as a basic building block for temperature compensation networks, the value of the fixed resistor is solely determined by the slope matching criterion. The gain of the circuit cannot be varied. To overcome this limitation an additional resistor can be added in series or in parallel to the thermistor. The two-element divider and two modifications with added resistors are presented in Fig. 7. Applying Thevenin's theorem it can be shown that the intrinsic shape of the compensation curve is not altered as long as the total network resistance "as seen by the thermistor" remains constant. This can be most easily seen in the circuit of Fig. 7b, where a series resistance has been added. Provided $R1 + R2 = R$ the current flowing through the divider in circuit Fig. 7b is the same as in Fig. 7a but the output voltage is now

$$U_o = U_i \times \frac{RT + R1 + R2}{R2},\qquad(17)$$

which compared to Fig. 7a is smaller by the factor $R2/R$. If used in the feedback loop of an operational amplifier, the gain will increase by $R/R2$. The opposite effect is achieved with the circuit of Fig. 7c. Again, the slope characteristic is not altered if

$$R = \frac{R1 \times R2}{R1 + R2}\;.\qquad(18)$$

Whatever method is chosen, automatic temperature compensation should never be completely relied upon. First of all, alignment of the compensation curve and the temperature dependence of the detector is never perfect (Chap. III.1) and deteriorates with increasing temperature range. Secondly, variation in membrane permeability may cause considerable error (Chaps. I.1, I.2). For the most accurate work, calibration should always be performed at the temperature of measurement.

6 Measuring Temperature and Atmospheric Pressure

Apart from the problem of automatic temperature compensation, temperature by itself is important as a physical variable. In most oxygen meters one of the temperature-sensing elements, normally used for compensation, is also used for temperature measurement by switching it into another circuit. Because of their large temperature coefficient of resistance, thermistors are very sensitive transducers and little signal amplification is needed, but the highly nonlinear response [Eq. (6)] is a very undesirable property for temperature measurement. By placing the thermistor in a voltage divider as discussed above, a reasonable linear response can also be achieved by choosing a suitable fixed resistance. A linearizing procedure for bridge circuits has been described [2] and computer programs for designing one- and two-thermistor networks have been given [1]. In digital data acquisition and processing systems highly accurate linearization can be performed by using a look-up table of the temperature-resistance dependence, but even then straightening of the thermistor signal before digitizing is advisable to keep the sensitivity constant at widely differing temperatures.

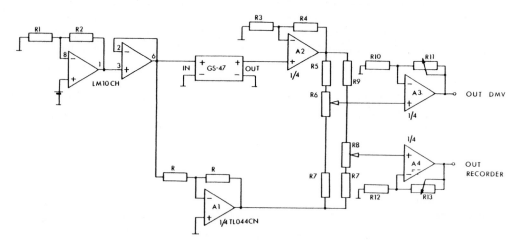

Fig. 8. Circuit schematic for an electronic barometer with the GS-47 pressure transducer; *R6* and *R8* scale offset adjustment; *R11* and *R13* span adjustment. The *LM 10* device provides a stable voltage for transducer excitation (5−10 V); this voltage is inverted by *A1* to get a negative voltage for the scale offset; *A2* is required for buffering the transducer output; *A3* and *A4* are noninverting adders for scale offset and span adjustment

Compared to temperature, normally occurring variations of atmospheric pressure at a given altitude have little effect on the accuracy of oxygen measurement and the deviation from the value calculated from the standard pressure-altitude relation (App. A) will hardly ever exceed $\pm$ 2.5%. Nevertheless, pressure is a factor affecting the calibration accuracy of a POS and automatic pressure compensation is of some importance for long term measurements, for instance in respirometry (Parts II, III). Since small and inexpensive electronic pressure transducers have become available, incorporating a barometer into an oxygen meter is practicable. In microprocessor controled detector circuits (Chap. I.11) compensation could then be performed automatically. A simple electronic barometer based on the GS-47 pressure transducer (GULTON S.C.D, 1644 Whittier Av. Ca. USA) is shown in Fig. 8. In this device the pressure membrane displaces the core of a differential transformer. This design has a better temperature stability than pressure sensors based on semiconductor strain gages.

References

1. Burke A (1981) Linearizing thermistors with a single resistor. Electronics 54:151–154
2. Forstner H, Ruetzler K (1971) Measurement of the microclimate in littoral marine habitats. In: Barnes H (ed) Oceanogr Mar Biol Annu Rev, vol VIII. George Allen Unwin Ltd, London, pp 225–248
3. Graeme GG (1973) Applications of operational amplifiers, third-generation techniques; the Burr Brown electronic series. McGraw-Hill, New York, p 225
4. Graeme GG, Tobey GE, Huelsman LP (1971) Operational amplifiers, design and applications. McGraw-Hill, New York, p 468
5. Hitchman ML (1978) Measurement of dissolved oxygen. J Wiley Sons and Orbisphere Corp, New York, p 255
6. LaForce RC (1976) Application of operational amplifiers to polarographic oxygen sensor systems. Appendix. In: Fatt I (ed) Polarographic oxygen sensors. CRC Press, Cleveland, Ohio, pp 255–263
7. Philbrick-Nexus Research (1969) Applications manual for operational amplifiers. Philbrick-Nexus Corp, Dedham, Mass, p 115
8. Vassos BH, Ewing GW (1972) Analog and digital electronics for scientists. Wiley Interscience, New York, p 405
9. Yellow Springs Instrument Co (1973) Instructions for YSI 5331 oxygen probe. Yellow Springs Springs Instrument Co, p 4

Chapter I.11 The Application of a Microprocessor to Dissolved Oxygen Measurement Instrumentation

I. Bals and J.M. Hale[1]

1 Introduction

The health of living species on the land and in the sea depends upon the presence of oxygen, and the avoidance of deterioration of perishable goods, or of corrodable metal parts depends upon the absence of oxygen. Hence there is often a need to check on oxygen concentrations in practice, and oxygen ranks high among the elements and chemical substances most frequently determined.

The conditions under which these measurements are made vary greatly. In some cases the oxygen activity is of interest, and in other cases it is oxygen concentration which is needed. The interesting concentration range is at least six decades wide. A wide variety of units are used in practice, including mg dm^{-3}, cm^3 dm^{-3}, μg at. dm^{-3}, percentage saturation, torr, kPa, mol dm^{-3}, and percentage by volume. Furthermore, different applications require different response times, sensitivities, stirring rates, stabilities, and so on.

No single analog instrument available in the past was really equal to the task of determining oxygen in all of these various situations. Modifications were made which resulted in a range of dedicated instruments, each of which answered the need of a small segment of the total spectrum of applications. But a microprocessor (MPU)-based instrument has the necessary versatility to match the variety of applications. Such an instrument is now available from Orbisphere Laboratories.

2 The Configuration of the Instrument

Figure 1 illustrates schematically the internal configuration of the Orpisphere model 2609 MPU-based instrument.

The polarographic oxygen sensor delivers a current to the instrument which is proportional to oxygen activity in the sample at constant temperature. Its sensitivity toward oxygen is temperature-dependent with a coefficient of about 3% per °C, but this

---

1 Orbisphere Laboratories, 3, Chemin de Mancy, CH–1222 Vésenaz, Geneva, Switzerland

Polarographic Oxygen Sensors (ed. by Gnaiger/Forstner)
© Springer-Verlag Berlin Heidelberg 1983

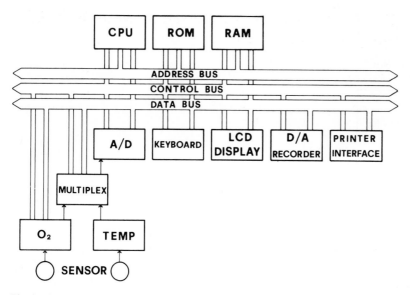

Fig. 1. The internal configuration of the Orbisphere model 2609 microprocessor based instrument

variation is reproducible and is therefore corrected by software. This current is continuously amplified and converted to a voltage.

The sensor also measures temperature by means of a high quality thermistor, and this measurement is used internally by the instrument to correct the oxygen measurement for temperature coefficient, as well as being available for display.

The multiplexer switches either the oxygen or the temperature signal to the measuring circuitry. The A/D converter is a monolithic 3 1/2 digit device which converts the incoming signals from the sensor to digital form, and stores them for interrogations by the MPU.

The MPU itself, labeled CPU in Fig. 1, is a RCA 1802 8 bit processor, which communicates with all of the devices under its command by means of three buses: a control bus, a data bus, and an address bus.

The program is stored in a ROM memory, of 2 1/2 k 8 bit words. The read/write memory is of the CMOS low standby drain type and is kept alive when the instrument is switched off. It stores constants and temporary data.

All remaining periferal devices are for communication between the user and the computer. There is a keyboard for input of data and for the specification of mode of operation, a liquid crystal display for output of results, a D/A converter and recorder output for continuous recording, and a printer interface to provide for a printed list of results.

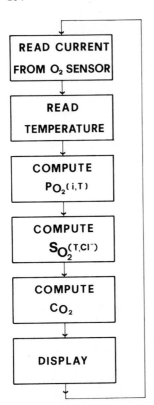

Fig. 2. The measurement loop of the control program

3 The Measurement Loop

A microprocessor-based instrument works in a totally different fashion from a conventional analog instrument.

Under program control it spends most of its time in a measurement loop which is illustrated in Fig. 2. One cycle round this loop requires about 1 s, and the loop is repeated endlessly, unless there is an interruption, as will be seen later.

During each cycle each of the steps in the boxes illustrated takes place. Thus the loop begins with an interrogation by the MPU, of the A/D converter, for an updated value of the current from the oxygen sensor. The A/D converter has its own clock, and performs anything up to 50 conversions per second. When the request arrives from the MPU, there is a delay until the conversion in progress is terminated, and the converter signals whether the precision of current measurement is optimized. If it is not, then the MPU directs a range change in the current measuring amplifier and the conversion is repeated. When the current measurement precision is at a maximum, the digital value is transferred to a memory register for use in subsequent calculations and circulation in the loop continues.

The MPU next switches the multiplexer to temperature measurement, and receives from the A/D converter a digitized value of the temperature in degrees contigrade. Again this value is stored in the Read/Write memory.

In the next step, the measured current is converted into oxygen partial pressure p_{O_2}. This conversion is based upon the following semiempirical equation:

$$p_{O_2} = K \, \exp{(\beta/T)} \times I_1, \tag{1}$$

where T is the absolute temperature, K and β are current and temperature-independent factors which are characteristic of the particular membrane used on the sensor, and I_1 is the current of the sensor. The appropriate magnitude of K and β are stored in RAM. Either one or both of these constants may be redetermined if desired whenever the sensor membrane is changed, by means of the calibration operations to be described in the next section.

The next step is to compute the proportionality coefficient between p_{O_2} and oxygen concentration. This is the solubility coefficient S_s, a function of temperature and chlorinity (App. A).

The oxygen concentration c_s in $cm^3 \ dm^{-3}$ is then just the product of the partial pressure and the solubility coefficient, and conversion into any other desired unit may be effected at this point by multiplication with the appropriate numerical scaling factor (App. A).

In the final step the O_2 concentration is transferred to the display and the loop repeated.

4 Changes in the Mode of Functioning

As previously stated, the instrument spends most of its time in a measurement loop, computing and recomputing the oxygen concentration and thereby continuously updating the displayed result.

If, however, the operator wishes to change the mode of functioning in any way, he touches a key on the keyboard and so interrupts the repetitive passage through the measurement loop. The sequence of ensuing events depends upon the identity of the key which was pressed, and there are three classes of behavior, as illustrated in Fig. 3.

Firstly, to change the displayed parameter — a choice of temperature, oxygen partial pressure, or oxygen concentration is offered — or to select some other unit for displayed oxygen concentration from the choice $cm^3 \ dm^{-3}$, $mg \ dm^{-3}$, or % saturation, a single key is pressed and return to the measurement loop is automatic.

Secondly, one has the possibility to specify the values of parameters not actually measured by the instrument, but which nevertheless modify oxygen concentrations in water. This applies to the barometric pressure $_bp$ and the chlorinity [Cl⁻] or salt content of the water. In this case, the instrument waits for the entry by means of the keyboard of the appropriate values of these parameters before returning to the measurement loop.

Finally, calibration again is a multi-key operation although for a different reason.

We require at least two different keys to be pressed in sequence for the recalculation of the scaling factor K in the formula for conversion of measured current into oxygen partial pressure, and two additional keys to be pressed for the recalculation of

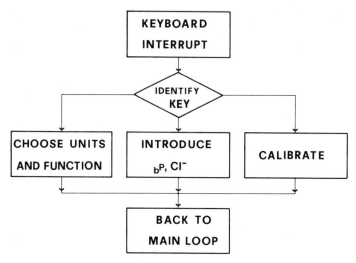

Fig. 3. Program flow following a keyboard interrupt

β the temperature coefficient. The reason for this approach is security, to avoid an accidental change in instrument sensitivity if the keyboard is touched unintentionally.

Calibration may be performed either in a titrated sample, or by reference to the air as a source of a known oxygen pressure. In the former case, the sensor is exposed to the titrated sample, the key C_{TITR} is pressed to interrupt cycling in the measurement loop, the oxygen concentration in the sample is entered by means of the keyboard into memory, and finally the key CAL is pressed to cause the actual recalibration calculation of K to occur.

Similarly, in the event that water-saturated air or air-saturated water is used as the calibration medium, the key C_{sat} is pressed first of all to interrupt cycling in the measurement loop, and to cause calculation of the oxygen concentration present in air-saturated water, in the chosen units, at the barometric pressure and chlorinity which were previously stored in the measurement loop. This concentration is displayed as an intermediate result, and is interesting in its own right since it enables comparison at any time between the actual concentration in a sample and the saturation value. This saturation value is displayed continuously until the key CAL is pressed, when calibration is completed (that is K is computed and stored in memory) and a branch back to the measurement loop occurs.

The key CAL has an effect only if pressed after the C_{TITR} or C_{sat} keys. This precaution ensures that the sensitivity of the instrument is not changed unknowingly due to accidental contact with the calibration key.

Recalculation of the temperature coefficient β is initiated in a similar fashion. The sensor is exposed to a calibration sample (either a titrated aqueous sample, or air-saturated water or water-saturated air) and the measurement loop is interrupted by pressing either the C_{TITR} or the C_{sat} key, whichever is appropriate. If necessary, the oxygen concentration in a titrated sample is entered, as before, by means of the keyboard. Finally the calculation of β is effected by pressing the key TEMP COEF. Certain precautionary tests are programmed into the calculation procedure to ensure that the temperature at which the previous CAL was performed is sufficiently different ($\Delta T \geqslant 5°C$) from the probe temperature at the moment of requesting the calculation of β, that adequate precision is achievable.

The operator of the instrument may choose not to redetermine the values of K and β when he changes the sensor membrane, since previously determined values of these constants are retained in memory even when the instrument is switched off. Provided that the membrane material and nominal thickness are unchanged, a maximum error of about ± 5% in the temperature range $0°–50°C$ is to be expected as a consequence of this action.

Accuracy may be improved by exposing the sensor to a calibration medium at any given constant temperature, and causing the recalculation of K as described above. The temperature coefficient retains its average value and is not recalculated by this procedure. This will result in a maximum error of 0.5% in later measurements at, or within 5°C of the calibration temperature, and a maximum 2% error at other temperatures.

Finally, recalculation of the temperature coefficient β will improve performance such that a maximum error is 1% in the temperature range $0°–50°C$ is to be expected.

Part II Laboratory Applications of Polarographic Oxygen Sensors

Chapter II.1   An Automated Multiple-Chamber Intermittent-Flow Respirometer

H. Forstner[1]

1 Introduction

Numerous types of respirometers, using a polarographic oxygen sensor (POS), have been described in the literature [1–4, 7–11] and this book also presents a selection of different designs (Chaps. II.1–II.9, III.3–III.5). Although the method is widely used, no commonly accepted design has so far evolved. A few types of respirometers are available commercially, but these are either custom designs, made to order for a research establishment, or simple closed-system respirometers for BOD or work with cell suspensions. There is no general purpose aquatic respirometer available that is comparable in distribution and scope to the WARBURG or the GILSON in the field of gasometric respirometry. One reason may well be that it is a relatively easy matter to put together a workable system involving a POS, especially for short-term experiments where a simple closed system suffices (Chap. II.6). On the other hand, there are cases where respirometer design has to be matched to a specific and difficult measuring task and the requirements of a particular organism. Mostly such studies are undertaken by researchers well versed in the use of POS, who are able and willing to design and build their own equipment. Nevertheless, in a wide range of applications, routine measurement of oxygen consumption is only part of a more comprehensive investigation, where the experimenter may have no experience in the use of POS and the design of a suitable respirometer. What is required in this case is a respirometer as a standard laboratory instrument.

That no generally accepted design has emerged may be due to the fact that, until recently, POS have been considered unreliable, at least in the hands of an inexperienced user. This is certainly no longer warranted for the present state-of-the-art sensors, if sensible procedures are followed (Chaps. I.1, I.2). Judging from the numerous examples where POS have been used, and from my own experience with such systems, it appears that the method has evolved sufficiently to be classed as a routine laboratory method. Consequently there is a need for a general purpose instrument, adapted for the most common tasks in respirometry, that can also be operated by the nonspecialist. The production of such a respirometer poses hardly any technical or mech-

1 Institut für Zoologie, Abteilung Zoophysiologie, Universität Innsbruck, Peter-Mayr-Str. 1A, A–6020 Innsbruck, Austria

Polarographic Oxygen Sensors (ed. by Gnaiger/Forstner)
© Springer-Verlag Berlin Heidelberg 1983

ical problems, the difficulty being in agreeing as to the "most common tasks", a fact which is reflected in the many different designs described in the literature.

In our department a design for a multiple chamber respirometer has evolved over the years, and has proved successful for measuring the respiration rates of diverse aquatic animals, both from marine and freshwater environments. The species investigated ranged from small shrimps (Chap. II.9), sea urchins, bivalve and gastropod mollusks, to various fishes, with a wet weight of 1–50 g organic tissue per animal chamber. Although this range represents only a small and arbitrary section of the animal kingdom, it is a fact that a considerable part of physiology is concerned with a few handy and readily available animals, large enough to fill a test tube and small enough to carry and care for.

The respirometer has been designed for completely automatic operation. It can be run as a stand-alone instrument, with output to a potentiometric recorder or a magnetic-tape station. The control logic also provides all necessary signals for interfacing to a microprocessor. Oxygen consumption in μmol kg^{-1} h^{-1} can then be calculated on-line during the experiment. The values can be displayed on a video terminal or a printer. Although compromises had of necessity to be made in the design, I believe that most of the requirements for a general purpose routine respirometer have been fulfilled.

2 Design Specifications

Although widely in use because of their simplicity, closed-bottle respirometers have serious limitations, and the drawbacks as compared to the flow-through method have been discussed in detail elsewhere ([6], Chap. II.3). Flow-through systems have a number of advantages: measuring is done in a constant environment and there is no accumulation of waste products; the POS signal is directly proportional to instantaneous oxygen uptake, furthermore the flow pattern in the animal chamber is directed and stable. The oxygen uptake rate is:

$$\dot{N}_{O_2} = \Delta c_s \times f = \Delta c_s \times (dV \times dt^{-1}), \tag{1}$$

where c_s is the oxygen concentration in the medium and f is flow. In a closed system oxygen uptake is determined from the decrease of oxygen concentration in the bottle

$$\dot{N}_{O_2} = (dc_s \times V) \times dt^{-1}. \tag{2}$$

Clearly the two equations are dimensionally identical, but the brackets indicate the different physical situation. In the first case V, the volume displaced, is a dynamic parameter which has to be actively maintained and is liable to fluctuations unless special care is taken (for instance by using a precision pump). In Eq. (2) V is the volume of the bottle, which for all purposes is fixed. Only time has to be measured accurately, which is much more simple than controlling flow, and is the main reason why closed systems are still preferred in less demanding applications.

In intermittent-flow respirometers, the constant volume principle is retained, but the basic disadvantages of the closed-bottle method are avoided. An intermittent-flow respirometer functions like a closed system during the measuring phase. After a fixed interval of time, or when oxygen tension has declined to a preselected level, flow is turned on until initial conditions are re-established (Chaps. II.5, III.4). The system is then closed again and the next measurement commences. In this way the physicochemical environment in the respirometer vessel is held nearly constant and the measuring cycle can be continued for as long as necessary. Evidently, true continuous measuring is not possible with an intermittent flow device and this has been cited as one of the drawbacks of the method [5]. But even in a continuous-flow respirometer time resolution is limited ultimately by time lags due to mixing effects. In most routine studies a relatively coarse time scale is adequate. To determine average values and to capture the influence of diel variation, measurements should in any case extend for at least a day. Measuring oxygen uptake at hourly intervals will give sufficient resolution in this situation and intermittent measuring is hardly a disadvantage. There is also an advantage: during flushing of the respirometer vessel, the oxygen uptake cannot be measured and the POS can be employed for other purposes. The sensor can be connected to another animal chamber or to a supply of air-saturated water. In this way flushing and measuring or calibration can be interleaved. The ratio between duration of the measuring phase and the measuring interval, i.e., the time after which measuring is repeated on the same unit, determines the number of chambers that can be accommodated. In fixing the length of the measuring phase, both biological and physical parameters have to be considered. To avoid the closed-bottle effect, oxygen depletion should not exceed a certain lower limit. What is permissible depends on whether the object of the study is more a regulator or a conformer, but as a rough rule p_{O_2} should not decrease by more than 20%. Laboratory stress should be kept to a minimum, which requires sufficient space for the organism in the respirometer vessel and a steady flow, fast enough to provide the necessary exchange, but not so fast as to be a stress factor by itself.

On the other hand, in order to increase the precision of the measurement, the difference in p_{O_2} should become as large as possible. The measuring phase should be long, so that the lags due to sensor time constant and delayed mixing become insignificant. Furthermore, in order to provide rapid circulation between animals and sensor and to provide homogeneous and near instantaneous mixing the flow should be vigorous. A compromise has to be found between these partially conflicting demands.

Using cylindrical animal chambers with a volume of $200-1000$ cm^3 flow rates of $0.2-0.5$ chamber vol/min gave good results. Under these conditions a linearly decreasing output is reached $1-3$ min after switching animal chambers. To improve the accuracy of the measurement, oxygen removal from the system is estimated by fitting a straight line to the decreasing p_{O_2} record. A minimum of 5 min of record is required to level out short-term variations, so that at least 8 min are required for the measuring phase. By fitting a regression line to discrete data points of the 5-min record, a difference in p_{O_2} of 1% could be measured with a precision of $\pm 0.05\%$ (95% confidence limits). Depending on the oxygen uptake rate of the organism studied and the oxygen solubility at the particular experimental temperature, a chamber volume must be selected to keep oxygen depletion between 1%-20% limits. To determine the necessary chamber volume, the expected range of O_2 uptake has to be estimated. The midrange value should produce an oxygen removal of 10% at the end of the measuring phase. The desired volume can be calculated by

$$V = 10 \, \frac{\dot{n}_{O_2} \, W \, t_m}{c_s} \, , \tag{3}$$

where $\dot{n}_{O_2}$ is the midrange oxygen consumption (μmol kg^{-1} h^{-1}), W is the average weight of the animals [kg], t_m is the length of the measuring phase and c_s is the oxygen in solution at the experimental temperature, salinity, and pressure. Simple as it may seem, selecting a suitably designed chamber is an important point in respirometric studies. The structure of the environment in the respirometer vessel will certainly affect the behavior of the animal and can result in widely differing oxygen consumption rates under otherwise identical conditions. Providing a proper substrate, support or shelter will help to reduce laboratory stress, but additional surfaces will increase bacterial respiration (Chap. II.9). It is difficult to lay down general rules, and conflicting requirements have to be balanced in each particular case.

The accuracy of the measurement (conformity to an absolute scale) is, of course, directly dependent on the accuracy of the sensor reading, and long-term drift may be a problem which can be avoided by calibrating the POS at intervals. Since the POS has to be switched between chambers in any case, this can be done without unduly complicating the design by connecting the sensor compartment once in each measuring cycle to a supply of equilibrated (air-saturated) water. A measuring cycle consists then of N measuring phases (where N is the number of animal chambers) and a calibration phase, equal in duration to a measuring phase. By selecting a measuring interval of 1 h, the time between successive measurements in the same chamber, up to six animal chambers can be accommodated. To make the chamber better accessible and to provide room for a flushing pump, the design presented here has been restricted to four chambers with a 12-min measuring phase. For the remaining 48 min in the 1-h measuring cycle, a small submersible centrifugal pump forces water from an external reservoir through the chambers (flushing phase). It is important to maintain approximately the same flow rate during the measuring cycle. Rapid flushing with a distinctly higher flow is a stress factor that causes erratic behavior and variable oxygen consumption rates. Since in the multiple chamber respirometer the duration of flushing is much longer than that of measuring, the flow optimized for measuring is also sufficient to ensure a return to constant initial conditions, and the throughput of the flushing pump can be adjusted accordingly.

The advantage of using a single POS for a number of respirometer vessels is offset to some extent by the need for a system of valves to sequence the different phases of the measurement cycle (i.e., measuring, flushing and calibration). The valve assembly is the critical and most complex part, and it is the key component determining a successful design. On the other hand, a multiple chamber design can be made very compact and only a moderately sized thermostat bath is needed.

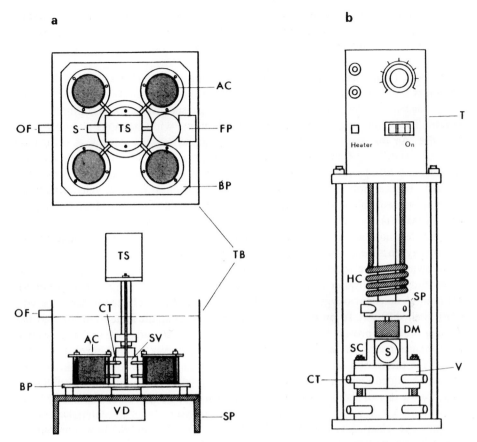

Fig. 1. a Top and side view of the respirometer system: *TB* thermostated bath, *BP* bottom plate, *AC* animal chambers, *CT* connecting tubes, *FP* flushing pump, *TS* thermostating and stirring unit, *S* sensor, *OF* overflow, *SV* sensor chamber and valve assembly, *VD* valve drive unit, *SP* support for thermostat bath. **b** Side view of central unit consisting of thermostat – stirrer and sensor – valve assembly: *T* thermostat, *SP* stirring pump, *HC* heating and cooling coils, *DM* drive magnet, *SC* sensor chamber, *V* valve

3 Description of the Respirometer System

An overall view of the respirometer system is shown in Fig. 1a. The central part consist of sequencing valve, sensor chamber, and the thermostating and stirring unit (Fig. 1b). This assembly is bolted to the bottom plate of the rectangular constant temperature bath. After two screws are loosened, this unit can be removed in one piece. The drive of the sequencing valve is situated externally, on the underside of the thermostat tank. The axle of the drive extends upward through the bottom plate and is sealed with two lip gaskets. The drive axle engages the axle of the sequencing valve with two asymmetrically placed pins. Thus the central part can be removed, but angular orientation of the two axles relative to each other is always and unambiguously maintained.

A rectangular plate, parallel to and about 2 cm above the bottom, supports the animal chambers and the flushing pump. This plate has a large central opening and adequate clearance at the outer rim, so that water can circulate freely around and below the chambers. The standard type of animal chamber consists of a cylindrical glass vessel and a 12-mm-thick perspex lid with an O-ring seal. The lid is clamped down by wing nuts on three rods extending from the bottom plate. Inflow and outflow are placed tangentially, near the bottom and the top respectively, for an optimum flow pattern. The connecting tubes are shaped to be in line with the inlets and outlets on the valve. An inexpensive alternative is all-glass preserving jars, with glass lid, rubber seal and metal-clamp fastening (Chap. II.9). Using a glass drill, two holes are made on the side, and fitted with rubber bulkhead seals through which stainless steel tubes are inserted. Another hole is made in the lid for removal of air bubbles. Once the system has been vented, the opening is closed with a small rubber stopper. Connections between chambers and valve are made with tygon tubing, over the ends of the glass or stainless steel inlet and outlet tubes, butting against each other to minimize the diffusion of oxygen. The flushing pump continuously circulates equilibrated water by way of the sequencing valve to all animal chambers except the one selected for measuring. When studying freshwater animals, this water can be drawn directly from the thermostat bath. A constant oxygen concentration in the bath is maintained by aeration with a coarse air diffuser. To prevent air bubbles from entering the system, the inlet of the flushing pump is provided with a bubble trap. A perspex tube, 10 cm long with an inner diameter of 5 cm and filled with perlon wool, serves for this purpose. On one side of the thermostat tank, at water level, there is an overflow tube. A small amount of water is added continuously; metabolic waste products are removed with the overflow. A flow rate of about 1 cm^3 min^{-1} g^{-1} animal (wet weight) is recommended [12].

Although the respirometer system is made of corrosion-resistant materials throughout (perspex, glass or stainless steel have been used), seawater should not be used as thermostating medium, because salt deposits will form on the heating coil. Equilibrated seawater must then be supplied from a separate vessel, which is also placed in the thermostat bath. The volume of this vessel must be at least twice the total volume of the animal chambers, and the size of the thermostat tank has to be increased correspondingly. An overflow arrangement for removal of wastes and a constant supply of seawater from a larger external reserve must also be provided. The respirometer consisting of animal chambers, sensor chamber, sequencing valve, flushing pump, and aeration vessel forms a completely sealed unit. The system can be removed as a whole and set up in a second tank, filled with seawater. The animals are inserted and the chambers closed. The system can now be transferred to the thermostat bath.

4 Sensor and Sensor Chamber

For routine applications in respirometry, a highly stable and temperature-compensated POS should be selected. In order to minimize problems due to leakage and to offset currents in connectors and at the input to the electronics, the output signal of the sensor should be of reasonable magnitude (10 μA at air saturation). Consequently the

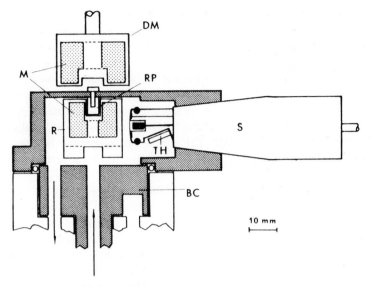

Fig. 2. Sensor chamber, cross-section and top view: *S* sensor, *TH* thermistor, *R* rotor; *M* magnet, *RP* rotor bearing pin, *BC* bottom of sensor chamber (= upper valve disk), *DM* drive magnet

cathode area of the POS should be large ($5-20$ mm^2). The POS should also have a correspondingly large anode surface and an ample electrolyte reservoir, to ensure constancy of the electrochemical parameters even during measurements extending for days or weeks.

Although oxygen uptake rates are generally measured at a constant temperature, short-term temperature variations, in the range of a few tenths of a centigrade, will occur on account of mixing delay and set point hysteresis, unless precision thermostating is used. Due to the pronounced temperature dependence of membrane permeability ($2\%-6\%$ per $°$C), even such small temperature changes produce a noticeable scatter in the output signal and the precision of the measurement is degraded. This source of error is removed if the sensor is properly temperature-compensated. For optimal compensation it is also important that the thermal time constants of the compensating elements and sensors are well matched, thus avoiding a phase lag. If this is not the case, although temperature compensation may be perfect at steady state, it is not effective for short-term fluctuations.

A well-compensated sensor will always produce the same current for identical values of p_{O_2}, even when measuring at widely differing temperatures. This is an important feature where a programmed temperature change is imposed during an experiment and obviates the need for range-switching in the sensor electronics.

A number of commercially available sensors are well suited for respirometer applications (ORBISPHERE, WTW, YSI) according to the specifications made above. The YSI 5750 nonstirring B.O.D. bottle probe was chosen, mainly because it has a conically tapered shaft which can be simply inserted into − and removed from − a matching bore in the side of the sensor chamber (Fig. 2). A watertight fit is made and no other sealing elements are required.

POS with a large area cathode are always diffusion limited (Chap I.1) and, depending on the particular parameters of the sensor (shape and area of cathode, membrane permeability), a certain minimum exchange of sample medium at the membrane surface is needed to obtain a reading which is both stable and representative for oxygen concentration in the sample. This can be achieved either by directional flow through the sensor chamber, normal or parallel to the sensor face, or by turbulent mixing in the chamber. In the latter case, current velocity at the sensor can be adjusted independently of bulk flow through the sensor chamber, and this is the method usually preferred for intermittent flow respirometers. The required stirring speed is determined by increasing the revolutions of the stirrer in increments until the output of the POS remains stable.

In a multiple-chamber respirometer, the sensor chamber must necessarily be separate from the animal chamber and water circulation has to be maintained between the two units. The stirring chambers used in continuous flow respirometers normally have inflow and outflow on opposite sides. The stirrer functions only to agitate the medium, while water exchange is maintained by a separate pump or by gravity flow. By placing the inlet axially and inserting the POS through the side of the stirring chamber, the functions of pump and stirrer can be combined in one element. With 20–60 revolutions per second flow rates from $2.5-7.5$ cm^3 s^{-1} can be achieved. A cylindrical vaned rotor, running on an axial stainless-steel-on-Teflon bearing is more reliable than a standard Teflon-coated magnetic stirring bar, which tends to get stuck occasionally. The rotor was made by molding a button magnet (ECLIPSE Nr. 822) in epoxy resin (CIBY Araldite D). By using the larger rotor (compared to the stirring bar), dead-space volume of the chamber is also reduced.

The stirrer is driven from the outside of the chamber by a matching drive magnet (ECLIPSE Nr. 823). The sensor chamber is placed on top of the sequencing valve (Fig. 4) and coaxially beneath the immersion circulator. The driving magnet is also embedded in expoxy resin and mounted on the axle of the circulator in the thermostat bath. This eliminates the need for a separate drive motor. Good magnetic coupling between the two magnets is essential for reliable operation. The gap between the poles of the specified magnets can be up to 6 mm, but should be as narrow as possible. This is achieved by making the top wall of the stirring chamber fairly thin, not exceeding 2 mm, to allow some clearance between this plate and the face of the magnets. If the chamber is made of perspex or a similar plastic, diffusion problems arise if the p_{O_2} difference across the wall is large. Oxygen from the thermostat bath diffuses into the chamber and apparent oxygen consumption is reduced. If the p_{O_2} difference is less than 15%, there is no noticeable effect. For critical applications the stirring chamber can be made of glass (Chap. III.4), but where the chamber is an integral part of the valve, it is more difficult to produce the required shape with this material.

5 The Sequencing Valve

The valve combination routing the flow between animal chambers, sensor and flushing pump is the heart of the intermittent-flow respirometer. The total valve assembly consists of two-ganged eight-position valves. A schematic exploded view is shown in Fig. 3 and a cut view of the assembled valve in Fig. 4. The symbols of the parts in these two figures will be referred to in explaining the valve and its function.

Each valve consists of a rotating valve disk (UD,LD) and a channel plate (UP,LP). The two valve disks are attached to the valve shaft (VS). A central bore in the shaft (VS) connects an opening in the lower valve disk (LD) with the stirring chamber. The connections leading to the animal chamber are on the sides of the channel plate. The outflow (UP) and the inflow (LP) of the flushing water supply are situated on corners. Stirring chamber (SC) channel plates (UP,LP) intermediate plate (IP) and bottom plate (VP) form the housing of the valve. The parts are held together by two threaded bolts. The upper valve disk (UD) is also the bottom of the stirring chamber.

The lumen of the valve is divided into two compartments which are completely separated during the measurement phase. The flow during the measurement phase can be traced in Fig. 4. Due to centrifugal action in the stirring chamber, water is forced down through the peripheral boring in the upper valve disk into the matching opening in the channel plate (UP) and out to the selected animal chamber. Water returns from the animal chamber through the corresponding connection in the lower channel plate (LD) down into the lower valve disk (LD) into the hollow drive shaft and up into the center of the stirring chamber. This completes the closed circuit. The system is sealed by O-rings between parts (SC,UP,LP), (UD,UP) and (LP,LD) as well as by the static O-ring seals between distributor disks and drive shaft.

The second compartment of the valve volume consists of 315° concentric channels in each of the valve disks (UD,LD) and a cavity in the lower (LP) and upper (UP) channel plate. The two cavities are separated by the intermediate plate. To keep down friction there is no seal here and a small amount of leakage is allowed. The upper and lower sides of each channel plate are connected by four holes in the intermediate positions between the openings to the animal chambers, and water can circulate freely between the cavity and the concentric channel in the distributor disk. Water from the flushing pump enters the inlet at the corner of the lower channel plate, tapping one of the vertical borings in the channel plate. Flow descends into the concentric channel and thence out to the animal chambers, except the one selected for measurement, which is sealed off from the system. The hole which connects to the center of the stirring chamber is in this position in the calibration phase: flow to the chambers ascends first into the cavity of the lower channel plate and then down through the three open connections in the plate into the concentric channel. Flow in the upper valve is routed the same way but in the opposite direction.

Although in the present design there are only four animal chambers, the valve has eight positions, the four additional intermediate positions serving for flushing the stirring chamber. After a measuring phase the valve disks advance 45° and are held there for 20 s while the drive of the Geneva mechanism completes a full turn. The connections to the stirring chamber are now in line with one of the intermediate borings, and water is forced through the stirring chamber, thus removing the oxygen-depleted medium from the previous measurement. During this time all connections to the animal chambers are open to the concentric channel and are supplied from the flushing pump. For calibration of the POS, this position is held for the full length of a measuring phase. To assist the pumping action of the stirring chamber, flow in this phase

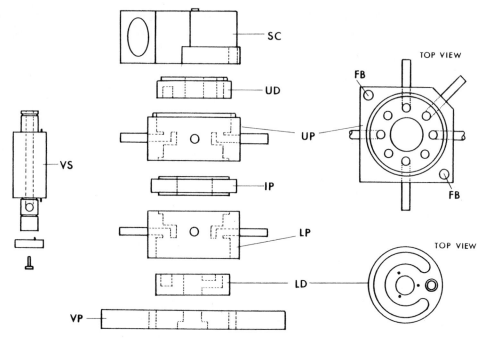

Fig. 3. Exploded view of the sensor chamber and valve assembly: *SC* sensor chamber, *UD* upper valve disk, *UP* upper channel plate, *IP* intermediate plate, *LP* lower channel plate, *LD* lower valve disk, *VS* valve shaft, *VP* valve bottom plate, *FB* fastening bolts

must come from the lower valve. This represents a certain disadvantage because flow in the animal chambers is necessarily reversed between measuring and flushing. In circular chambers this has not been found to cause any noticeable disturbance to the animals, as long as the flow rate is held approximately constant.

6 Valve Drive

The valve drive has been placed outside, and is fixed to the bottom plate of the thermostat tank, so that access to the respirometer system from above is unobstructed. The details of the drive unit are shown in Fig. 5a. Accurate angular positioning of the drive shaft, corresponding to the eight positions of the valve, is achieved with a 45° (angle) Geneva mechanism. A Geneva mechanism is a special type of gear which transforms continuous revolutions of a drive shaft into intermittent motion of the the driven wheel. Every revolution of the drive shaft produces a fixed angular displacement of the driven wheel, but unlike a cogwheel gear which moves continuously, position is held as soon as the driving pin clears the slot in the driven wheel. Once the driving pin is disengaged the Geneva mechanism is self-locking, due to the halfmoon-shaped extension on the drive shaft. The sequence of operation is shown in Fig. 5b. Although the mechanism may look complicated, tolerances are not in fact critical.

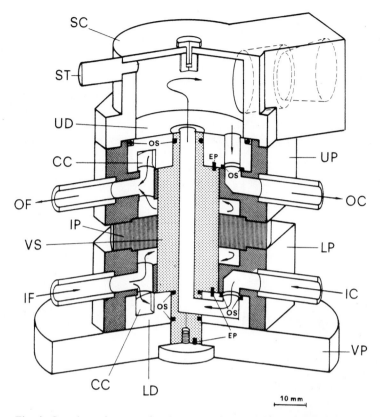

Fig. 4. Cut view of sensor chamber and valve assembly; rotor and sensor not shown: *SC* sensor chamber, *ST* vent stopper, *UD* upper valve disk, *UP* upper channel plate, *OF* outflow to reservoir, *OC* outflow to animal chamber, *CC* circular channel in valve disk, *IP* intermediate plate, *LP* lower channel plate, *IF* inflow from flushing pump, *IC* inflow from animal chamber, *LD* lower valve disk, *VS* valve shaft with central bore, *EP* engaging pin between valve shaft and valve disks, *OS* O-ring seals, *VP* valve bottom plate

Dimensions of the two parts are given in Fig. 5c. The radially slotted wheel is made of perspex, the driver of brass. The parts can be made on a lathe and a milling machine using standard workshop techniques. Geneva mechanisms are also available from suppliers of precision mechanical components (for instance: PIC, POB 335, Benrus Center, Ridgefield Ct., 06867 USA, Catalog. No. EU-4).

The drive mechanism is actuated by a 220 V split-pole motor through a 1:1000 reduction gear, completing two revolutions per minute and developing a torque of 150 N cm (Heidolph-Elektro KG., D–8420 Kelheim/W. Germany, Type 602.106.0001). The driver of the Geneva mechanism is mounted directly on the output shaft of the gearbox.

The position of the valve drive is monitored by three microswitches (Fig. 5a). One switch (SW 1) is actuated by the driver pin of the Geneva mechanism in the locked position. The other two are operated by a cam-wheel on the driven axle. There are five cams on the upper plane (SW 2), four at 90°, corresponding to the valve positions

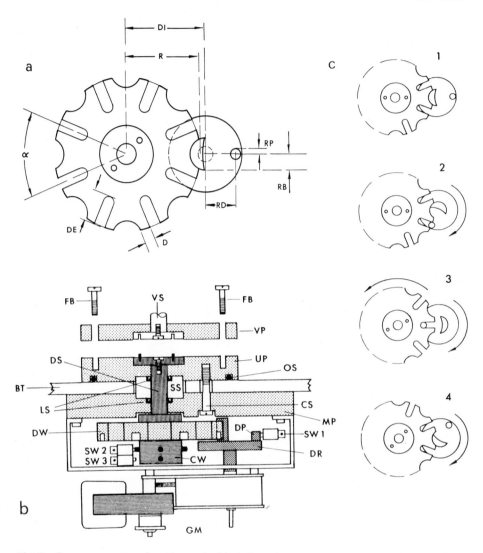

Fig. 5a. Geneva movement, face view and critical dimensions (length in mm): α stepping angle (α = $360°/2$ N, where N is the number of animal chambers), R radius of driven wheel (for $45°$ movement 80 mm is recommended), DI distance between drive axle and driver, $DI = R \times \sqrt{1 + \mathrm{tg}^2\,\alpha}$, RD radius center of drive pin to center of driver axle, $RD = R \times \mathrm{tg}\,\alpha$, RP radius of driver pin, DE depth of drive slots on driven wheel, $DE = RD + RP + 0.5$, RB should be chosen to make $D > 2$ mm.
b Side view of the drive unit, upper part in cross-section to show the arrangement of seals and engagement with valve shaft: VS valve shaft, VP valve bottom plate, FB fastening bolts for central unit, UP upper part of drive shaft seal, DS drive shaft, OS O-ring seal, LP lip seal, SS seal seat, BT bottom of thermostat bath, CS clamping screw (1 of 3 shown), MP mounting plate for drive unit, DW driven wheel of Geneva movement, DR drive wheel, DP drive pin, $S1$, $S2$, and $S3$ microswitches, CW cam-wheel, GM gear motor. **c** Geneva movement, sequence of operations: *1* locked position, *2* drive pin engaging, *3* advancing, *4* drive pin disengaged, driven wheel locked again

where an animal chamber is connected to the sensor, and one in a 45° intermediate position for the calibrating phase. The lower plane (SW 3) has only one cam coincident with the calibrating phase. If the valve is in one of the working positions the switches are open.

To realize a simple stand-alone controller, the state of SW1 and SW2 are combined in a logical OR function with the output of a timing device, to start and stop the drive motor. The output pulse from the timer must extend until SW1 has closed. Then SW1 and SW2 will keep the motor running until the next position of the valve has been reached. Alternatively, the switches can be used as input flags for a microprocessor, which in turn controls the motor. SW1 being open indicates the calibrating phase and can be used for synchronizing the measuring cycle.

7 Calculating Oxygen Consumption

Oxygen consumption is calculated from the recorded values of POS output and the oxygen solubility under experimental conditions (Chaps. I.2, I.11, App. A). While temperature and salinity are usually controlled factors, barometric pressure may change during an experiment and should therefore be monitored. Changes of 1 kPa (10 mbar) are not uncommon during 24 h and will introduce an error of about 1%. With the specific design parameters, decrease in p_{O_2} is always essentially linear and the slope can be determined by fitting a straight line (Fig. 6). If the output is recorded on a potentiometer recorder, this line is fitted by eye. A chart speed of 300 mm h^{-1} is recommended. Two points on the fitted line, intersecting scale marks on the chart paper, are selected and the horizontal distance (l) between them is measured. In order to ensure sufficient resolution, l should be from 100–150 mm. Taking now the vertical distance (h) between the intersections and the value recorded during calibration (100%), both in units of scale marks, oxygen uptake rate (μmol h^{-1}) in the system is

$$\dot{N}_{O_2} = \frac{h}{100\%} \times \frac{\text{chart speed}}{l} \times c_s \times V. \tag{4}$$

If the values are stored digitally, the slope is determined by least-squares regression and confidence limits can also be calculated. POS output should be converted with 10-bit resolution, but since p_{O_2} will never drop to less than 80% of the starting value, an 8-bit converter and a zero-supression are adequate (Chap. I.10). With a 12-min measuring period, and discarding the first 3 min of the record, a sampling interval of 30 s gives 18 data points for fitting the regression. This is normally sufficient and will give a significant regression down to 0.25% decrease in p_{O_2}. In calculating the regression equation, one measuring interval is taken as a unit increment of the independent variable and the estimate of slope b is then the decrease in p_{O_2} per interval. The oxygen uptake rate (μmol h^{-1}) is now

$$\dot{N}_{O_2} = \frac{b}{100\%} \times \frac{3600}{t_s} \times c_s \times V, \tag{5}$$

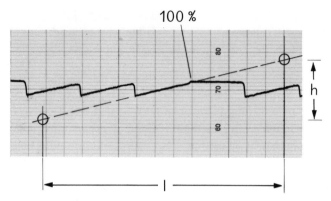

Fig. 6. Evaluating oxygen uptake from a strip-chart recording, actual recording shown: chart-speed is 120 mm h^{-1}, l = 132 mm, 100% = 72 (scale marks), h = 16 (scale marks)

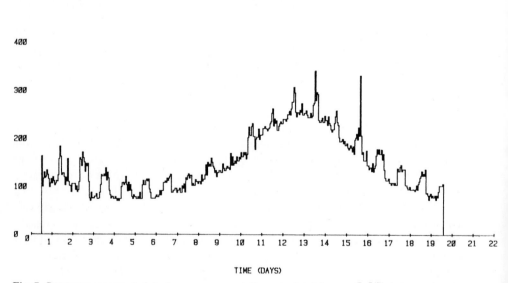

Fig. 7. Computer generated plot of oxygen consumption calculated from values stored on magnetic tape. Experimental temperature was varied sinusoidally, with a period of 16 days. Mean temperature was 20°C, amplitude ± 5°. Although the experiment was performed in darkness, a persisting diel activity can be seen; time marks at midnight (goldfish, weight 6 g)

where t_s is the sampling interval in seconds. If a zero-supression has been used, the fixed offset has to be added to the stored 100% values before they enter the calculation. Dividing now by animal weight (kg) gives the weight specific oxygen uptake rate in μmol kg^{-1} h^{-1}. Wet weight is generally used. It has been shown to be just as good a measure as dry weight or total nitrogen content [5] and the determination is less complicated.

8 Accuracy of the Measurement

Statistical analysis has shown that, with one size of animal chamber and allowing a maximum oxygen depletion of 20%, a resolution of 1:400 can be achieved. Independently thereof the accuracy of measurement, that is the conformance to an absolute scale, is determined by the accuracy of the sensor calibration. The value measured during the calibration phase provides this reference [Eq. (5)]. It is not necessary to know the actual starting concentration in the chamber which, due to the respiratory uptake of the animal, is always somewhat below saturation. This is in fact an advantage because the formation of gas bubbles is avoided. Accuracy depends then on how well inflowing water is equilibrated with a gas of known oxygen content (usually air). Errors caused by long-term drift of the POS signal are eliminated by recalibrating the sensor in each measuring cycle. The effect of short-term fluctuations is removed by fitting a regression line. To check the calibration, saturation in the thermostat bath from which the water supply for the respirometer system and for calibration of the sensor is drawn, was compared against a carefully equilibrated standard (Chap. I.2); using a POS no measurable difference was found. The test was performed with a tank volume of 40 dm^3 and with a total oxygen consumption of 800 μmol h^{-1}. As result of the high turbulence and the consequent efficient mixing in the thermostat bath a near perfect equilibrium is achieved. When temperature and atmospheric pressure are known, oxygen equilibrium concentration in μmol dm^{-3} (or mg dm^{-3}) can be calculated with the equations given in Appendix A. With regard to the controlling variables temperature and atmospheric pressure, the numerical accuracy of these equations is better by two orders of magnitude than is required for respirometric work. Normally temperature is held constant during an experiment, but in order to maintain accuracy, precise thermostating is needed; a temperature change of only 0.25°C will change oxygen equilibrium concentration by approximately 0.5%.

Atmospheric pressure is more of a problem, because it is an external variable, subject to random fluctuations. Maximum diel pressure changes, generally encountered, are likely to cause an error of the order of 1%, and although this may be tolerable in some studies, a continuous pressure record is recommended in long-term studies. If a barograph is not available in the laboratory, pressure recordings can be acquired from a meteorological station near the site. In the case where the measurement parameters are processed on-line, or stored in numeric form for later off-line processing, an electronic barometer is useful (Chap. I.10). The output signal of such a barometer can be digitized along with temperature and oxygen and all input variables for calculating oxygen consumption are then available.

Fortunately, the error resulting from uncertainty about the true value of c_s, is a proportional error. That is, whatever the magnitude of the calculated slope, the error is always a certain percentage of the calculated oxygen consumption. Consequently, the resolution of the measurement is little affected. Also it could be shown that this proportional error can be kept quite small. Analyzing a series of 84 100%-readings (each obtained by averaging 12 measurements during the calibration phase) and accounting for variations in atmospheric pressure, it was found that accuracy could be maintained to better than ± 0.2% (95% confidence limits).

Acknowledgment. This work has been supported by the Fonds zur Förderung der wissenschaftlichen Forschung, Project Nr. 2939.

References

1. Bielawski J (1961) The use of a solid platinum electrode for continuous recording of the rate of respiration. Comp Biochem Physiol 3:261–266
2. Courtney WAM, Newell RC (1965) Ciliary activity and oxygen uptake in *Branchiostoma lanceolatum* (Pallas). J Exp Biol 43:1–12
3. Davies SP (1966) A constant pressure respirometer for use with medium sized animals. Oikos 17:108–112
4. Eriksen C, Feldmeth CR (1967) A water current respirometer. Hydrobiologia 29:495–504
5. Fry FEJ (1971) The effect of environmental factors on the physiology of fish. In: Hoar WS, Randall DJ (eds) Fish physiology, vol VI. Academic Press, London New York, pp 1–87
6. Kamler E (1969) A comparison of the closed-bottle and flowing-water methods for measurement of respiration in aquatic invertebrates. Pol Arch Hydrobiol 16:31–49
7. Klekowski RZ, Kamler E (1968) Flowing-water polarographic respirometer for aquatic animals. Pol Arch Hydrobiol 15:121–144
8. Lenfant C (1961) A method for measuring V_{O_2} and V_{CO_2} of very small sea animals. Appl Physiol 16:768
9. Oertzen JA, Matzfeld V (1969) Eine Apparatur zur kontinuierlichen Respirometermessung an marinen Organismen. Mar Biol 3:336–340
10. Pattee E (1962) Methodes de mesure du metabolisme respiratoire chez les animaux aquatiques. Hydrobiology 14:40–56
11. Schramm W (1966) Kontinuierliche Messung der Assimilation and Atmung mariner Algen mittels der elektrochemischen Sauerstoffbestimmung. Helgol Wiss Meeresunters 13:275–287
12. Sprague JB (1973) The ABC's of pollution bioassay using fish. In: Cairns J Jr, Dickson KL (eds) Biological methods for the assessment of water quality, vol 528. ASTM Special Publ, Philadelphia, pp 6–23
13. Zeuthen E (1953) Oxygen uptake as related to body size in organisms. Q Rev Biol 28:1–12

Chapter II.2   The Application of Polarographic Oxygen Sensors for Continuous Assessment of Gas Exchange in Aquatic Animals

J.P. Lomholt and K. Johansen[1]

1 Introduction

Most aquatic invertebrates, as well as fish, depend on active ventilation of water for irrigating respiratory surfaces. These may be specialized in the form of gills or ctenidia, or be unspecialized parts of the general body surface. Movement of water may occur by active muscular pumping or by ciliary action. Typically, the passage of water across respiratory surfaces occurs in discrete channels and most commonly unidirectionally.

For the purpose of analyzing respiratory gas exchange, it becomes important to confine the ventilatory current for measurement of both volume ventilated and gas composition of the expired water. This paper reports on the application of polarographic oxygen sensors (POS) for continuous measurement of O_2 tension in expired water from fishes and selected invertebrates. Electromagnetic flow-measuring techniques have been used in combination with the POS for evaluation of total external gas exchange. Traditionally such analysis has depended on sequential sampling of water for analysis of water O_2 content by the Winkler method (App. D), or measurement of p_{O_2} in water by injection of sampled water into cuvettes holding POS (Chap. II.10). Such sample measurements of expired water gas composition may often give a distorted picture of normal gas exchange, since natural breathing movements may be periodic or phasic, giving intermittent values of O_2 tension in expired water of indeterminable value unless related to ventilatory changes.

We will describe some experimental arrangements, which we have used to record continuously the inspired as well as expired oxygen tension and hence the extraction of oxygen from the ventilatory current in fishes and invertebrates like decapod crustaceans and polychaetes. Additionally, these arrangements have permitted simultaneous and direct recording of the magnitude of the ventilatory current.

The equipment consists of a standard Radiometer p_{O_2} sensor (E 5046) mounted in a thermostatted cuvette (D 616) and connected to a Radiometer PHM 71 or 72 acid-base analyzer. The sensor cuvette is placed alongside the experimental aquarium, a few centimeters below the water surface. Water to be analyzed for O_2 tension is passed to the sensor cuvette by gravity through polyethylene cannula (Clay Adams PE 100).

---

1 Department of Zoophysiology, University of Aarhus, DK–8000 Aarhus C., Denmark

Polarographic Oxygen Sensors (ed. by Gnaiger/Forstner)
© Springer-Verlag Berlin Heidelberg 1983

This material is sufficiently impermeable to allow the passage of water of zero oxygen tension without any measurable rise in tension, even if the passage time extends to 1–2 min.

A further check on the validity of the p_{O_2} readings of water passed through polyethylene catheters was made by setting the POS at zero by injecting an O_2-free solution (sodium sulfite and borax solution) into the cuvette. Subsequently, water deoxygenated by N_2 equilibration was passed through the polyethylene catheter into the cuvette without any readable change in the zero setting. This checking procedure excluded the possibility that O_2 molecules could pass through the catheter walls and distort the true water p_{O_2} values at the water flow rates employed.

By way of a three-way stopcock and a catheter equal in length to the sampling catheter the cuvette holding the POS is connected to a beaker placed in the experimental aquarium and containing water of the experimental temperature. During calibration the stopcock is switched and air-saturated water passes the POS at the same flow rate as during sampling from the animal.

In experiments involving rapid changes of the inspired oxygen tension two sensors should be employed, one of which is used for continuous monitoring of ambient p_{O_2}. If measurements are carried out at conditions of stable inspired p_{O_2}, one sensor will suffice.

The ventilatory flow of water passing the respiratory exchange surfaces is recorded by a Statham SP 2202 electromagnetic flowmeter.

2 Applications on Fish

Obtaining samples of exhaled water from undisturbed and unrestrained fish presents various problems. If exhaled water is sampled via a single catheter in the branchial cavity, uncertainty may exist as to whether the sample represents mixed exhaled water [1, 2]. On the other hand, attaching rubber membranes to animals to confine the expired water current in order to obtain truly mixed expired water, may place considerable restraint on the animal.

In a study of ventilation and gas exchange in the carp (*Cyprinus carpio*), we have employed a respiration mask (Fig. 1) made from thin flexible rubber [6]. The mask is tailored to the fish on the basis of a cast of the fish head, made by rapid impression of the head in dental cement in the way described in detail by Glass et al. [3]. The mask can be equipped with electromagnetic flowmeter probes at the exit orifices extending from each of the opercular openings. Catheters for continuous water sampling were similarly placed well inside the outflow extensions from the opercular openings (Fig. 1). The success of this procedure depends on the special morphology of the fish head and the magnitude of breathing movements. In the carp the movements of the mouthparts involved in breathing are rather small, whereas in other species, like trout, breathing involves large movements of the jaws. In such cases a mask technique is difficult to apply because it will restrict normal breathing movements. When useful, the mask will permit continuous sampling of truly mixed expired water for monitoring of oxygen tension as described above.

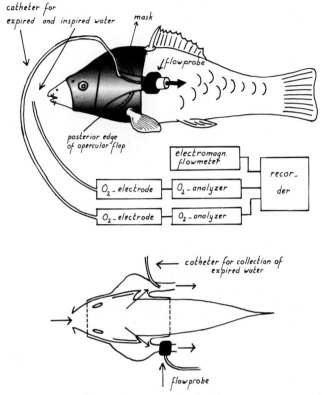

Fig. 1. Carp equipped with rubber mask permitting channeling of respiratory water current for measurement of ventilation by an electromagnetic flow probe and sampling of exhaled water [6]

This experimental arrangement was used on the carp to study how gas exchange parameters were affected by acclimation to hypoxic water [6]. It was demonstrated that not only were values for O_2 extraction from the respiratory water current higher than those hitherto reported for fish, but hypoxia acclimation resulted in a further elevation of O_2 extraction when compared with values from normoxic fish studied during acute exposure to hypoxic water (Fig. 2). These significant findings most certainly would have escaped notice if water O_2 tension had not been continuously monitored in combination with concurrent measurements of water ventilation.

A similar procedure was employed to study respiratory adjustments of the flounder (*Platichtys flesus*) to hypoxic water [5]. In this case advantage was taken of the fact that during undisturbed ventilation the fish rests buried in the sediment. Expired water was sampled continuously via a catheter attached inside the outflow orifice of a funnel placed in the sediment over the upper operculum (Fig. 3). Gill ventilation was recorded concurrently by attaching a flow probe to the neck of the collecting funnel. Alternatively, flow could be recorded from a separate funnel placed over the mout of the fish (Fig. 3). The latter procedure excludes the possibility that a portion of the water current may exit from the lower opercular opening. If care was taken that the edges of the funnels were buried in sediment, no water leaks around the edges could be detected.

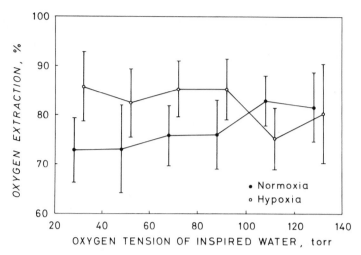

Fig. 2. Percentage extraction of oxygen from the ventilatory water current in normoxia- (*filled circles*) and hypoxia- (*open circles*) acclimated carp as a function of inspired oxygen tension. Each *point* is the average of 30–60 determinations in five different fish. *Bars* are 1 SD. Except at the highest level of oxygen tension average values for the two acclimation groups are significantly different (t-test, $P < 0.001$) [6]

This was checked by placing small amounts of a concentrated dye immediately above the sediment surface close to the edge of the funnel.

The main finding of the *Platichtys* study, directly dependent upon the experimental arrangement described above, was that hypoxia acclimation of the flounders allowed them to maintain an O_2 uptake rate in hypoxic water twice the value for flounders acutely exposed to hypoxia (cf. Chap. II.3). The higher level of O_2 uptake in the hypoxia-acclimated flounder compared to the acutely hypoxia-exposed fish resulted from an ability to maintain the same O_2-extraction in spite of a higher level of ventilation. The recording of these differences depended on the simultaneous monitoring of

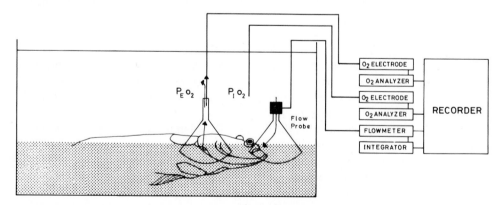

Fig. 3. Schematic illustration of a flounder in the experimental aquarium showing arrangement used for measurement of inspired ($p_{O_2}^i$) and expired ($p_{O_2}^e$) water O_2 tensions as well as inspired water flow [5]

inspired-expired O_2-tensions and concomittant ventilation changes during hypoxic exposure.

The funnel technique as applied to the flounder has the advantage that no equipment needs to be attached to the animal itself, but its application naturally is limited to burrowing animals.

3 Applications on Invertebrates

Placements of masks or other structures confining respiratory water currents in invertebrate animals are easily done if the species studied possesses an exoskeleton, such as most decapod crustaceans, shell-bearing mollusks, or tubiculous polychaetes. Similarly, burrowing forms, like many holothurians and polychaetes, can be studied without contact disturbance with the animal itself.

Figure 4 pictures the brachyuran crab, *Cancer magister,* equipped with a mask molded for an optimal fit after casting the area surrounding the mouth parts in dental cement. The mask can be glued to the exoskeleton by various fast-setting glues or be attached by anchoring bands (sutures, rubber bands) to protruding parts on the exoskeleton. A POS was placed with the sensing end inside the mask screening the well-mixed exhaled water before it exited through an electromagnetic flow probe giving a continuous record of the ventilated volume. Other catheters could be placed at appropriate sites for sampling of arterial and venous blood as indicated in Fig. 4. Crabs subjected to the procedure allowed samplings and continuous recordings of information needed for a complete gas exchange characterization of the animal [4]. The same approach has recently been applied on four other species of crabs including *Carcinus*

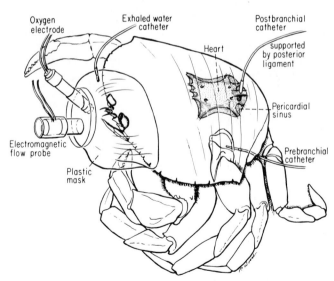

Fig. 4. Schematical representation showing placement of catheters for blood sampling and the mask containing the oxygen sensor and the electromagnetic flow probe [4]

maenas, weighing only about 20 g, thus demonstrating the applicability of the technique also to small species.

Among tubiculous polychaetes, species practicing unidirectional ventilation of a burrow are ideally suited for gas exchange analysis by the described method. Preliminary experiments on the polychaete *Nereis diversicolor,* weighing as little as 0.2 g, allowed simultaneous measurements of ventilation and O_2 extraction, yielding values for O_2 uptake corresponding well with those obtainable by closed respirometry.

For small species having a low ventilation, the direct passage of exhaled water from the expired current may not be possible, since a certain flow rate through the O_2-sensor cuvette is necessary for continued, stable sample flow. In such cases the application of a funnel for confinement of the expired water current in burrowing forms will act as a reservoir, giving average values of O_2-extraction when the funnel water is intermittently siphoned by gravity or aspirated by a pump through the O_2 cuvette.

Another important limitation of the technique relates to the dimensions of tubes and other flow channels interposed in the natural ventilatory current set up to connect the latter to the flow and p_{O_2} measuring devices. The propagation of ventilatory currents in all aquatic animals, whether they employ active muscular pumping or ciliary activity, is powered by very small pressure gradients, often less than 1 cm of H_2O (0.1 kPa). This fact implies that normal ventilatory currents are subject to very low values of resistance. If the arrangements of masks and tubes connecting the measuring equipment significantly alter the resistance to water flow, distorted values of ventilation may be recorded, leading in turn to nonrepresentative values of O_2-extraction. Because of this limitation, the largest possible flow probes and channeling devices should be used. However, since the flow probes are velocity sensors, the size of the probe diameter must be selected as a compromise between the requirement for a low resistance and a velocity high enough to detect pulsatile flow. In the described applications the added resistance imposed by connecting channeling devices in direct series with the natural ventilatory current have been small and negligible judged from pressure buildups less than 1 mm H_2O. By comparison the pressure gradient across the gills of fish is typically more than an order of magnitude greater.

In summary we see the following principal advantages of using continuous recordings of ventilatory flow and water O_2 tensions in aquatic animals:

a) The simultaneous and continuous monitoring of ventilatory flow and inspired and expired water p_{O_2} permits detection of phasic changes in O_2 extraction and gas exchange.

b) When ambient water p_{O_2} (inspired) varies as during hypoxic exposure, the time course of compensatory changes in ventilatory flow and O_2 extraction can be evaluated.

c) A minimum of physical disturbance of the experimental animal is required. Attachment of masks or connecting vessels imposes a minimum of restraint, allowing unhindered breathing movements and for many species data may be obtained during locomotory activity, like swimming.

d) For burrowing species no physical contact with the animal is required.

e) Calibration of POS, as well as flow probes, can be done easily and rapidly in situ.

References

1. Davis JC, Watters K (1970) Evaluation of opercular catherization as a method for sampling water expired by fish. J Fish Res Board Can 27:1627–1635
2. Garey WF (1967) Gas exchange, cardiac output and blood pressure in free-swimming carp (*Cyprinus carpio*). Ph D Dissertation, State Univ New York Buffalo
3. Glass ML, Wood SC, Johansen K (1978) The application of pneumotachography on small unrestrained animals. Comp Biochem Physiol 59A:425–427
4. Johansen K, Lenfant C, Mecklenburg TA (1970) Respiration in the crab, *Cancer magister.* Z Vergl Physiol 70:1–19
5. Kerstens A, Lomholt JP, Johansen K (1980) The ventilation, extraction and uptake of oxygen in undisturbed flounders, *Platichtys flesus.* (Responses to hypoxia acclimation). J Exp Biol 83: 169–179
6. Lomholt JP, Johansen K (1979) Hypoxia acclimation in carp — How it affects O_2-uptake, ventilation and O_2-extraction from water. Physiol Zool 52(1):38–49

Chapter II.3   The Twin-Flow Microrespirometer and Simultaneous Calorimetry

E. Gnaiger[1]

1 Introduction

Continuous long-term monitoring of respiratory rates in aquatic organisms is only possible in open-flow systems providing controlled environmental conditions during the experiment. This is a basic requirement for many topics in ecophysiological and applied research, and there is an apparent need for more detailed respiratory studies with organisms of different size and under a wide range of conditions. The most important aspects are: metabolic adaptation and acclimation to environmental factors (Chap. II.2); functional relations of oxygen uptake and locomotory activity (Chap. II.5) and growth; internal rhythms (Chap. II.7); quantification of respiratory energy loss complementing assimilation and production measurements in energy budget studies (Chap. III.4); sublethal effects of environmental contaminants; biological oxygen demand in water quality control.

Microrespirometers are commonly closed, thus not permitting extended experimental periods without the interference of uncontrolled variables. Methodological difficulties arise since the mere miniaturization of an ordinary respirometry system does not produce a functional microrespirometer. Due to the unfavorable volume-to-surface ratio in animal chambers with a volume less than 0.5 to 1 cm^3, increasing attention has to be paid to problems of oxygen diffusion and bacterial growth. The respiratory rate per unit volume (closed systems) or throughflow (open systems) is likely to decrease with size of the experimental animal, since reduction of volume or of throughflow entails new limitations of system functions. Consequently, an improved sensitivity and stability of the measuring and recording system is required. The twin-flow microrespirometer was developed to meet the need for a new apparatus for automatic and continuous monitoring of oxygen uptake in small aquatic animals. This instrument combines the following features:

a) Long-term stability and high precision. This is achieved by use of two polarographic oxygen sensors (POS) alternately switched to the measuring or the calibration position (twin-flow principle). Thus automatic control of baseline stability is possible without interruption of the oxygen record.

1 Institut für Zoologie, Abteilung Zoophysiologie, Universität Innsbruck, Peter-Mayr-Str. 1A, A–6020 Innsbruck, Austria

Polarographic Oxygen Sensors (ed. by Gnaiger/Forstner)
© Springer-Verlag Berlin Heidelberg 1983

b) High sensitivity and time resolution. Miniaturization of the stirring chambers and minimal dead volume reduce the response time. A good time resolution of metabolic patterns is achieved even at low rates of throughflow, when the system is run at its lower limit of detection (3 nmol O_2 h^{-1}).

c) Controlled environmental conditions. The open-flow system permits long-term recording of oxygen uptake at constant levels of oxygen concentration. Parameters of the medium can be altered during the experiment (e.g., in toxicological studies) without interfering with the continuous record of respiration.

d) Flexibility. Every type of animal chamber can be used in conjunction with the twin-flow system which therefore can be adapted for several types of aquatic organisms and for the simultaneous monitoring of additional physiological parameters (e.g., locomotory activity).

e) Simultaneous calorimetry. The twin-flow microrespirometer was specifically designed to provide a constant flow regime even during calibration. This is a basic prerequisite for simultaneous respirometry and calorimetry in an open-flow system.

Under aerobic and balanced physiological conditions, excellent agreement between indirect (respirometric) and direct calorimetry was observed over short as well as long periods of time (cf. Chap. II.4). In these situations accurate calculation of respiratory energy dissipation is possible (App. C). Under behavioral, environmental, and toxicological stress, however, total energy dissipation (aerobic and anoxic processes) can be considerably higher than indicated by aerobic respiration. Simultaneous direct and indirect calorimetry not only open new perspectives in ecological bioenergetics, they assist in exploring mechanisms of metabolic energy expenditure.

2 Choice of the Proper Respirometer

At the present stage of technology, the choice for the construction or — if available — for the purchase of a particular type of respirometer is predominantly governed by economic considerations. The biological arguments for or against the various systems have been widely discussed and the technical problems have been solved in principle even for the more extravagant fields of application such as in situ measurements in the deep sea (Chap. III.5) or for space biology [20]. However, the capital investment involved is unrealistic for standard laboratory applications. On the other hand, inexpensive and simple respirometric techniques may result in scientific and economic failure if the hours wasted on unsatisfactory experiments are taken into account. Such costly experience stimulated the perfection of the twin-flow respirometry system for the automatic long-term monitoring of oxygen uptake in very small macrofauna or a reasonable number of meiofauna (< 1 mg dry weight of biomass).

2.1 Gasometric Methods

The most sensitive microrespirometers (10^{-3} to 10^{-5} μmol O_2 h^{-1}) are modifications of the Cartesian diver technique [29, 49] which is becoming increasingly important in studies of unicellular species and meiofauna ($<$ 0.1 mg wet weight). In the case of the electromagnetic diver [33], recording is automatic and continuous. Its application requires high skill and practice, especially if metabolic effects of low p_{O_2} levels are studied [32]. The water volume in the diver has to be kept small to avoid diffusion errors. When confined in a drop of water zooplankton and other active animals are restricted in their locomotory behavior: some copepod species die within 2 to 10 h in the Cartesian diver [21]. A premortal increase in metabolic rate followed a plateau, the physiological interpretation of which may be complex, although the level of oxygen consumption agreed with that measured in an open-flow system [21].

If the experimental oxygen uptake rates are in the order of 1 μmol O_2 h^{-1}, Warburg and Gilson respirometers may be considered for studies of small aquatic animals. However, a comparison of methods makes evident that shaking the respirometer flask disturbs the animals and influences metabolic rates (Table 1). While the average rate is doubled in aquatic oligochaetes by shaking, it decreases in other animals [48]. Nonwetting species such as *Daphnia* are caught in the surface film of the water in shaken Warburg flasks. Good agreement between oxygen uptake under these highly abnormal conditions with that in a closed bottle respirometer [44] suggests cautious interpretation of results obtained with both of these methods (see below). Since agitation distors the respiratory response to other environmental factors in various species [42], the Warburg method cannot be recommended for monitoring the physiological effect on animals of aquatic pollutants.

In nearly all gasometric respirometers, the gas phase in equilibrium with the medium containing the organism is scrubbed free of carbon dioxide. Although alimination of CO_2 has no detectable effect on *Lumbriculus* [14], it decreases the metabolic rate of other species [45, 48].

2.2 Open Versus Closed Respirometric Systems

Analysis of dissolved oxygen circumvents the problem of CO_2 absorption and p_{O_2} equilibration between the aqueous and gaseous phase. Without much instrumental expense and with some practical skill, the micro-Winkler method (App. D) may be used in a simple closed-bottle respirometric technique. Disturbance of the animals by transfer into a respirometer usually results in greatly increased oxygen uptake rates during the first hours of the experiment [21, 25, 45]. This may explain a large fraction of the difference in rates obtained with the closed-bottle and open-flow respirometric method [8, 25]. The continuous record of oxygen uptake by POS aids substantially in interpreting the results.

Various additional factors acting in concert may obscure the physiological interpretation of measurements made in closed systems: because oxygen is gradually depleted in the course of the experiment, respiration decreases in animals whose respiration rate is oxygen-dependent. At the same time, the rate may be unstable due to starvation. Continuous monitoring by POS of p_{O_2} down to oxygen depletion in closed systems reveals a response to the rapid change in oxygen content [34]. This response may, however, be masked by other effects varying with time in an uncontrolled manner [13]. Due to the fact that in closed respirometers no sustained measurements are

Table 1. Comparison of manometric (Warburg) respirometry involving shaking of the animal chamber (A) and methods without mechanical disturbance of the animals (B–C); B: closed system respirometry with continuous or discontinuous p_{O_2} analysis by Winkler or POS; C: microcalorimetric flow method, the oxygen equivalent of heat production was estimated at $1 \text{ mW} \triangleq 8 \text{ } \mu\text{mol O}_2 \text{ h}^{-1} \triangleq 0.18 \text{ cm}^3 \text{ O}_2 \text{ h}^{-1}$ (App. C). Test animals: *Tubifex tubifex* (T.t.) and *Lumbriculus variegatus* (L.v.) (Oligochaeta). Experimental temperatures (θ_{exp}) and individual wet weights ($_wW_{exp}$) are given. Dry weight is $0.17 \times {_w}W$

Method	Species	θ_{exp} (°C)	$_wW_{exp}$ (mg)	$\dot{n}_{O_2}$ (20°C) μmol O$_2$ h^{-1} g^{-1} Range[a]	Mean[a]	Mean[b]	Ref.
A	T.t.	18–20	?	9.2–13.8	11.9		[22]
A	T.t	17–19	?	12.9–24.6	18.2		[31]
A	L.v.	20	?	6.2– 9.0	7.6		[28]
A	T.t.	25	1.9	12.7–21.6	16.6	11.0	[11]
A	T.t.	20	2.5	23.4–25.6	24.5	17.4	[40]
A	T.[c]	20(?)	?	11.8–24.6	18.8		[47] [g]
A	L.v.	20	10/1	10.4–18.4[d]	14.4	10.4	[g]
A	T.t.	20	7.8	4.5– 6.3	5.4	5.0	[7]
				9.4–19.9[e]	14.7	11.0	
B	T.t.	19	?	6.7– 7.4	7.1		[g]
B	T.t.	14	6	8.6– 9.5	9.0	7.9	[g]
B	T.t.	20	10/1	4.4– 8.1[d]	6.2	4.4	[g]
B	T.t.	20	2.4	3.2– 4.1	3.9	2.7	[4]
B	T.t.	15	1.8	5.1– 7.4	6.5	4.2	[g]
B	T.[f]	25	5	10.0–11.6	10.8	9.1	[g]
C	L.v.	20	10	5.4– 8.3	6.9	6.9	[14]
				5.2– 9.2[e]	7.2	5.9	

[a] $\dot{n}_{O_2}$ (20°C) is the weight specific rate of oxygen consumption at 20°C, calculated as

$$\dot{n}_{O_2} \text{ (20°C)} = \dot{n}_{O_2} \times \theta_{exp} \times Q_{10} \times \frac{20-\theta_{exp}}{10}; Q_{10} = 2.2$$

[b] $\dot{n}_{O_2}$ (10 mg) is the specific rate of oxygen consumption corrected for temperature (20°C) and individual weight ($_wW_{exp}$) and expressed for a standard individual of 10 mg $_wW$

$$\dot{n}_{O_2} \text{ (10 mg)} = \dot{n}_{O_2} \times {_w}W_{exp} \times (10 \text{ mg}/_wW_{exp})^{(b-1)}; b = 0.75$$

[c] *Tubifex* and *Limnodrilus* in mixed samples
[d] Calculated rates for individuals of 10 mg and 1 mg according to the published weight-rate relationship
[e] 95% confidence limits of the mean
[f] *Tubifex templetoni*
[g] For references see [14]

possible at constant low oxygen concentration, ecologically meaningful information on physiological functions in a low oxygen environment remains elusive (see also Chap. II.2). Furthermore, the experimental oxygen regime is unidirectional in closed systems as opposed to fluctuating environmental conditions in nature, which can be simulated only in open flow respirometers.

Users of open-flow respirometers mention also the disturbing effect of accumulating excretory products on oxygen uptake, while users of closed systems ascribe little significance to this effect. These attitudes reflect the lack of experimental proof for either argument. Addition to the medium of 2 mmol dm^{-3} acetate, an endproduct excreted by several anoxic invertebrates, did not significantly influence the aerobic and anoxic rate of heat dissipation of *Lumbriculus* (unpublished observation). However, the role of accumulating excreta in supporting a high rate of bacterial respiration has also to be considered in closed respirometers.

As the pitfalls of closed systems become increasingly recognized, intermittent flow respirometers are coming into wider use ([1], Chaps. II.1, II.5), although a glance through the literature shows that for zooplankton the closed bottle method is still most commonly relied upon.

3 System Design Parameters of the Twin-Flow Respirometer

Just as the choice of a POS represents a compromise between stability, sensitivity, stirring requirements, and response time (Chap. I.1), so does any particular design of a respirometer involve certain compromises regarding (1) the sensitivity, (2) time resolution of oxygen uptake rates, (3) a more or less disturbing environment for the experimental animal, and (4) ease of construction and operation. The decisions are dictated by the specific application, by the desired accuracy, and by keeping costs low.

In describing the twin-flow microrespirometer, I will briefly specify the main arguments underlying the construction of the system. Their importance will become more readily apparent in the discussion of system functions and in some case studies. All constituent parts of the respirometer in contact with the experimental medium are made of stainless steel or glass with the exception of the microvalves, the coating of the stirring magnets, and the Teflon stoppers of the animal chamber. Plexiglas and other easily machinable materials with a high oxygen solubility and permeability must be avoided. High adaptability of every system component and exchangeability of the independent functional parts determined the design. While it proved almost impossible to reduce the cost of the apparatus without considerable detriment to functional parameters, the most expensive component — the time lost due to methodological weaknesses — can be eliminated for a wide range of applications beyond those mentioned in this article.

3.1 The Flow Regime

Many applications of POS in conjunction with flow respirometers in studies on small aquatic animals have been described ([3, 8, 9, 21, 23, 26, 30, 37, 39, 43], Chap. II.7). Calibration of the POS and measurement of oxygen uptake are usually achieved by an alternating water flow to the oxygen sensor, either directly from a water reservoir or after passage through the animal chamber. If calibrations are made only at the beginning and at the end of an experimental run, then the accuracy of the measurements is

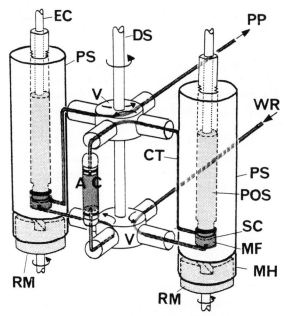

Fig. 1. The twin-flow microrespirometer. The flow regime is shown with the POS on the right hand side in the calibration position. The medium is pumped from the water reservoir (*WR*) to the first stirring chamber (*SC*) and POS by the lower four-way valve (*V*). The water flows upward through the stainless steel capillary tube (*CT*) along the side of the stainless steel POS sleeve (*PS*) and is directed by the upper four-way valve (*V*) to the animal chamber (*AC*) with flow direction downwards. Flow through the second stirring chamber with the POS in measuring position is symmetrical. The outlet of the upper valve is connected to the peristaltic pump (*PP*) by a Teflon tube. *DS* drive shaft, *EC* electrode cable, *MF* magnetic follower, *MH* stainless steel magnet housing, *RM* rotating magnet

limited by the nonlinear drift of the sensor signal. Intermittent calibration in these respirometers results in a discontinuous record of oxygen uptake and interrupts the flow of water through the animal chamber in most instruments. During calibration, the change in water current and the gradual oxygen decline may disturb the animals and interfere with internal rhythms.

These disadvantages do not apply to the twin-flow respirometer (Fig. 1). Due to the symmetrical arrangement of the two POS and measuring chambers, a continuous flow is maintained in a constant direction through the animal chamber (AC) and stirring chamber (SC) both during measurement and calibration. This is, in fact, the crucial requirement for simultaneous calorimetry, since the slightest change in pressure and flow through the animal chamber disturbs the calorimeter baseline. As the two four-way valves (V) are switched at the same time, the sensor previously in the calibration position (when water from the reservoir enters the stirring chamber directly) is set into the measuring position (water passes the other POS and the animal chamber before flowing into the stirring chamber), simultaneously the other POS is switched to the opposite position (Fig. 6). The POS switched to the measuring position shows, after some latency at an initial plateau, an overshoot which must not be mistaken for a fluctuation in oxygen consumption (Fig. 8, POS 2). It is roughly proportional to the oxygen up-take rate (cf. Figs. 6 and 8) and is due to water from the stirring chamber (previously in the measuring position) and the capillary tubes connecting this chamber to the two valves, passing through the animal chamber for a second time (Fig. 1). Therefore a compromise has to be made between the advantage of frequent calibrations (every 2 h) with intermittent interruptions of the continuous record of oxygen uptake, and a long measuring and intercalibration interval (e.g., 12 h) with less certainty as to the calibra-tion factor of the measuring POS.

Microvalves with minimum volumes (< 90 mm^3), as used in gas or liquid chromato-graphy, are commercially available. In the present system two four-way valves (Pharmacia, LV4) are connected head to head. A central bore through the plug of the upper valve contains the drive shaft, which provides an easily accessible means of switching the two valves simultaneously (Fig. 1) either by hand or by an electronically timed motor. As the Pharmacia connections are vulnerable to damage by the stainless steel capillary tubes, there is always the danger of leakage and high diffusion rates. Hamilton valves (HVP 86779) provide a promising alternative, but no bore can be drilled through their plugs.

A LKB 10200 Perpex peristaltic pump with gear ratios of 1:200 or 3:250 is em-ployed to keep the flow rate constant at chosen intervals between 3 and 25 cm^3 h^{-1}. The observed variability of maximally $\pm 0.7\%$ may be largely due to errors in the mea-surement of the flow rate. With care taken to exclude evaporative loss of the water col-lected in a measuring cylinder, the long-term stability of flow rates within experiments was found to be 0.1%. Due to aging of the silicon rubber tubings, the flow rate chang-ed much more between experiments. The connection of the pump to the outflow of the respirometer renders the precaution of oxygen diffusion into the peristaltic tubings unnecessary.

3.2 The Open-Flow Microrespirometer-Calorimeter

For simultaneous direct calorimetric and indirect respirometric measurement of metabolic activity, the twin-flow microrespirometer can be connected to an open-flow calorimeter (Fig. 2). At present, the best-suited commercial calorimeter for monitoring small animals is the LKB-2107 flow sorption microcalorimeter [13, 14]. The geometry of the heat detector system sets a restrictive upper size limit for the test animals. Optionally the LKB system can be equipped with a 0.5 cm^3 pyrex flow sorption chamber and a 1.4 cm^3 stainless steel flow ampoule. The latter, however, is not recom-mended for very accurate work due to problems associated with the standard calibra-tion procedure [46]. The flow-through vessel is contained in an aluminum block, and in the case of the flow sorption chamber the gold outlet capillary tube is coiled to en-sure perfect heat exchange in the calorimeter detector. Thus, any heat effect in the animal chamber builds up a temperature gradient across the thermopiles, which in turn gives rise to a voltage signal while the heat dissipates into the constant temperature heat sink of the calorimeter. The ratio of the measured potential and the heat dissipat-et by the animals or by an electrical resistor (calibration heater) is the static gain value of the system, amounting to 0.055 μV μW^{-1} [13] or roughly 6 μV per 1 μmol O$_2$ h^{-1} consumed in aerobic metabolism.

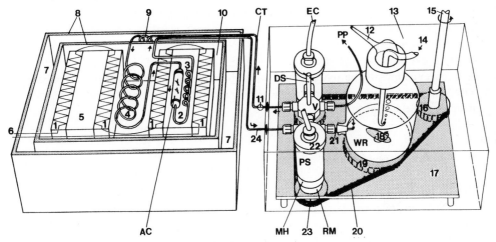

Fig. 2. The open-flow respirometer-calorimeter: the twin-flow microrespirometer (cf. Fig. 1) connected to the LKB-2107 flow sorption microcalorimeter. *AC* animal chamber, *CT* capillary tube, *DS* drive shaft for switching the two four-way valves, *EC* electrode cables of the POS, *MH* magnet housing of the bottom assembly of the stirring chamber, *PP* outlet Teflon tubing connected to the peristaltic pump, *RM* rotating magnet driving the magnetic follower in the POS-stirring chamber, *V* four-way microvalve, *WR* water reservoir for p_{O_2} equilibration. *1* thermopile, *2* heat detector, *3* detector heat exchanger (gold *CT*), *4* internal heat exchanger (gold *CT*), *5* static reference detector, *6* thermovoltage transducer cables connecting to the Keithley 150B microvolt ammeter, *7* constant temperature air bath of the calorimeter, *8* thermal insulation, *9* external heat exchanger (gold *CT*), *10* heat sink (aluminum block), *11* three-way microvalve connecting the stainless steel *CT* from the POS in calibration position with the gold *CT* to the animal chamber, *12* gas outlet, *13* constant temperature water bath of the respirometer, *14* gas line from the gas mixing pump to the water reservoir, *15* drive shaft connected to the synchronous motor, *16* gear wheel driving the toothed rubber belt, *17* lower mounting plate of the gear wheels and rotating magnets [the upper mounting plate (Fig. 3) is not shown here], *18* magnetic stirring bar in the water reservoir, *19* bracing wheel with impeller magnet, *20* rubber drive band, *21* stainless steel *CT* from the water reservoir to the POS, *22* retaining nut fixing the POS sleeve to the upper mounting plate (Fig. 3), *23* gear wheel (below mounting plate) driving the rotating magnet (above mounting plate), *24* capillary connection from the animal chamber to the POS in measuring position

The original Teflon inlet and outlet tubings of the calorimeter had to be replaced by gold capillary tubes to prevent gaseous diffusion and inhibit bacterial growth [13]. A three-way microvalve interpolated between the respirometer and the calorimeter inlet (Fig. 2) is advantageous during exchange of the animal chamber and for cleaning the apparatus. Inflow water is thermally equilibrated by passage through the external and internal heat exchangers before flowing through the animal chamber in the heat detector. The static reference detector compensates for all symmetrical disturbance effects of external temperature fluctuations. A symmetrical arrangement of the gold inlet and outlet capillaries is therefore necessary since the metal tubes enhance thermal leakage between the insulated heat sink and the calorimeter thermostat, and asymmetries result in diminished baseline stability. Since the drift of the calorimeter baseline as observed with high signal amplification (3 μV or about 5 μW recorder full scale) is correlated with the dynamic properties of the necessarily nonideal thermostat, mathematical baseline correction procedures can be employed [27]. The flow of water through the heat detector shifts the baseline by some constant value as long as the flow rate and the viscosity of the medium are constant. Rapid switching of the respirometer four-way valves for POS calibration produces no disturbing effect whatsoever. Between experiments the oxygen sensors can be removed from the stirring chambers for electrolyte renewal and application of new membranes without disassembling any other part of the apparatus.

3.3 POS and Stirring Chamber

Various types of POS may be used with the twin-flow respirometer. A choice has to be made between small and large cathodes, in other words between stirred and unstirred POS (Chap. I.1). In preliminary experiments small-cathode sensors (Radiometer, E 5048/0) were sensitive to pressure and electrical interferences in a stopped-flow system. Hence a large-cathode POS (YSI 5331) was chosen and equipped with a minimum volume stirring chamber (Fig. 3). This sensor produces a current of 0.09 to 0.13 μA (air-saturated water at $12^\circ C$ with the YSI standard membrane) which is easily converted into a voltage (R_L 100 kΩ, Chap. I.10) and displayed with a potentiometric compensation recorder. It proved practical to use the 300% zero suppression and set the calibration point at air saturation to about 90% of recorder full scale. The full scale deflection is then roughly equivalent to a change of 25% air saturation, and even high oxygen consumption rates should not reduce the oxygen content in excess of this limit, which would require resetting of the suppression range. After about 2 weeks of continuous operation, the signal of the YSI sensor usually becomes noisy and drifts off. This indicates that the electrolyte and membrane should be replaced. If the anode has turned dark it should be cleaned with a 25% ammonia solution.

Within the stirring chamber (Fig. 3) a strong and stable water current is generated at the POS membrane. At the same time the chamber acts as a centrifugal pump and exerts a pressure in the direction of throughflow (Fig. 4). The disk-shaped magnetic stirrer (Radiometer D 4030) displaces the major part of the volume of the stirring chamber and leaves just enough space to avoid resistance to the throughflow. At high rotation speeds, the minimum dead volume of the measuring chamber is also important to prevent wobbling of the stirrer.

Even at 30 rotations per second a stirring effect (0.06% per rotation/second) was observed. Therefore a synchronous motor (Papst, SZ 62.75-4-173 DeM/K-B357) is employed and the two rotating magnets (Fig. 1) are driven via a toothed rubber belt at a constant stirring speed of 25 rotations per second ($\stackrel{\triangle}{=}$ 1500 rpm). A third rotating magnet is cast in an epoxy gear wheel which acts as the bracing wheel for the drive band and is of larger diameter than the other gear wheels. This magnet agitates the magnetic stirring bar in the water reservoir (Fig. 2) to enhance gas equilibration of the water (Chap. I.2) during changes in experimental p_{O_2}. The gear wheels and drive magnets are mounted on stainless steel shafts which rotate in waterproof ball bearings. The stirring assembly also provides vigorous mixing of the constant-temperature water bath which is essential for precise thermoregulation.

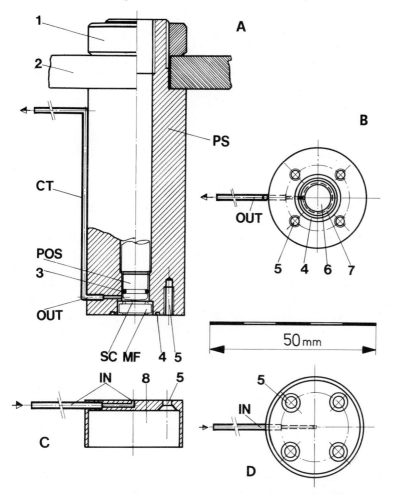

Fig. 3A–D. The stirring chamber. Cut view and cross-section of the POS sleeve (**A, B**) and the bottom assembly (**C, D**) respectively. *1* retaining nut, *2* Perspex mounting plate, *3* O-ring of the POS serving as a seal of the stirring chamber (*SC*), *4* O-ring groove for sealing the stirring chamber between POS sleeve and bottom assembly, *5* threaded hole for fixing the bottom assembly to the POS sleeve, *6* central bore of the POS sleeve, *7* stirrer guide, *8* drive magnet housing. Other symbols as in Figs. 1 and 2

An additional advantage of the stirring chambers warrants consideration. If the lower four-way valve (Fig. 1) is switched while the other valve is kept in a constant position, then the stirrer on the right-hand side maintains a flow through the animal chamber. In this way the open-flow system is changed into a closed-flow respirometer. The water flow is high at 25 rotations per second, despite the resistance of the capillary tubings and valves, and may be regulated by a needle valve interpolated between the four-way valve and the animal chamber. An open-flow current respirometer can be made by inserting a third stirring chamber (without POS) to produce a short-circuit flow (cf. Fig. 4).

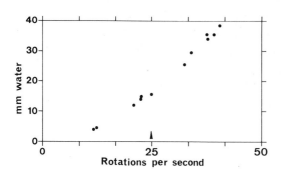

Fig. 4. Pressure exerted by the stirring chamber of the POS as a function of stirring speed as measured by the rise of the water column in a Teflon capillary tube (1 mm water = 9.80 Pa)

For some applications, however, rigorous stirring of the medium may not be feasible. The bottom assembly of the measuring chamber (Fig. 3C,D) may be used in conjunction with any other POS sleeve providing an O-ring seal. With a micro-cathode sensor (e.g. Chap. I.7) no magnetic stirrer is required. Then the unidirectional flow regime, irrespective of valve position, is especially advantageous, since the central bore in the bottom assembly leads the throughflow directly to the central cathode of the POS and minimizes oxygen depletion at the membrane. This is important even with some POS which are specified as "stirring-insensitive".

3.4 The Animal Chamber

The twin-flow respirometer can be connected to any type of animal chamber. The latter has to be adjusted in shape and in size to the experimental animal and to the desired range of flow rates. The animal chamber may also provide facilities for additional simultaneous measurements and may even be remote, such as in simultaneous direct and indirect calorimetric measurements (Fig. 2). The design of the animal chamber, however, is not as simple as it might at first appear to be. Every construction involves a compromise between minimum restriction of the animal and time resolution of oxygen uptake rates, since the volume and the pattern of water exchange in the animal chamber and the flow rate comprise the key factors determining the time response characteristic of a flow respirometer (Table 2).

Table 2. Interdependence of throughflow and volume of the animal chamber relative to animal size in determining the system specifications of a flow respirometer

| | | Volume | | |
		Large	Small	General
Flow	Low	Poor time resolution, large p_{O_2}-gradients	Reasonable time resolution	High accuracy
	High	Reasonable time resolution susceptible to animal behavior	Good time resolution, well-defined p_{O_2}	Low accuracy
	General	Unrestrictive	Restrictive	

A

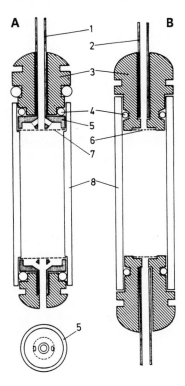

B

Fig. 5A,B. Animal chambers. **A** pyrex chamber of the LKB-2107 flow sorption microcalorimeter, 0.5 cm³ volume. **B** similar glass chamber with the same inner dimensions for direct connection to the microrespirometer. *1* gold capillary tube, *2* stainless steel capillary tube, *3* Teflon cap, *4* O-ring, *5* gold nut, *6* Nickel grid (as used in electron microscopy), *7* nylon gauze, *8* glass tube

Generally two ideals may be approached for obtaining an optimum time resolution: (1) a capillary tube chamber without mixing, where a laminar flow of water extends over the whole cross-section; and (2) a chemostat vessel where the rate of mixing is high relative to the throughflow. In the latter case the chemostat theory applies and backcalculation of actual rates is possible using mathematical correction factors for the time course of oxygen concentration at the outflow (see below). The turbulence required for ideal mixing and the damping of the signal increase with increasing volume of the animal chamber, thus a small volume respirometer provides a more accurate time resolution. Both ideals can only be approached to some extent and are restrictive or disturbing to most animal species.

In the applications discussed below, the 0.5 cm³ pyrex chamber of the LKB flow sorption microcalorimeter was used for simultaneous calorimetric and respirometric measurements (Fig. 5A). A similar glass tube chamber (Fig. 5B) served for independent applications of the twin-flow microrespirometer. However, for the simultaneous measurement of locomotory activity of microzooplankton a plain glass chamber of the same inner dimensions with side walls of stainless steel had to be constructed. This ensures an undistorted optical image of the animals (Gnaiger and Flöry, in prep.). These chambers allowed free locomotory movement of planktonic copepods.

In ecological bioenergetics, the aim is to determine oxygen uptake under natural, unrestrained conditions. The resolution of rapid changes in respiratory rate is then impossible and less important. In such studies large unstirred chambers provide the animals with as natural an environment as possible, and time averages of oxygen consumption can be measured accurately despite the uncertainty which prevails regarding short-term fluctuations.

4 Precision and Accuracy in Static and Dynamic Analyses

The accuracy of oxygen uptake measurements with the twin-flow microrespirometer depends on several variables, last but not least on the skill of the experimenter. These variables will be discussed on the basis of experience gained under experimental conditions considered to be difficult, i.e., with factor combinations resulting in low respiratory rates (0.01 to 0.1 μmol O_2 h^{-1}), especially at low p_{O_2} (down to 4 mbar = 0.4 kPa or 2% air saturation).

Experiments were performed in a constant temperature room ($17 \pm 1°C$) with relative air humidity of 30% to 65%. The respirometer was immersed in a water bath of $6 \pm 0.02°C$. The animal chamber was placed either in the water bath or for simultaneous heat dissipation measurements in the calorimeter at an identical temperature. The biomass in the experiments ranged from 0.7 to 5 mg dry weight. Millipore-filtered (0.45 μm) tap water was used as the experimental medium and was pumped at a flow rate of 5.7 cm^3 h^{-1} through the system. The water reservoir was continuously aerated or equilibrated with a gas mixture at a gas flow rate of about 8.5 dm^3 h^{-1} (Wösthoff gas mixing pump, 1SA 27/4). Since humid room air served as one component of the gas mixture, a special correction factor was employed for the calculation of p_{O_2} in solution (App. B).

4.1 Calibration Factor

The barometric pressure, $_bp$, was frequently recorded and interpolated. Hence the oxygen concentration of thermostated water entering the POS chamber in calibration position, c_{in}, is

$$c_{in} = S_s \times p_{O_2},$$ (1)

where S_s is the solubility of oxygen at the experimental temperature [μmol dm^{-3} kPa^{-1}], and p_{O_2} is the experimental partial pressure of oxygen [kPa] (App. A and B). The calibration factor of the POS in terms of partial pressure, F_p [kPa μA^{-1}], is

$$F_p = \frac{p_{O_2}}{I_{in} - I_r},$$ (2)

where I_{in} is the electrode signal corresponding to the calibration p_{O_2} at time t, and I_r is the residual (zero) current of the sensor [μA]. I_r amounted to 0.5 to 1.5% of the aerobic signal and can be estimated by the intercept of an I_{in}/p_{O_2} regression in experiments with several p_{O_2} levels. However, the variability of the "oxygen current", $I_{O_2} = I_{in} - I_r$, contributes to the variance of the intercept. In 10 experiments lasting for 1 to 3 days with 4 p_{O_2} levels (100%, 10%, 5%, 2% air) the 95% confidence interval of

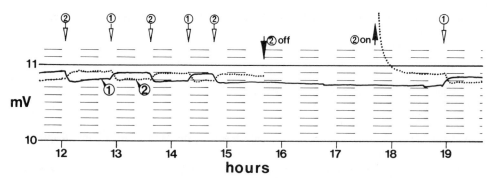

Fig. 6. Blank oxygen consumption run (12°C, air-saturated water, f = 5.9 cm³ h⁻¹) recorded on a two-channel potentiometric recorder (10 mV ≙ 0.1 μA). Switching from measuring to calibration and vice versa is shown by *arrows* and the *corresponding numbers* indicate which POS is switched into calibration position. The polarizing voltage (0.8 V) of POS 2 in calibration position was intermittantly switched off (*black arrow off*), and on (*black arrow on*)

the intercept (= extrapolated I_r) ranged from 0.1% to 1.7% of the electrode current in air. This uncertainty affects calculations of aerobic respiratory rates by less than 1%, but becomes increasingly important at low p_{O_2}. Zero calibrations in Na-sulfite solutions made before mounting the POS in the respirometer are not recommended. In many cases these were considerably higher than calibrations made in the respirometer with pure nitrogen, where I_r was extremely stable once the steady state was reached. The instability of the calibration factor can therefore be attributed mainly to the variable oxygen current at constant I_r: frequent calibrations at experimental p_{O_2} are appropriate to cancel out this source of error (Fig. 6). Since the barometric pressure changes slowly with time, it is convenient to combine all changes in a concentration-based calibration factor, F_c [μmol dm⁻³ μA⁻¹],

$$F_c = \frac{c_{in}}{I_{in} - I_r} .$$ (3)

The signal of the POS in measuring position, I_{out}, is then simply converted to the oxygen concentration of the medium leaving the system, c_{out},

$$c_{out} = F_c \times (I_{out} - I_r) .$$ (4)

The calibration factor will vary depending on the type of POS, the quality of the membrane and O-ring fit (Chap. I.1), ageing and poisoning of the cathode, anode and electrolyte (Chaps. I.2, I.5), and on bacterial growth on the membrane (Chap. II.9). In long-term tests from 40 to 120 h, the stability of the calibration factor was better than ± 2.5% under aerobic conditions. Occasionally following long anoxic periods, however, the signal of one POS did not stabilize for many hours or reached a steady state 15% higher than previously, while the other POS remained stable. This aberrant behavior remained unexplained (cf. Chap. I.4, control of membrane). Even for highly reliable POS, a sufficient stability < 0.2% can be achieved only with the aid of frequent calibrations of the undisturbed sensor, which is possible in the twin-flow system.

4.2 Oxygen Reduction Ratio, Blank Oxygen Consumption, and Oxygen Diffusion

The change in the amount of oxygen per unit time, $\dot{N}_{O_2}$ [μmol h^{-1}] is proportional to the flow rate, f [dm^3 h^{-1}], and the change in oxygen concentration,[2]

$$\frac{\Delta N_{O_2}}{\Delta t} = f \times (c_{in} - \bar{c}_{out}), \tag{5}$$

where $\bar{c}_{out}$ is the mean oxygen concentration in the POS chamber in measuring position during the time interval Δt. The time interval chosen has to be small relative to changes in barometric pressure, but large as compared to irregular fluctuations in c_{out} (see below). The oxygen reduction ratio, R_{O_2}, relates the concentration change to the input oxygen concentration,

$$R_{O_2} = \frac{c_{in} - \bar{c}_{out}}{c_{in}} = \frac{\bar{I}_{in} - \bar{I}_{out}}{\bar{I}_{in} - I_r}, \tag{6}$$

whence Eq. (5) becomes

$$\dot{N}_{O_2} = R_{O_2} \times c_{in} \times f. \tag{7}$$

S.D.(c_{O_2}), the relative standard deviation of the mean of consecutive steady-state measurements of c_{in} or c_{out}, was 0.12%. This results in a sensitivity of 3 nmol h^{-1} at a flow rate of 6.10^{-3} dm^3 h^{-1} (see Table 3). For comparison, the sensitivity of the Cartesian diver is 0.05 nmol h^{-1} [28] and the sensitivity of the flow microcalorimeter is 2 μW [13], equivalent to about 16 nmol O_2 h^{-1}.

Table 3. Relative standard deviation of oxygen uptake measurements [%] as a function of R_{O_2} and S.D.(c_{O_2}). From [37]

R_{O_2}	S.D.(c_{O_2}) [%]			
	0.1	0.5	1	2
0.05	2.8	14.0	28.0	55
0.10	1.3	6.7	13.0	27
0.15	0.9	4.4	8.7	17
0.20	0.6	3.2	6.4	13
0.25	0.5	2.5	5.0	10

The calculation of respiratory rates of the experimental animals, however, is complicated by the side effects of blank oxygen consumption and oxygen diffusion, so that confidence limits of animal respiration cannot be ascertained with the same accuracy. The oxygen consumption by the relatively large cathode of the POS is about 1 nmol h^{-1} under aerobic conditions, which is below the detection limit (Fig. 6). The aerobic blank oxygen consumption rate also could not be fully explained by bacterial

---

2 Although a negative value should be assigned to the amount of oxygen removed within a system, the convention of reporting oxygen uptake rates as positive values is adopted here for convenience. Hence oxygen diffusion into the system (which should be positive) is negative in this context

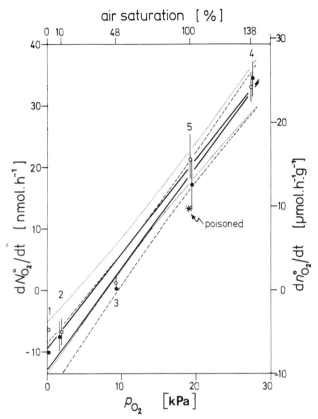

Fig. 7. Blank oxygen consumption as a function of p_{O_2} in experiment RJC1 after removing the animals from the animal chamber. The changes in oxygen are expressed as absolute rates, $dN^{\ominus}_{O_2}/dt$, and as $dn^{\ominus}_{O_2}/dt$ which is the blank rate relative to the dry weight of the animals whose oxygen uptake was recorded for 95 h (see Fig. 12). The blank rates were observed from 97 to 377 h in the following sequence of air saturation values: 100-48-138-100-48-10-0-100. The negative sign indicates net oxygen diffusion into the system. *Open* and *closed circles* are the means, and the *dotted* and *broken lines* are the 95% confidence limits of the two regression lines for POS 1 and POS 2 respectively [$\dot{N}^{\ominus}_{O_2}$ (POS 1) = $-9.52 + 1.530 \times p_{O_2}$; $\dot{N}^{\ominus}_{O_2}$ (POS 2) = $-13.21 + 1.649 \times p_{O_2}$). The *bars* represent the standard deviations of the means. The number of readings before and after calibration is indicated. The *arrow* points to the blank oxygen consumption obtained 20 h after poisoning with formaldehyde (4%)

respiration, as antibiotics or poisons exerted only a slight effect on the blank rate (Fig. 7). In 18 experiments with planktonic copepods the blank averaged 22 ± 5 nmol h^{-1}, but reached extraordinarily high values above 50 nmol h^{-1} in other cases. Hence the blank has to be determined and subtracted from the total rate for every experiment to achieve a precision better than 5% for an animal respiratory rate of 100 nmol h^{-1}. In studies of the rate/p_{O_2} relationship the blank rates have also to be determined at several p_{O_2} levels. Below 30%–50% air saturation oxygen diffusion into the system exceeded the chemical and biochemical processes of oxygen removal in blank runs. A least-squares regression gives a good estimate of the blank rate, which is a linear function of p_{O_2} (Fig. 7). By subtraction we obtain the specific rate of oxygen consumption, $\dot{n}_{O_2}$ [μmol h^{-1} g^{-1}], by the experimental animal(s) with total weight, W,

$$\frac{\Delta n_{O_2}}{\Delta t} = (c^{\ominus}{}_{out} - \overline{c}_{out})\frac{f}{W},\tag{8}$$

where $c^{\ominus}{}_{out}$ is the oxygen concentration of the outflowing medium corresponding to the blank rate determined at or calculated for the experimental p_{O_2}.

When the water bath was thoroughly flushed with nitrogen, the apparent diffusion rate at $p_{O_2} = 0$ could be reduced to zero. Without an isoperibolic oxygen jacket, diffusion rates averaged -12 ∓ 6 nmol h^{-1} in the last 8 experiments in conjunction with the calorimeter, corresponding to $c^{\ominus}{}_{out} = 2$ μmol dm^{-3} or 0.6% air saturation. With less experience, however, diffusion rates of -10 to -30 nmol h^{-1} were typical for the respirometric system alone using Pharmacia valves.

4.3 Two POS for Twin Measurements

So far the evaluation of the precision of respiratory rate measurements has been restricted to the analysis of the signal of a single sensor. With the twin-flow system, however, two independent measures of the oxygen concentration ratio are obtained at intervals (Fig. 8). Agreement between these twin measurements tends to confirm the above analysis. The absolute difference between the blank oxygen consumption or diffusion rates as measured by the two sensors was 2.5 ± 2.2 (S.D.) nmol h^{-1} in 25 control runs at p_{O_2} from 0 to 280 mbar (0–28 kPa, 0%–140% air saturation). The difference may in part be due to slight asymmetries in the sites of oxygen consumption and diffusion. Therefore, the rates of the blank obtained for each sensor were subtracted from the respective total consumption rate, although the regressions for the two POS were not significantly different (Fig. 7). This comparison indicates that an estimated sensitivity of 3 nmol h^{-1} for a single sensor system may be on the optimistic side, while the sensitivity is improved by taking the mean of the two respiratory rates [Eq. (8)] obtained with the twin-flow system.

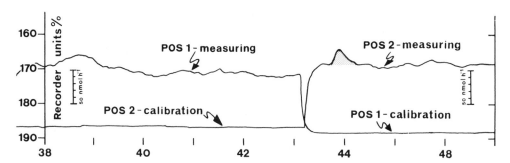

Fig. 8. Recorder traces of the two POS before and after switching the four-way valves of the twin-flow microrespirometer, in an experiment with 40 *Cyclops abyssorum* in the calorimeter chamber (experiment RJC1, 100% air saturation, see also Fig. 12). The hatched area is due to a second passage of some water through the animal chamber and stirring chamber after switching the POS into measuring position

   Another at least equally important advantage of the twin-flow microrespirometer is the provision of objective criteria for the discrimination between normal and irregular system function; one may reject "bad data" on two levels: (1) If the calibration factor varies in a manner which renders the correction by interpolation impossible, then the data are bad for a functionally obvious reason. (2) If the twin measurements of respiratory rate disagree, then something must have been wrong for not necessarily obvious reasons. A third rejection criterion is provided by simultaneous calorimetry: scepticism is warranted if the ratio of oxygen consumption/heat dissipation shifts irregularly (see below). Objective discrimination between reliable and unreliable data is a potent means of reducing variability due to intermittent instrumental malfunction. This considerably enhances the significance of long-term experiments, especially as these cannot be repeated very often for reasons of time.

4.4 Response Time and Correction for Instrumental Lag

The dynamics of respiratory rates are a sensitive indicator of an animal's physiological state. Hence the accurate resolution of short-term patterns of oxygen uptake may be just as significant in analytical tests as the comparison of average oxygen uptake levels [18]. However, due to the inertia of a respirometer, true instantaneous rates may differ considerably from the apparent rates which are directly obtained from the untreated oxygen records (Fig. 8). This has usually been neglected [12]. The magnitude of the time delay and damping of *internal* changes in respiratory rate is not directly obvious, while after *external* changes of p_{O_2} the equilibration time of a respirometer is evidenced by the nearly exponential time course of the oxygen records (Fig. 9). The transient response of a respirometer reflects the dynamic properties of the various constituent parts which determine the over-all time behavior of the system (Fig. 9). On the basis of this information, suitable means can be contrived to alleviate the problem of inertia, where possible, by improving the physical characteristics of the most sluggish system component, or otherwise by correcting the observed values with tested mathematical models.

   The response time, $\tau_{99\%}$, of the YSI sensor varied from 60 to 90 s, and steady state was reached within 200 to 400 s after transfer from air into Na_2SO_3 solution at room temperature. Old sensors should be discarded, if, despite a tight membrane fit and cleaning and regenerating the anode, equilibration times become very long (Fig. 9, curve 1). In the stirring chamber (Fig. 3) the response time of the POS is not limited by the thickness and the diffusion coefficient of the membrane and electrolyte layer (Chap. I.1), but depends primarily upon the flow rate, and on the mixing and the volume of water in the measuring chamber and the inlet flow tube. The 99% response time after switching the four-way valves ranged from 15 to 30 min, apparently increasing with a larger step change in p_{O_2} (Fig. 8, POS 2; Fig. 9, curve 2). A transit time elapses from the initial p_{O_2} change in a defined section of the respirometer (water reservoir, valve, first stirring chamber or animal chamber) until the POS starts to react. For simplicity this purely additive delay in the system's transfer function was always eliminated in the various expressions of time behavior.

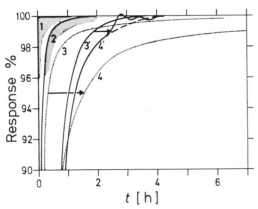

Fig. 9. The transient response to external p_{O_2} changes of functional components of the respirometer at different levels of integration at a flow rate of 5.7 cm³ h⁻¹. *1* POS after transition from air to Na-sulfite solution; the response of defective sensors is delayed (*shadowed area*), *2* POS in the stirring chamber after switching from measuring position (86% air saturation) to calibration position (100% air saturation). A delay of the response (*shadowed area*) indicates the trapping of gas bubbles in the flow system. *3* and *3'* POS in calibration position after changing the p_{O_2} of the gas equilibrating the 0.4 dm³ water reservoir. *3* change from air to pure nitrogen at a high gas flow rate. *3'* change from 10% O_2 in N_2 (= 48% air) to 10% air in N_2 at a lower gas flow rate. *4* and *4'* POS in measuring position corresponding to curve 3 and 3' respectively; both curves are corrected for the transit time due to the flow path from the first to the second stirring chamber (0.60 h with the animal chamber in the calorimeter). *4* the pyrex chamber in the calorimeter contained 40 dead specimens of *C. abyssorum* after poisoning (experiment RJC2, see Fig. 15); the O_2 concentration change was −326 μmol dm⁻³ during the first hour. *4'* the animal chamber in the calorimeter contained 40 active specimens of *Cyclops* (experiment RJC1, see Fig. 12); the O_2 concentration change was −120 μmol O_2 dm⁻³ during the first hour; irregularities are due to fluctuations in oxygen consumption by the animals

The instantaneous events taking place at the inflow of the stirring chamber can be calculated from the recorded values and from the exponential transient response curves (Fig. 9) by employing mathematical deconvolution methods [2, 38]. However, the response time of the POS in the measuring position depends predominantly upon the turbulence and the volume of water in the animal chamber which exceeds the volumes of the stirring chambers by a factor of about 20. In an unstirred animal chamber not only the geometry, but also the activity, position, and size of the animal(s) determine the effective dead volume and the distribution of throughflow. This precludes the use of invariant time constants for the resolution of the immediate respiratory response to an externally varied oxygen regime, since the physical effects of animal behavior are highly variable. This is illustrated by the different time lags between curves 3 and 4, and 3' and 4' in Fig. 9 (arrows) where equilibration is considerably enhanced by the locomotory activity of the live animals. Attainment of the final equilibrium and the form of the response curve also depend on the magnitude of the external p_{O_2} variation: only after large concentration changes does the amount of back diffusion from or into the oxygen-permeable materials exceed the limit of detection.

With this in mind, a pessimistic picture emerges regarding the applicability of simple mathematical models to improve the time resolution. However, these *internal* oxygen changes are an order of magnitude below the *external* transient changes analyzed in Fig. 9. They never exceeded 25 μmol dm^{-3} during 1 h and were usually below 6 μmol dm^{-3} in a 0.3-h period (Fig. 17), whence higher-order exponential terms (necessary to describe curves 4 and 4' in Fig. 9) become less significant. Fry [12] considered a first-order model as applied to the mixed reactor (chemostat) to correct for the lag in open-flow fish respirometers [10]. Equation (5), including the lag term, then becomes

$$\frac{dN_{O_2}}{dt} = \frac{dV}{dt} \times (c_{out}^{\ominus} - c_{out}) + V \times \frac{dc_{out}}{dt} , \tag{9}$$

where $dV/dt = f$ [dm^3 h^{-1}], V [dm^3] is the homogeneously mixed volume of the respirometer, and dc_{out}/dt [μmol h^{-1}] is the instantaneous rate of change in oxygen concentration of outflow water. The ratio of the system's storage and flow parameter is the time constant, τ [h],

$$\tau = \frac{V}{f} = \frac{V}{dV} \times dt, \tag{10}$$

or the period whereafter the response has completed 63% [35].

None of the simplifying assumptions inherent in Eq. (9) holds for a flow microrespirometer. Especially in conjunction with the flow calorimeter, the capillary connections from animal to measuring chamber contain a significant volume of water (2–3 cm^3) in which oxygen fluctuations are damped due to the parabolic current profile and diffusive oxygen exchange between different water layers in laminar flow. Additional terms are required for the description of such a cascade system [2, 34]. But with an unstirred animal chamber the primary problem hinges on the changing and nonlinear effect of animal behavior on the transfer function of the whole system. Poor mixing in the animal chamber reduces the effective volume and the first-order time constant [Eq. (10)], and hence the outflow water transmits oxygen changes in the flow path more quickly. But at the same time a quite unpredictable high-order system is generated due to the slow and irregular interchange with the residual dead volume. This makes the significance of any refined deterministic model questionable. A first-order approximation was therefore considered, where all response characteristics (high-order, nonlinear) are lumped in a single, empirical time constant [35]. While according to Eq. (10) and using the volume of the animal chamber the theoretical time constant is calculated as 0.09 h, the transient response as obtained with active animals in the respirometer-calorimeter suggests a time constant ranging from about 0.2 to 0.3 h (Fig. 10). This is the period corresponding to the break frequency above which the amplitude ratio (the static gain value) and the phase response of the output signal fall off relative to the real fluctuations in oxygen uptake. With inactive animals this period is quite different, and high frequency fluctuations of the output signal may be artifacts due to movements only. Thereforth apparent oxygen consumption rates were averaged over 0.33 h intervals and no better time resolution was attempted.

The simultaneous measurement of heat dissipation and oxygen consumption provides the unique possibility of "calibrating" the transfer function of the twin-flow microrespirometer relative to the response time of the flow calorimeter in different experimental situations. The calorimeter time constant was determined at 120 to 150 s

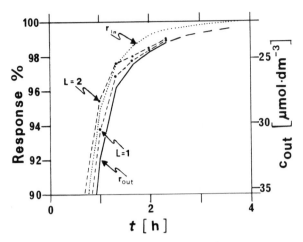

Fig. 10. Correction of time lag in the transient response to external p_{O_2} changes. The observed input transient (*dotted line*, r_{in}) and the observed output of the POS in measuring position (*full line*, r_{out}) correspond to curves 3' and 4' respectively of Fig. 9. The *dotted lines* are the corrected r_{out}-curves, r_{cor}, according to the equation [cf. Eq. (11)]

$$r_{cor} = r_{out,t} - \frac{L}{\Delta V} \times (r_{out,t-1} - r_{out,t}) \ .$$

Two boundary values for the lag factor, L [cm^3], were inserted as shown in the figure. The lumped first-order time constant is $L/(\Delta V/\Delta t)$; above the 97% response the significance of high-order terms becomes evident. The time constant is calculated by inserting r_{in} instead of r_{cor} in the above equation and solving for L

by electrical calibration and was not corrected for. The validity of considering a first-order correction for the respirometer can already be gauged from the generally high correlation coefficients of the untreated calorimetric and respirometric data. With more than one significant time constant relative to the calorimeter response, much larger deformations of the oxygen uptake record would be expected. However, the correlation coefficient could be optimized (Fig. 11) by adjusting the storage term or lag factor in the following difference equation [cf. Eq. (9)],

$$\frac{\Delta N_{O_2}}{\Delta t} = \frac{\Delta V}{\Delta t} \times (c_{out}^\Theta - \overline{c}_{out,t}) + L \times \frac{(\overline{c}_{out,t-1} - \overline{c}_{out,t})}{\Delta t} \ , \tag{11}$$

where Δt [h] is the time interval, $\overline{c}_{out,t-1}$ and $\overline{c}_{out,t}$ [μmol dm^{-3}] are the average oxygen concentrations of outflow water in successive time intervals, t−1 and t, respectively, and L [dm^3] is the empirical lag factor (Fig. 10). The lag factor as adjusted relative to the calorimeter response will be indexed as L_c. The calibration of L_c (Fig. 11) rests on the assumption that the averages of the calorimeter output per 20 min accurately reflect fluctuations in aerobic metabolism, and that peaks of heat dissipation are not due to intermittently activated anoxic processes. The effects of likely deviations from this assumption on estimating L_c have to be critically examined for each situation (see below). The magnitude of the lag factor (the effective storage volume of the respirometer) estimated in the transient response analysis (Fig. 10) and the optimization of L_c with

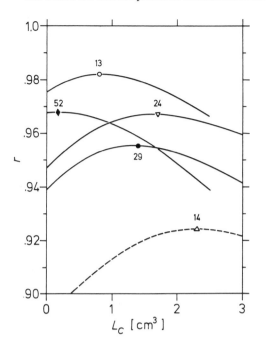

Fig. 11. Variation of the lag factor, L_c [Eq. (11)], for the construction of optimum curves of the correlation coefficient, r, between simultaneous calorimetric and respirometric measurements averaged over 0.33 h periods. *Open symbols* with one specimen of *Salvelinus alpinus* in the post-hatching stage under aerobic (*circle*) and hypoxic conditions (see Fig. 18 to 20). *Closed symbols* with 40 specimens of *Cyclops abyssorum* in the adult and cope-podite V stage under aerobic (*circle*) and hypoxic conditions (see Figs. 15 to 17). The number of subsequent 20-min intervals for each experimental situation is given in the figure

different animal species and numbers (Fig. 11) agree well. This suggests the suitability of the approach. Actual rates of the test animals used may be expected to fall in the range of corrected rates corresponding to lag factors between 0.5 and 1.5 cm³ [Eq. (11)]. However, no statistical confidence intervals can be estimated. Standardized mixing of the animal chamber, digital data acquisition, and more realistic models are necessary to improve the above analysis.

5 Applications of the Twin-Flow Microrespirometer

In order to avoid giving the impression of an exaggerated emphasis on methodological details (which have too often been disregarded), we shall briefly inquire into some of the applications of the twin-flow microrespirometer. Case studies are presented to illustrate the practical advantages of the design and to demonstrate the importance of careful data analysis.

5.1 Oxygen Dependence in Open-Flow Respirometry

The physiological response of a population of *Cyclops abyssorum* from Kalbelesee (see Chap. III.1) to a hypoxic environment was investigated in long-term experiments. Oxygen uptake as a function of p_{O_2} and time after collection is shown in Fig. 12. The respiratory rate decreased exponentially independent of oxygen concentration. Only

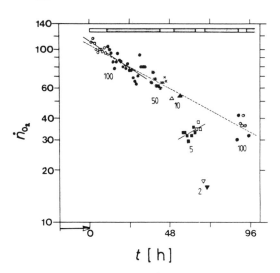

Fig. 12. Oxygen uptake, $\dot{n}_{O_2}$ (μmol h^{-1} g^{-1} dry weight), of *Cyclops abyssorum* as a function of time, t [h], and externally varied p_{O_2} (numbers in % air saturation) experiment RJC1: 1.36 mg total dry weight, 6°C; constant flow of 5.75 cm^3 h^{-1} through the respirometer-calorimeter; see also Figs. 7 and 8). The *arrow* on the time axis indicates the period between sampling the animals from lake Kalbelesee and the beginning of the experiment. The *broken line* is the least-squares interpolation of initial and post-hypoxic rates at air saturation. The *full lines* indicate the calculated exponential slopes of aerobic decrease and hypoxic increase, respectively, of oxygen consumption. The flow regime determining the successive periods in measuring and calibration position of a POS is shown by the *light* and *dark bar* on the top and by different symbols for the hourly averaged rates

after the transition from 10% to 5% air saturation of the inflow water did oxygen uptake immediately drop to half of the preceding rate. During the subsequent recovery period, oxygen consumption gradually approached the same level as that extrapolated from the initial aerobic rates, and nearly abolished the oxygen-induced change. Short-term acclimation to severe hypoxia is also seen in Fig. 15. These trends, uniquely resolved in an open-flow system, draw attention to the misleading picture which would inevitably result from this experiment if a closed respirometer were used (Fig. 13). In a closed system of large volume, the instantaneous starvation effect on the diminishment of oxygen uptake is obscured by the apparent dependence on the simultaneously reduced oxygen content. With a small volume and consequently rapid oxygen depletion, however, the acclimatory response to the varied oxygen regime remains elusive.

5.2 Hypoxia, Respiration and Activity

Oxygen uptake and locomotory activity were measured as a function of p_{O_2} and laboratory stress with 40 *C. abyssorum* in a 0.5 cm^3 plain glass animal chamber (Gnaiger and Flöry, in preparation). Active animals moved freely in the water column, but showed an oxygen-dependent tendency to attach to the walls of the animal chamber, while inactive specimens lay motionless on the bottom. The distribution of animals be-

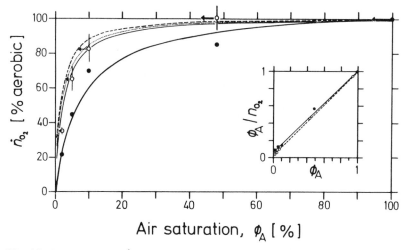

Fig. 13. Oxygen uptake, $\dot{n}_{O_2}$ (in % of the aerobic rate), as a function of air saturation, ϕ_A, and linearization of the $\dot{n}_{O_2}/\phi_A$ plot (*inset*) (experiment RJC1, see Fig. 12). *Full line* the average of aerobic rates (12 h before the external p_{O_2} change) is taken as the static aerobic reference value, and the inflow water represents the experimental conditions. *Other lines* the rate/time regression of aerobic oxygen consumption (Fig. 12, dotted line) served as a time correction for the aerobic reference value, and the p_{O_2} of inflow water (*right line with arrows*), outlfow water (*left line*), or the mean of the two (*middle line*) was used in the calculations

tween water column (N_c), walls (N_w), and bottom (N_b) was observed during prolonged periods of constant controlled oxygen conditions. From this distribution an index of relative activity,

$$\frac{N_c + f_w \times N_w}{N_c + N_w + N_b} \tag{12}$$

could be contrived, where $N_c + N_w + N_b$ is the sum of individuals in the respirometer, and f_w is a weighting factor for the number of animals intermittently attached to the walls, which expresses their state of activity relative to the unrestricted activity of free-swimming animals. f_w can also be considered as an expression for the artificial effect of confinement of planktonic animals in a small volume chamber. It was derived by linearizing the oxygen consumption/activity relationship at various p_{O_2} levels between 100% and 2% air saturation (Fig. 14). The intersect of this plot with the ordinate agrees well with the relative activity observed under anoxic conditions, while the negative intersect with the aerobic respiration axis hints at the level of anoxic metabolism in sustaining physiological functions in the low p_{O_2} range.

5.3 Simultaneous Respirometry and Calorimetry

The twin-flow microrespirometer was primarily designed for the establishment of a simultaneous direct and indirect calorimetric method (Fig. 2), and herein appears its most promising and innovative field of application. Within the pratical context of this

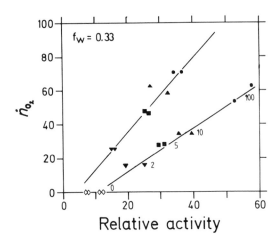

Fig. 14. Oxygen uptake, $\dot{n}_{O_2}$ (μmol h^{-1} g^{-1} dry weight), as a function of locomotory activity and air saturation (numbers in %) in two experiments with *Cyclops abyssorum* (40 individuals) at different levels of "laboratory stress" (Gnaiger and Flöry, in prep.)

book, however, I want to point out some of the problems encountered in operating simultaneously, over several days, and near the limit of detection, two instruments, both of which are dependent upon the continuous function of a number of mechanical and electronic components (Fig. 15). However, unique advantages are provided by the simultaneous measurement of both oxygen uptake and heat dissipation: an erroneous result with one apparatus is very likely to become apparent by comparison with the other. The second case study with salmonid fish larvae (Fig. 18) demonstrates the fascinating results that remunerate the considerable effort and expense invested. Some speculative comments suggest the broad spectrum of questions which can be solved or, even more important, may be unexpectedly introduced, by the application of the open-flow microrespirometer-calorimeter.

In one series of experiments the calorimeter baseline did not stabilize due to evaporative heat loss in the heat detector. Only after the time course of oxygen uptake and heat dissipation coincided was there evidence of the final diminution of this interfering effect (Fig. 15). In the same experiment 12 out of 40 *C. abyssorum* were egg-carrying females. Nauplii were washed out as they hatched, and some were observed in the collecting cylinder. Nauplii are intolerant to anoxia (from hour 68 to 87, Fig. 15). Upon return to aerobic conditions, the apparent oxygen uptake increased steadily, while heat dissipation even decreased after an initial overshoot. This implausible discrepancy motivated the application of antibiotics in detecting possible bacterial decomposition of dead nauplii trapped in the capillary system. As expected, the respirometric signal inflected immediately, to converge with the calorimeter output after 27 h. This suggests that the bacterial interference was by this time largely under control. The disturbed pattern of heat dissipation indicates a significant effect of the antibiotics on the animals. Much valuable information is revealed even in such a "faulty" simultaneous experiment: although the drift of the calorimeter baseline was uncontrolled but steady (during the first 48 h in Fig. 15), the fluctuations in heat dissipation could be compared quantitatively with the instantaneous respiratory rates during different sections of the experiment. There was a remarkably close correlation between the regular fluctuations in heat dissipation and oxygen uptake (Fig. 16). This

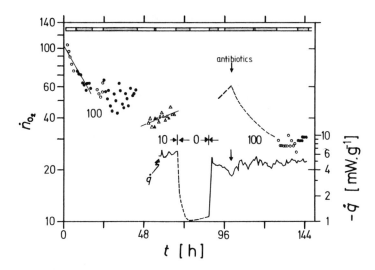

Fig. 15. Oxygen uptake, $\dot{n}_{O_2}$ (μmol h^{-1} g^{-1} dry weight), and heat dissipation, $\dot{q}$ (mW g^{-1} dry weight), of *Cyclops abyssorum* as a function of time, t [h], and externally varied p_{O_2} (experiment RJC2: 1.418 mg total dry weight; 6°C; at constant flow of 5.65 cm^3 h^{-1}). See Fig. 12 and text for further explanations

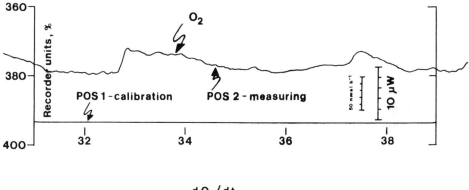

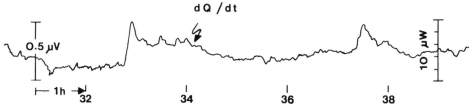

Fig. 16. Recorder traces in a simultaneous calorimetric and respirometric experiment showing the activity pattern of a group of 40 specimens of *Cyclops abyssorum* (experiment RJC2, see Fig. 15). The p_{O_2} record of the POS in measuring position is shifted backwards relative to the calorimeter power-time curve by 11 min to correct for the transit time in water flow from animal to stirring chamber

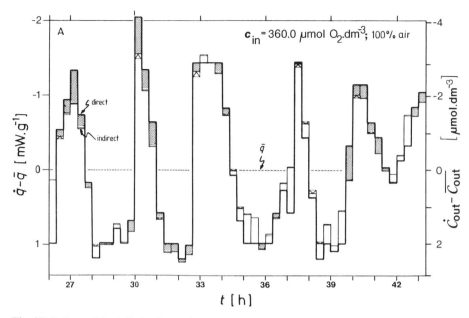

Fig. 17. Pattern of heat dissipation and oxygen consumption expressed as the deviations from the mean value, $\bar{q}$, during the period shown (experiment RJC2, see Figs. 15 and 16). *Full line* 20-min averages of the direct calorimetric rate corrected for baseline drift. *Thin line* 20-min averages of oxygen uptake expressed as heat dissipation on the basis of the theoretical oxycaloric equivalent of -450 kJ (mol O_2)$^{-1}$ (App. C); the corresponding changes in oxygen concentration of outflow water are indicated on the right hand side. Changes within 20 min did not exceed 5 μmol O_2 dm^{-3}. The difference in oxygen content of inflow and outflow water averaged 16.9 μmol dm^{-3} (= 4.7% of air saturation). The mean oxygen consumption was 51.3 μmol O_2 h^{-1} g^{-1} dry weight (calculated $\bar{q}$ = 6.41 mW g^{-1})

suggests an almost exclusively aerobic mechanism sustaining spontaneous activity peaks of *C. abyssorum* under these conditions (Fig. 17), as well as in 10% air saturation (Fig. 11). This in turn was the prerequisite for estimating the likely magnitude of the average lag factor of the respirometer for these periods of observation (Fig. 11).

The effect of varied p_{O_2} on one individual fish embryo (*Salvelinus alpinus*) was investigated with the respirometer-calorimeter (Fig. 18). The eggs of this salmonid species are capable of maintaining a stable level of anoxic heat dissipation over many hours and show a linear dependency of metabolic rate on oxygen up to air saturation as measured with the LKB flow microcalorimeter [13, 16]. After hatching, however, oxygen dependency was only apparent after a sudden fall in p_{O_2}, but within one day the normoxic level of metabolism was reestablished (Fig. 18) (cf. Chap. II.2). Simultaneous direct and indirect calorimetry brings about the resolution of two distinct phases in this acclimatory response to lowered p_{O_2}. (1) Anoxic compensation (24 to 44 h): Here aerobic sources of metabolic energy expenditure (App. C) are augmented by anoxic processes compensating for the drop in metabolic rate as indicated by the increased ratio of measured and calculated heat dissipation. In the calculation the experimentally determined oxycaloric equivalent of aerobic metabolism was used. (2) Conservative (anabolic) compensation (48 to 52 h): Here the ratio of measured

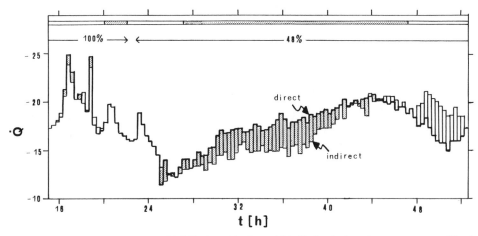

Fig. 18. Simultaneous direct (*thick line*) and indirect (*thin line*) calorimetric measurement of heat dissipation, $\dot{Q}$ [μW], of one individual *Salvelinus alpinus* as a function of externally varied p_{O_2} and time of exposure (at 8°C, constant flow rate of 5.67 cm³ h⁻¹; 5.11 mg dry weight; see also Figs. 19 and 20). The oxygen consumption under normoxia (p_{O_2} = 19.5 kPA, 0 to 22 h) averaged 157 nmol O_2 h⁻¹. The experimental oxycaloric equivalent averaged −451 kJ (mol O_2)⁻¹. After the reduction of oxygen content to 48% air (p_{O_2} = 9.3 kPa) two phases of short-term acclimation became apparent: *anoxic compensation* (25 to 43 h, the *hatched area* indicates the anoxic part of heat dissipation); and *conservative compensation* (44 to 53 h, the *open area* indicates the difference between the expected and observed rate of heat dissipation). The *light* and *dark bar* on the top indicates measuring and calibration positions of the two POS. No estimates of respiratory rates could be obtained during the p_{O_2} transition period (23 to 25 h)

and calculated heat dissipation (App. C) fell below the aerobic value and hence far below the mean ratio for the preceding anoxic compensation period (Fig. 18). This indicates the apparent activation of coupled reactions in which part of the energy that would otherwise be expected to dissipate as heat is conserved. The interesting question arises as to the fate of the accumulated anoxic end product lactic acid [19]. Reutilization in aerobic respiration and coupled glyconeogenesis could theoretically be the explanation [17].

The subsequent test with antibiotics indicated that no bacterial contamination had occurred in this experiment. The animals displayed a highly elevated rate of heat dissipation with large contributions from anoxic sources after exposure to 0.2 g dm⁻³ streptomycin- and neomycinsulfate (Fig. 20, see also Chap. II.9). The simultaneous recorder traces of the calorimeter and respirometer during recovery from this stress are shown in Fig. 19, and the results of the whole experiment are summarized in Fig. 20. It is also interesting to note the similarity of the lag factor of the respirometer calculated for the periods of standard aerobic metabolism and recovery from the drug-induced stress (Fig. 11). For the stress period itself, however, fitting of the respirometer lag factor relative to the *calorimeter* response produces an artifact, as the *physiological* response of the animal is complicated by anoxic processes and cannot be separated from the instrument's response time (Fig. 11).

Contrary to every expectation, further reduction of p_{O_2} to 19% air saturation did not induce a continued activation of anoxic pathways. In fact, as early as 6 h after the

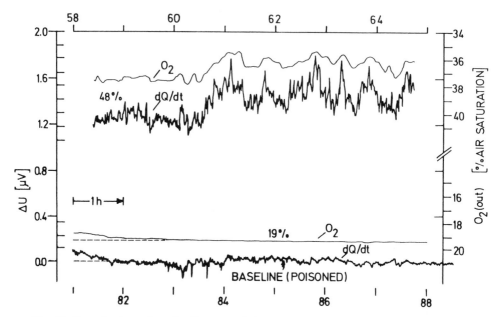

Fig. 19. Recorder traces in a simultaneous calorimetric and respirometric experiment with *Salvelinus alpinus*. A section of the period at 48% air saturation (p_{O_2} = 9.28 kPa) is shown on the top (58 to 65 h). The calorimeter baseline and the outflow oxygen value after poisoning at 19% air saturation of inflow water are shown below (81 to 88 h) (see also Fig. 18)

oxygen transient, a physiological state was observed comparable to the conservative compensation in the above terminology (Fig. 20). The pattern of oxygen consumption leveled off to a rather smooth line, which indicates cessation of locomotory activity and hence a specific reduction of dissipative metabolism under apparently severe hypoxia. Were net-anabolic processes still going on?

Our present understanding of energetic mechanisms involved in coping with variations of the environmental oxygen regime is largely restricted to a descriptive approach in whole animal studies. Oxygen uptake measurements direct interest toward the various expressions of *critical* p_{O_2} [24], without reflecting the most significant metabolic changes that may occur as a result of channeling the energy flow through glycolytic pathways. On the other hand, direct calorimetric results on their own do not provide any information as to the *limiting* p_{O_2} at which anoxic metabolism is switched on to supplement the diminishing aerobic energy source [13, 40]. Besides the aerobic or anoxic *source* of biochemical energy (ATP, electrochemical potentials), external changes in p_{O_2} will also affect the *fate* of this energy: Little is known about the p_{O_2} dependence of *conservative* metabolism, where net-anabolic reactions associated with growth or energy storage reduce the heat change below the value calculated from oxygen consumption using the traditional oxycaloric equivalent (App. C) (Figs. 18, 20). The ratio of observed and expected heat dissipation forms the experimental basis for expressing the "caloric efficiency" [15] and hence quantifying the share of conservative metabolism in the total rate. Again, in invertebrates the rate of *dissipative* metabolism as a function of p_{O_2} [15] has rarely been analyzed in terms of its func-

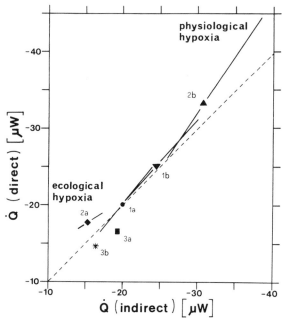

Fig. 20. Relation of direct and indirect calorimetry in *Salvelinus alpinus* during different environmental and physiological conditions. *Hatched line* expected relation on the basis of the theoretical oxycaloric equivalent of -450 kJ $(\text{mol O}_2)^{-1}$. The *symbols* indicate the mean rates during specified periods and the *full lines* are the corresponding regressions (Bartlett's method of best fit) using the lag factors, L_c (see Fig. 11). *1a* normoxic period (see Fig. 18); *2a* anoxic compensation period under *environmental hypoxia* (48% air saturation, see Fig. 18); *2b* anoxic compensation period under *physiological anoxia* (stress period after addition of antibiotics; 48% air saturation). After normalization of the metabolic activity the values returned again to the region of correspondence (*1b*); *3a* conservative compensation period at 48% air saturation (see Fig. 18); *3b* at 19% air saturation after period 1b

tional components such as locomotory activity and maintenance (Fig. 14). Direct and indirect calorimetric experiments in open-flow instruments as well as activity measurements and biochemical determinations [19] are required to improve our insight into the regulatory mechanisms underlying the apparently confusing relationship between oxygen conformity and hypoxia tolerance. Not only do studies of the influence of oxygen on animal energetics satisfy the needs of environmental physiologists, but serve equally the exploration of basic principles in cellular physiology.

The merits of direct calorimetry are on the verge of being recognized in the context of ecophysiological energetics ([13–20, 41], Chap. II.4). In constructing energy budgets we still rely upon our dubious belief in theoretical values for the caloric equivalent of oxygen consumption for most animals, and even these calculations were combined with previously unrecognized errors (App. C).

The contribution of anoxic metabolism during hypoxia and in sustaining high levels of physical activity can be deduced from an increasing bulk of biochemical evidence, while the integrated economy of glycolytic and gluconeogenetic processes and the energetic cost of regulatory mechanisms in anoxic-aerobic transitions can only be poorly understood without the crucial test by direct and indirect calorimetry. However, most comparative biochemists are reluctant to ponder over the significance of calorimetric and respirometric analyses, and experimental as well as conceptual difficulties of alignment direct such studies into the realm of interdisciplinary research [19]. While progress has been made in the thermochemical analysis of anoxic and aerobic metabolism [15], the possible effect of anabolic processes on the heat changes accompanying aerobic metabolism has not yet even been considered in energy budget calculations for animals. A lesson in microbiology [7], however, suggests elucidating the question of caloric efficiency of animal growth in simultaneous calorimetric and respirometric experiments. Besides tracking the net enthalpy changes that accompany aerobic respiration under various physiological conditions [13, 14], these studies are likely to throw new light on existing theories of biological thermodynamics [6, 15, 36].

Acknowledgments. This work was supported by the *Fonds zur Förderung der wissenschaftlichen Forschung in Österreich,* projects no. 2919 and 3917. Improvements have been achieved in a further development of the twin-flow microrespirometer in cooperation with M. Ortner. This instrument is now commercially available.

References

1. Atkinson HJ, Smith L (1973) An oxygen electrode microrespirometer. J Exp Biol 59:247–253
2. Belaud A, Trotter Y, Peyraud C (1979) Continuous evaluation of P_{a,O_2} in fish: Recording and data processing. J Exp Biol 82:321–330
3. Bishop J (1976) A continuous recording, differential respirometer for a closed, flowing seawater system. Oikos 27:127–130
4. Brinkhurst RO, Chua KE, Kaushik N (1972) Interspecific interactions and selective feeding by tubificid oligochaetes. Limnol Oceanogr 17:122–133
5. Brkovic-Popovic I, Popovic M (1977) Effects of heavy metals on survival and respiration rate of tubificid worms. II. Effects on respiration rate. Environ Pollut 13:93–98
6. Calow P (1977) Conversion efficiencies in heterotrophic organisms. Biol Rev 52:385–409
7. Dermoun Z, Belaich JP (1979) Microcalorimetric study of *Escherichia coli* aerobic growth: Kinetics and experimental enthalpy associated with growth on succinic acid. J Bacteriol 140:377–380
8. Dries RR, Eschweiler L, Theede H (1979) An improved equipment for continuous measurement of respiration of marine invertebrates. Kieler Meeresforsch 4:310–316
9. Edwards RW, Learner MA (1960) Some factors affecting the oxygen consumption of *Asellus*. J Exp Biol 37:706–718
10. Evans DO (1972) Correction for lag in continuous-flow respirometry. J Fish Res Board Can 29:1214–1216
11. Fowler JD, Goodnight CJ (1965) The effect of environmental factors on the respiration of *Tubifex*. Am Midl Nat 74:418–428
12. Fry FEJ (1971) The effect of environmental factors on the physiology of fish. In: Hoar WS, Randall DJ (eds) Fish physiology. Academic Press, London New York, pp 1–98
13. Gnaiger E (1979) Direct calorimetry in ecological energetics. Long-term monitoring of aquatic animals. Experientia Suppl 37:155–165
14. Gnaiger E (1980) Energetics of invertebrate anoxibiosis: Direct calorimetry in aquatic oligochaetes. FEBS Lett 112:239–242
15. Gnaiger E (1980) Das kalorische Äquivalent des ATP-Umsatzes im aeroben und anoxischen Metabolismus. Thermochim Acta 40:195–223

16. Gnaiger E (1980) Direct and indirect calorimetry in the study of animal anoxibiosis. A review and the concept of ATP-turnover. In: Hemminger W (ed) Thermal analysis, vol II. ICTA 80. Birkhäuser, Basel Boston Stuttgart, pp 547–552

17. Gnaiger E (1980) The enthalpy of growth – a thermodynamic analysis. Abstr 4th Int Symp Microcalorimetry Appl Biol Univ Coll Wales, Aberystwyth

18. Gnaiger E (1981) Pharmacological application of animal calorimetry. Thermochim Acta 49: 75–85

19. Gnaiger E, Lackner R, Ortner M, Putzer V, Kaufmann R (1981) Physiological and biochemical parameters in anoxic and aerobic energy metabolism of embryonic salmonids, *Salvelinus alpinus.* Eur J Physiol Suppl 391:R57

20. Gnaiger E, Tiefenbrunner F (1982) Microcalorimetry for continuous monitoring of biological processes. In: Message M (ed) Cell and molecular biology in space. Elsevier North Holland, Amsterdam (in press)

21. Gyllenberg G (1973) Comparison of the Cartesian diver technique and the polarographic method, an open system, for measuring the respiratory rates in three marine copepods. Commentat Biol 60:3–13

22. Harnisch P (1935) Versuch einer Analyse des Sauerstoffverbrauchs von *Tubifex tubifex* Müll. Z Vergl Physiol 22:450–465

23. Hart RC (1980) Oxygen consumption in *Caridina nilotica* (Decapoda Atyidae) in relation to temperature and size. Freshwater Biol 10:215–222

24. Herreid CF (1980) Hypoxia in invertebrates. Comp Biochem Physiol 67A:311–320

25. Kamler E (1969) A comparison of the closed-bottle and flowing-water methods for measurement of respiration in aquatic invertebrates. Pol Arch Hydrobiol 16:31–49

26. Kanwisher J (1959) Polarographic oxygen electrode. Limnol Oceanogr 4:210–217

27. Kaufmann R, Gnaiger E (1981) Optimization of calorimetric systems: Continuous control of baseline stability by monitoring thermostat temperatures. Thermochim Acta 49:63–74

28. Kirberger C (1953) Untersuchungen über die Temperaturabhängigkeit von Lebensprozessen bei verschiedenen Wirbellosen. Z Vergl Physiol 35:175–198

29. Klekowski RZ (1971) Cartesian diver microrespirometry for aquatic animals. Pol Arch Hydrobiol 18:93–114

30. Klekowski RZ, Kamler E (1968) Flowing-water polarographic respirometer for aquatic animals. Pol Arch Hydrobiol 15:121–144

31. Koenen ML (1951) Vergleichende Untersuchungen zur Atmungsphysiologie von *Tubifex tubifex* M. und *Limnodrilus claparedeanus* R. Z Vergl Physiol 33:436–456

32. Lasserre P, Renaud-Mornant J (1973) Resistance and respiratory physiology of intertidal meiofauna to oxygen-deficiency. Neth J Sea Res 7:290–302

33. Lovtrup S (1973) The construction of a microrespirometer for the determination of respiratory rates of eggs and small embryos. In: Kerkut G (ed) Experiments in physiology and biochemistry, vol VI. Academic Press, London New York, pp 115–152

34. Mangum C, Winkle Van W (1973) Responses of aquatic invertebrates to declining oxygen conditions. Am Zool 13:529–541

35. Milsum JH (1966) Biological control systems analysis. McGraw-Hill Electronic Science Series, New York, 466 pp

36. Morowitz HJ (1978) Energy flow in biology. Academic Press, London New York, 344 pp

37. Nagell B (1975) The open-flow respirometric method: precision of measurement in general and description of a high precision respirometer for aquatic animals. Int Rev Gesamten Hydrobiol 60:655–667

38. Nimi AJ (1978) Lag adjustment between estimated and actual physiological responses conducted in flow-through systems. J Fish Res Board Can 35:1265–1269

39. Olson TA, Rueger ME, Scofield JI (1969) Flow-through polarographic respirometry for aquatic animals. Hydrobiologia 34:322–329

40. Palmer MF (1970) Aspects of the respiratory physiology of *Tubifex tubifex* in relation to its ecology. Zool London 154:463–473

41. Pamatmat MM (1978) Oxygen uptake and heat production in a metabolic conformer (*Littorina irrorata*) and a metabolic regulator (*Uca pugnax*). Mar Biol 48:317–325

42. Pattee E (1965) Sténothermie et eurythermie les invertébrés d'eau douce et la variation journaliére de tempépature. Ann Limnol 1:281–434
43. Platzer I (1967) Untersuchungen zur Temperaturadaptation der tropischen Chironomidenart *Chironomus strenskei* Fittkau (Diptera). Z Vergl Physiol 54:58–74
44. Richman S (1958) The transformation of energy by *Daphnia pulex.* Ecol Monogr 28:273–291
45. Rohde RA (1960) The influence of carbon dioxide on respiration of certain plant-parasitic nematodes. Proc Helminthol Soc Wash 27:160–164
46. Spink C, Wadsö I (1976) Calorimetry as an analytical tool in biochemistry and biology. In: Glick D (ed) Methods in biochemical analysis, vol 23. Wiley-Science, New York, pp 1–159
47. Whitley LS, Sikora RA (1970) The effect of 3 common pollutants on the respiration rate of tubificid worms. J Water Pollut Contrib Fed 42:R57–R66
48. Wightman JA (1977) Respirometry techniques for terrestrial invertebrates and their application to energetics studies. NZJ Zool 4:453–469
49. Zeuthen E (1950) Cartesian diver respirometer. Biol Bull 98:303–318

Chapter II.4   Simultaneous Direct and Indirect Calorimetry

M.M. Pamatmat[1]

1 Introduction

Much of our knowledge of respiration rates is derived from measurements of oxygen uptake, one method of indirect calorimetry. The amount of oxygen consumed per unit time is often calculated from the decrease in partial pressure of oxygen in an enclosed air or water volume and then converted to equivalent heat production by using Ivlev's [5] oxycalorific coefficient of 4.8 cal cm^{-3} O_2 consumed ($= 450$ kJ mol^{-1}). This average conversion factor was stoichiometrically determined by Ivlev and subsequently shown to be valid for fully aerobic living organisms [6, 7]; it can actually vary between 442 when the substrate is fat or protein and 477 kJ mol^{-1} when carbohydrate is oxidized [5]. For further discussion and new calculations of the oxycaloric equivalent see Appendix C.

When interested solely in the respiration rate of fully aerobic organisms, one may simply measure oxygen uptake, avoiding direct measurement of metabolic heat production which is difficult to do even with modern instruments. First, however, we need to know when and under what conditions organisms are fully aerobic. The fact that an individual is taking up oxygen does not mean that it is fully aerobic. It could have an energy demand that is greater than its rate of oxygen uptake and consequently build up an oxygen debt. Then following a period of partial anaerobiosis it could have an elevated rate of oxygen uptake until its oxygen debt has been repaid. If we assume that values of 442 to 477 kJ mol^{-1} O_2 consumed mean that the organism is fully aerobic, then it must be partly anaerobic, building up an oxygen debt when this ratio exceeds 477 and repaying an oxygen debt when the value drops below 442 kJ mol^{-1}.

Our present knowledge does not allow us to infer from continuous measurements of oxygen alone whether an organism is in one of the above-mentioned states or another. Simultaneous direct and indirect calorimetry is required to improve our understanding of fluctuations in metabolism. Direct calorimetry permits continuous determination of total metabolism as the experimental animal switches back and forth between fully aerobic and anaerobic or intermediate states while simultaneous measurements of oxygen monitor the changing rates of oxygen uptake: the combined results

1 Tiburon Center for Environmental Studies, San Francisco State University, P.O. Box 855, Tiburon, CA 94920, USA

Polarographic Oxygen Sensors (ed. by Gnaiger/Forstner)
© Springer-Verlag Berlin Heidelberg 1983

tell to what degree the animal becomes anaerobic and later repays its oxygen debt (cf. Chap. II.3).

Several species of mussels, clams, and other benthic organisms are facultative anaerobes [1, 4, 11]. When bivalves voluntarily shut their shell valves for any reason and stop ventilating they cease taking up oxygen and may remain anaerobic for many hours. This shell-valve-closing behavior has been regarded as a disadvantage when using these animals to test sublethal metabolic effects of pollutants since the change in their metabolic rate could no longer be accurately detected in terms of oxygen uptake [2]. In fact, it is of considerable interest to determine the lowest concentration of noxious substances that will cause these animals to shut their shells. While the shortcoming of oxygen measurements is obvious, direct calorimetry, on the other hand, could detect not only when the test animal actually shuts its valves, but also the effect of the test chemical on the organism's aerobic and anaerobic metabolic rate.

2 Oxygen Measurements

The partial pressure of oxygen inside respiration chambers during direct calorimetry has been measured with a Radiometer PHM72Mk2 Digital Acid-Base Analyzer equipped with a Radiometer E5046 polarographic oxygen sensor (POS) [7]. This instrument has a lower detection limit of 0.01 kPa p_{O_2}. It has a full response time to a step change in p_{O_2} of 4 min. When transferring the oxygen sensor back and forth during calibration between atmosphere (for air calibration) and sulfite solution (for zero p_{O_2} adjustment), successive readings in atmosphere or sulfite solution could be off by as much as 0.3 kPa. After prolonged continuous usage, the zero setting could also drift by as much as 0.3 kPa. Not knowing whether the drift is linear or not, no correction has been attempted in the calculations of oxygen uptake.

Oxygen uptake, ΔV_{O_2} [cm^3 STPD], is calculated using the following equation:

$$\Delta V_{O_2} = (p_{O_2,i} - p_{O_2,f}) \times V \times 273°C/[(273°C + \theta) \times 101.32],$$

where $p_{O_2,i}$ and $p_{O_2,f}$ [kPa] are initial and final p_{O_2}, V is the gas volume of the respirometer chamber, and θ is the experimental temperature [°C]. The maximum possible error in the calculated oxygen uptake is less than 2%.

3 Direct Calorimetry

The metabolic heat production of benthic macrofauna has been measured with twin heat-flow calorimeters [3, 7–9]. The instrument has been described as a wattmeter [10]. In fact, it merely produces thermopile voltage in proportion to the heat flow through its semiconductor sensors. Hence, in using it for measuring metabolism it is crucial to eliminate or compensate all other causes of heat flow, including fluctuations in ambient temperature which would result in temperature differentials and unwanted heat flow within the calorimeter.

TOP X-SECTION

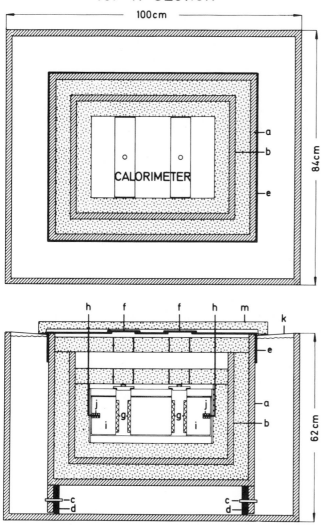

LONGITUDINAL SECTION

Fig. 1. Setup of double-twin heat-flow calorimeter using a water bath for environmental control. The calorimeter is contained inside two plywood boxes (*a* and *b*). Each box is lined with aluminum sheets and 5-cm thick styrofoam. Box *a* is coated with fiberglass inside and outside and immersed in the water bath; it is held by pins (*c*) to the bottom of the water bath, which has two strips of 2-cm thick PVC (*d*). Box *a* is capped with 5-mm thick aluminum heat shield (*e*) whose sides extend into the water. The openings of the aluminum heat shield are covered with plates (*f*) of the same thickness. The opposing sensors (*g*) are parted during insertion and removal of samples by pulling on a stout monofilament nylon line (*h*) attached to sliding blocks (*i*), thus compressing springs (*j*). When the line is released the springs push the sliding blocks and the sample is held in place between opposing sensors. The water temperature is maintained by an immersion cooler and heater-circulator. In addition, water is circulated by two submersible pumps on opposite sides of the water bath. The water bath is covered with a sheet of polyvinyl plastic (*k*) which is taped to the aluminum heat shield. The top styrofoam cover (*m*) rests on this polyvinyl plastic

The present setup uses a water bath as environmental control system (Fig. 1). The calorimeter has been operated at water bath temperatures from $5°$ to $30°C$, which were as much as $10°$ below to $3°C$ above room temperature. Room air temperature fluctuations of $± 2°C$ have no detectable effect on baseline stability. Long-term baseline drift and fluctuations do occur and are continuously monitored to determine appropriate corrections in the power-time curves [8]. Baseline drift and fluctuations become worse as sample thermal mass increases, but these are minimized by carefully balancing the mass of sample in the experimental chamber and that of a dummy sample in the compensation chamber [7].

The present instrument has a calibration constant of 2.16×10^{-5} W μV^{-1} ($= 5.16 \times 10^{-6}$ cal s^{-1} μV^{-1}). When signals are steady or changing slowly thermopile voltages are directly proportional to heat production rates, but not during periods of rapid signal fluctuations, as often observed in normal power-time curves of macrofauna, because of instrument lag. Nevertheless, the area under the curve over a time period represents the total amount of heat produced during that period, and area divided by time represents the average heat production rate during that period. Areas are integrated by planimetry. Overall accuracy of metabolic heat production estimates is believed to be consistently better than 90% and often considerably higher than this. The calorimeter's full response time to a step change in power output varies from less than 20 min to about 2 h, depending upon the mass, specific heat, and thermal conductivity of the medium in which a resistance heater is immersed. For example, with a small resistance heater inside a large clam measuring 50 mm long $\times$ 35 mm thick, switching an electric current on or off shows a full response time of 2 h. However, the change in power output from a resistance wire inside a dead clam does not accurately simulate the response time of the instrument to metabolic changes in a living clam. This problem is still under investigation.

4 Combining Direct and Indirect Calorimetry

Direct and indirect calorimetry may be combined in a variety of ways, in each case the respiration vessel being inside a calorimeter:

a) experimental animal in a closed vessel, either in air or water, whose p_{O_2} is monitored continuously with a POS inside the vessel,

b) experimental animal in a flow-through vessel, either in air or water, whose p_{O_2} is measured continuously at the inflow and outflow (Chap. II.3), and

c) experimental animal in a closed vessel, either in air or water, which is sampled and flushed periodically and its p_{O_2}, p_{CO_2}, etc. determined by any suitable method.

In the first setup, the POS must not produce measurable heat and the wire leads must not conduct a significant amount of heat into or out of the calorimeter. The choice of respiration vessel volume is dictated both by the constraints of the calorimeter and the metabolic rate of the test organism. Since the calorimeter may take as long as 4 h to regain thermal equilibrium after being opened to place the respiration vessel inside, an appropriately large volume of air or water must be used to minimize the change in p_{O_2} until useful thermopile signals can be obtained.

A flow-through respiration vessel is better than a closed system because it avoids unidirectional change in p_{O_2} and accumulation of excretory metabolic products (Chap. III. [3]). However, there is a limit on the rate of water flow that would not affect the calorimeter's baseline stability and calibration constant. Additional complications will arise as the respiration vessel volume increases and water mixing becomes necessary.

A suitable compromise between continuous flow-through and closed systems is intermittent flow. Small-diameter inlet and outlet tubings will allow periodic sampling and flushing of the respiration medium without opening the calorimeter. Each sampling will produce a transient disturbance in the heat-flow measurement, even if only from the differential pressure effect, which, however, will be of a much shorter duration than that caused by opening and placing a sample inside the calorimeter.

5 Case Studies

In the following experiments with *Modiolus demissus* in air, the mussels were first scrubbed with a wire brush under running tap water and then placed wet inside an epoxy-spray-painted and oven-baked tin canister. The entire POS body was inside the respiration vessel, its coaxial wire leads going through a small hole in the can and the hole was sealed with Dow Corning Silicone rubber sealant. In one experiment the can also had two 2-mm Tygon tubings leading outside the calorimeter. In each case the canister was sealed with a wide Neoprene rubber band. A two-channel Rustrak recorder running at 6.35 mm h^{-1} was quite satisfactory for long-term monitoring except for the drawback of having unequal scale markings, designed to linearize the galvanometer needle swing, which makes planimetry tedious.

In experiment 1, mussel 1 (Fig. 2) showed a steady basal rate of heat production from the time the calorimeter stabilized until the 11th h. Heat production increased at this time, accompanied by a rapid drop in p_{O_2}. Repeatedly thereafter as heat production rate (thermopile voltage) decreased, p_{O_2} leveled off and as heat production increased rapidly, e.g., at the 25th, 37th, and 71st h, there was a sudden drop in p_{O_2}. During 35 h when this mussel produced 143 J of heat it consumed 249 μmol O_2. The ratio of 574 kJ mol^{-1} O_2 indicates that the mussel was partly anaerobic. If calculated over the entire 72 h shown, the value would be even higher as the mussel showed no oxygen uptake for 30 h.

In experiment 2, mussel 2 (Fig. 2) showed a different pattern of changing heat production rates with time, but like mussel 1 the increases in heat production at the 21st and 78th h were accompanied by a rapid fall in p_{O_2}. Following a relatively sharp drop in heat production, thermopile voltage rose at the 10th h but this rise was not accompanied by a corresponding drop in p_{O_2}. Apparently, the rapid drop in p_{O_2} occurs only after a relatively long preceding period of lowered heat production. Mussel 2 showed a ratio of 475 kJ mol^{-1} O_2 over a 48-h period.

Another mussel that had been kept at 5°C in air for 1 week before being placed in the calorimeter at 20°C showed a characteristically different power-time curve while p_{O_2} dropped exponentially (mussel 3, Fig. 3). It is noteworthy that this individual was

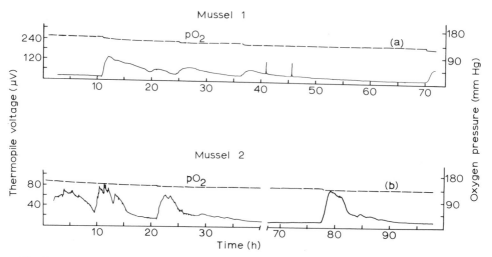

Fig. 2. Power-time curves of *Modiolus demissus* at 20°C and simultaneous oxygen measurements during direct calorimetry. O_2 uptake was calculated from occasional readings of the digital display of the Radiometer Analyzer which gave finer resolution than the stripchart recording of its analog output. Mussel 1 produced 143 J of heat during the first measurable 35 h, with p_{O_2} decreasing from 20.69 to 18.08 kPa for a total O_2 uptake of 249 μmol O_2. The ratio of heat production to O_2 uptake was 576 kJ mol^{-1}. Mussel 2 produced 108 J of heat during the first measurable 48 h, with p_{O_2} decreasing from 20.69 to 18.31 kPa for a total uptake of 227 μmol O_2. The ratio of heat production to O_2 uptake was 475 kJ mol^{-1}. Sudden drops in p_{O_2} are seen when the power-time curves rise rapidly from a basal level, especially when this lasts many hours

never seen to open its shells from the time of its collection. At 20°C its power-time curve showed sharp fluctuations of about ± 5 μV with a periodicity of 20–30 min, without the large decline shown by the other two, without the leveling off to basal levels and the rapid increases. Its mean heat production rate decreased linearly with time. The ratio of total heat production for 48 h to oxygen uptake was 484 kJ mol^{-1}. This mussel was gaping when its respiration vessel was opened and its shells remained open in spite of handling. When placed in aerated seawater at 20°C it recovered its normal irritability and closed its shells when disturbed.

Mussel 3 was returned to the calorimeter after 1 week in aerated water. Its second power-time curve (Fig. 3b) is different from its first, but the average heat production rate changed little (1.95×10^{-3} W vs. 1.78×10^{-3} W). The ratio of its heat production to oxygen uptake during the second run was 471 kJ mol^{-1}. After flushing the respiration vessel with air, the ratio of heat production to oxygen uptake during the next 72 h was 453 kJ mol^{-1}. Flushing the chamber with air (Fig. 3b, 3c) had no noticeable effect on the metabolic trend. Following reestablishment of thermal equilibrium of the calorimeter, mussel 3 showed the same level of heat production as immediately before flushing (Fig. 3b) or a smooth continuation of the decreasing trend already established before flushing (Fig. 3c). The cycle appears to be independent of the oxygen pressure.

Finally, mussel 4 (Fig. 3d) showed the same fluctuations in its thermogram as the other three, but when part of the mid-ventral shell was chipped off to make a ≈ 1-cm^2 opening, the cycle of intermittent heat bursts and decline to basal level disappeared.

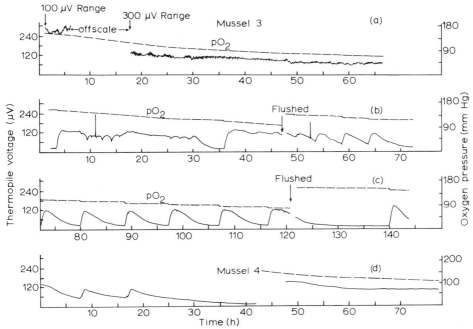

Fig. 3a–d. Power-time curves and simultaneous p_{O_2} measurements for mussel 3 at 20°C: **a** after 1 week at 5°C, **b** after 1 week in aerated water at 20°C, **c** continuation of **b**. The chamber was flushed with air at the times indicated, raising p_{O_2} from 13.36 to 19.19 kPa the first time and from 11.67 to 20.66 kPa the next. The metabolic cycle began to appear only after 45 h. Raising p_{O_2} had no apparent effect on the trends of the power-time curve, considering the interruptions caused by injection of air and consequent disturbance of thermal equilibrium. **d** Power-time curve of mussel 4 with shells intact during the first 42 h, and subsequent to breaking off pieces of the mid-ventral shell valves to make a permanent opening. Like the other three individuals it showed fluctuations which disappeared following the opening of the shell valves. With shell valves open p_{O_2} decreased smoothly

Continuous p_{O_2} measurements with the shells broken open showed a slow exponential decrease similar to that of mussel 3 following acute temperature change (Fig. 3a). The ratio of heat produced to oxygen uptake during 16.3 h was 477 kJ mol^{-1}.

The mussel 4 experiment points to the complete closing and opening of the shells as the immediate cause of the stepwise decreases in p_{O_2}. Shell closure traps several milliliters of air inside the mantle cavity. The mussel probably remains aerobic after closing its shells until all oxygen in its mantle cavity is used up. As the oxygen inside the mantle cavity is depleted the mussel's metabolic rate decreases to basal level. The steady basal rate for hours, and lasting over 2 days sometimes, probably represents anaerobic metabolism. When the shell valves subsequently open anoxic air mixes with the rest of the air in the canister and causes a sharp drop in p_{O_2} reading.

The smooth exponential decrease in p_{O_2} in the case of mussel 3 following acute temperature change from 5° to 20°C (Fig. 3a) indicates that its shells were gaping throughout the 66-h period. The sharp fluctuations in its power-time curve may be the result of more or less rhythmic muscular contractions.

It is now evident that the POS should have been placed inside the mussel's mantle cavity, or a second one placed there. As it was, the POS inside the can failed to continuously measure the true oxygen uptake by the mussels when their shells were closed. When p_{O_2} reading was stable (O_2 uptake was apparently zero) there was actually some oxygen uptake until the air inside the mantle cavity became anoxic. The present measurements by themselves do not show when the mussels actually became fully anaerobic. However, their steady basal metabolic rates per unit weight are the same as that of anaerobic clams and mussels [8, 9]. It is believed that they actually had zero O_2 uptake for long periods while producing heat anaerobically and during such periods the ratio of heat production to oxygen uptake was infinitely high.

The differences in p_{O_2} readings when the mussels were open (p_{O_2} values were decreasing, such as those used in the calculations of heat production to $\dot{O}_2$ uptake ratios) still reflect actual oxygen uptake. The calculated ratio for mussel 1 (576 kJ mol^{-1} O_2) shows that this species could remain predominantly anaerobic for extended periods and oxygen uptake measurements alone could underestimate its actual metabolic rate. Furthermore, the wide range of metabolic pattern and large amplitude of fluctuations in the metabolic cycle of *Modiolus demissus* in air raise questions about the usefulness of measurements lasting only several hours. On the other hand, the observed ratios of 453 to 484 kJ mol^{-1} O_2 signify that even though the mussel becomes anaerobic periodically it eventually fully oxidizes its substrate, including the metabolic end products formed during anaerobiosis.

These case studies illustrate that oxygen uptake measurements alone suffice for many investigations on invertebrate metabolism, but combining direct and indirect calorimetry is necessary to understand certain metabolic patterns.

Acknowledgment. Supported by U.S. National Science Foundation Grant OCE77-08634.

References

1. Chen C, Awapara J (1969) Intracellular distribution of enzymes catalyzing succinate production from glucose in *Rangia* mantle. Comp Biochem Physiol 30:727–737
2. Davenport J (1977) A study of the effects of copper applied continuously and discontinuously to specimens of *Mytilus edulis* (L.) exposed to steady and fluctuating salinity levels. J Mar Biol Assoc UK 57:63–74
3. Gnaiger E (1979) Direct calorimetry in ecological energetics. Long term monitoring of aquatic animals. Experientia (Suppl) 37:155–165
4. Hammen CS (1976) Respiratory adaptations: invertebrates. In: Wiley M (ed) Estuarine processes, vol I. Uses, stresses, and adaptation to the estuary. Academic Press, London New York, pp 347–355
5. Ivlev VS (1934) Eine Mikromethode zur Bestimmung des Kaloriengehalts von Nährstoffen. Biochem Z 275:49–55
6. Kleiber M (1962) The fire of life. An introduction to animal energetics. John Wiley, New York, 454 p
7. Pamatmat MM (1978) Oxygen uptake and heat production in a metabolic conformer (*Littorina irrorata*) and a metabolic regulator (*Uca pugnax*). Mar Biol 48:317–325

8. Pamatmat MM (1979) Anaerobic heat production of bivalves (*Polymesoda caroliniana* and *Modiolus demissus*) in relation to temperature, body size, and duration of anoxia. Mar Biol 53: 223–229

9. Pamatmat MM (1980) Facultative anaerobiosis of benthos. In: Tenore KR, Coull BC (eds) Marine benthic dynamics. The Belle W. Baruch Library in Marine Science, no 11. University of South Carolina Press, Columbia, pp 69–90

10. Wadsö I (1974) A microcalorimeter for biological analysis. Sci Tools 21:18–21

11. Zwaan A de, Wijsman TCM (1976) Anaerobic metabolism in Bivalvia (Mollusca). Characteristics of anaerobic metabolism. Comp Biochem Physiol 54B:313–324

Chapter II.5   An Automated, Intermittent Flow Respirometer for Monitoring Oxygen Consumption and Long-Term Activity of Pelagic Crustaceans

L.B. Quetin[1]

1 Introduction

Recently there has been an increase in interest regarding the effects of environmental parameters such as temperature, pressure, oxygen, pH, and salinity on the oxygen consumption of pelagic animals [1, 3, 4, 6, 8, 11]. In these studies polarographic oxygen sensors (POS) were used with closed respirometers rather than continuous-flow systems to avoid problems of flow control and the need for a second POS in the system. There are, however, two major limitations to closed respirometry systems: (1) oxygen consumption rates cannot be measured over extended periods of time; and (2) the partial pressure of oxygen to which the animal is exposed cannot be maintained at a constant level.

An intermittent flow respirometer is a compromise between continuous flow respirometers and closed respirometers. Such a system retains some of the simplicity and low cost of closed respirometers without limiting the duration of the experiment due to depletion of oxygen. A modification of a closed respirometry system and activity monitor [9] was used to simultaneously measure the oxygen consumption rate and activity of the deep-sea pelagic crustacean *Gnathophausia ingens* at different temperatures and oxygen levels [7]. This system uses the output of a POS and timed controllers to maintain the partial pressure of oxygen in the respirometer within chosen limits and to record periodically the animal's oxygen consumption rate. Activity is recorded continuously as pleopod beats per minute. The pleopods provide the forward thrust for swimming in large pelagic crustaceans, which has a large effect on the metabolic rate of the animal. Measuring this activity allows more detailed examination of the effect of environmental parameters on oxygen consumption.

2 The Respirometry System

The respirometry system consists of four functional parts: (1) the temperature-regulation equipment, (2) the respirometer and POS, (3) the pump system that circulates

1 Marine Science Institute, University of California at Santa Barbara, Santa Barbara, CA 93106, USA

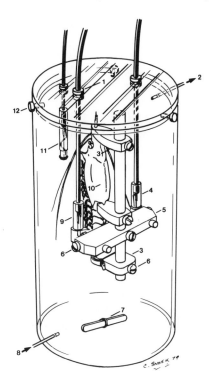

Fig. 1. The mysid *Gnathophausia ingens* held to record pleopod activity in the intermittent flowing water respirometer. *1* rubber seal, *2* water outlet, *3* clamp, *4* photoresistor, *5* holder, *6* set screw, *7* stirring bar, *8* water inlet, *9* light-emitting diode (LED), *10* animal, *11* POS, *12* Plexiglas pin

water through the respirometer, and (4) the control equipment that maintains the oxygen in the respirometer within a set range.

2.1 Temperature Regulation

Temperature is regulated in the respirometer to ± 0.1°C using a Forma-Temp, Jr. circulating water bath which circulates chilled freshwater through the waterjacket on the outside of the glass reservoir. The glass reservoir contains 12 cm^3 of filtered seawater which is deep enough to cover the respirometer completely.

2.2 Respirometer and POS

The respirometer is a Plexiglas cylinder 7 cm in diameter, 19.5 cm long, and 4 mm thick (Fig. 1). A Plexiglas plate is cemented to one end and the other end is sealed by a close-fitting, removable top which is inserted into the cylinder, sealed with silicone stopcock grease and fixed by three small Plexiglas pins (Fig. 1, 12). Two leads from the activity monitor and one from the oxygen sensor penetrate the top through three 1-cm holes which are sealed with rubber stoppers.

The inlet port of the respirometer is 1 cm from the bottom and the outlet port 1 cm from the top on the opposite side. The ports are 1-mm holes into which 1-mm

I.D. vinyl tubing is inserted. The tubing extends 0.5 m from the outlet port and emp-
ties into the seawater reservoir containing the respirometer. The tubing to the inlet
port forms a long cooling coil. The small size of the ports and the long length of tubing
extending from them prevent any effective oxygen exchange between the respirometer
and the surrounding aerated seawater when the circulating pump is turned off. Therefore, with the pump off the respirometer acts as a closed system. This feature simplifies the system by eliminating valves at the inlet and outlet ports (cf. Chap. II.1).

Self-constructed POS are used to measure the p_{O_2} in the respirometer (Chap. I.8).
A magnetic stirring bar in the bottom of the respirometer provides sufficient stirring.
The magnetic stirrer is placed directly below the respirometer, reservoir and water
bath.

2.3 Pump System

The respirometer is submerged in the reservoir containing filtered, aerated seawater.
This keeps the respirometer at a constant temperature and is also the source of aerated
seawater which circulates through the respirometer (Fig. 2). Water is drawn through
aquarium tubing from the reservoir, pumped through a mixing pump attached above
the seawater reservoir, through an air trap, and then through a 2.5-cm coil of 1-mm I.D.
vinyl tubing (cooling coil) before entering the respirometer. The mixing pump (Rupp
Industries, Inc., Bellville, Ohio, USA) was used at a pumping rate of 100 cm^3 per min.
Both the air trap and cooling coil are immersed in the seawater reservoir. The air trap
is simply an inverted 30-cm^3 test tube with a two-hole rubber stopper inserted into

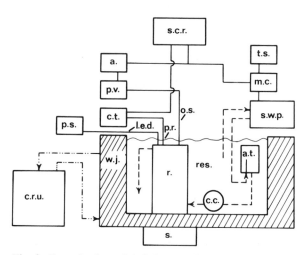

Fig. 2. General schematic of the experimental appartus. *Arrows* point in the direction of water
flow. Symbols are as follows: – – – seawater flow, · · – · · chilled freshwater fow, —— electrical
connections, *a.* amplifier, *a.t.* air trap, *c.c.* cooling coil, *c.r.u.* circulating refrigeration unit, *c.t.*
cardiotachometer, *l.e.d.* light emitting diode, *m.c.* millivolt controller, *o.s.* oxygen sensor, *p.r.*
photoresistor, *p.s.* power source for l.e.d., *p.v.* polarizing voltage for o.s., *r.* respirometer; *res.*
reservoir, *s.* magnetic stirrer, *s.c.r.* strip-chart recorder, *s.w.p.* seawater mixing pump, *t.s.* timed
switch, *w.j.* water jacket

the open end. Two glass tubes were inserted through the stopper, one ending halfway up the inside of the test tube and the other ending flush with the rubber stopper. Water enters the trap through the long glass tube and releases any air bubbles to the top of the trap.

2.4 Oxygen Regulation

The mixing pump is controlled by two mechanisms: (1) a pH-mV controller (type 45E, Chemtrix Inc., Hillsboro, ORE, USA) using input from the POS, and (2) a pre-set timed switch (model T101, Intermatic Inc., Spring Grove, ILL, USA). The timed switch is necessary because the mV controller available is only able to maintain the oxygen in the respirometer within limits too narrow to obtain an adequate estimation of oxygen consumption. By switching the mixing pump on or off the mV controller maintains the oxygen level in the respirometer within ± 0.20 kPa. The timed switch works independently of the p_{O_2} in the respirometer to swtich the mV controller and therefore the mixing pump on or off at fixed intervals set at the beginning of the experiment.

In a similar system a relay circuit energized by a reed switch mounted on the recorder is used instead of the mV controller [10]. When the oxygen level in the respirometer reaches a preset low p_{O_2} the system is opened to replenish oxygen up to the preset high p_{O_2}. The control unit can be adapted and optimized variably to meet the demands of the special experimental situation.

3 The Activity Monitor

The activity monitor is composed of a rod, on which the animal is held in place by clamps, and a light-emitting diode (LED)-photoresistor combination. The photoresistor and LED are each encapsulated in epoxy cement (TRA CON 2113) and soldered to a length of co-axial cable. A 2.5 cm^3 plastic syringe body sealed at the needle attachment is used as a mold. The photoresistor and LED are centered in their respective molds and then covered with epoxy cement until at least 5 mm of the outer covering of the co-axial cable is immersed. This ensures that the LED and photoresistor are waterproof. After curing, the LED and photoresistor are carefully filed to their final shape. The face of each is highly polished, using pumice. Plexiglas rods (6.4 mm) are attached at right angles to the photoresistor and the LED to fit into the 6.8 mm holes of the holder (Fig. 1, 5).

All other parts of the activity monitor are made of Plexiglas. The dimensions given in the following description pertain to an activity monitor used for an 8–10-cm long pelagic mysid. For animals of different sizes and morphology the system would have to be modified. The clamps for keeping the animal in a fixed position, and the activity monitor have a 10-mm hole in the top center through which a 152-mm long, 9.5-mm diameter rod slides (Fig. 1). A 4.8-mm nylon set screw tightens against the rod to keep each sliding piece in a fixed position.

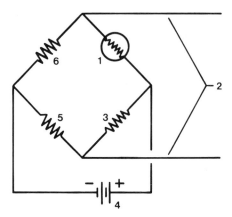

Fig. 3. Wheatstone bridge circuitry used with the photoresistor to record the pleopod activity of *Gnathophausia ingens*. *1* photoresistor, *2* output, *3, 5* two matched resistors (we used 22 K), *4* power supply, *6* variable potentiometer similar to the resistance of the photoresistor

The morphology of *Gnathophausia ingens* makes it very easy to attach the animal to the rod using three clamps. Two clamps, each with a small horizontal hole for insertion of either the rostrum or posterior dorsal spine, hold the anterior part (carapace) of the animal to the rod. The animal is attached to the rod by fixing the first clamp in position and sliding the rostrum through the hole as far as possible. The second clamp is then moved forward and the posterior dorsal spine inserted into the hole in the clamp. This clamp is then fixed in position. Before the abdomen of the animal is secured, the LED-photoresistor combination must be attached to the rod. A taut rubber band is attached flush with the bottom edge of the third clamp. The tail of the animal is inserted between the two sides of the rubber band. Care must be taken when adjusting this clamp to ensure that the abdomen is not stretched unnaturally away from the carapace causing abnormal activity.

The LED and photoresistor can be rotated, moved vertically, and moved along the length of the animal, and are positioned as close as possible to a pair of pleopods without interfering with their movements. In this instance, the distance between the LED and photoresistor is 25 mm, and the light beam from the LED is interrupted by the beating of a single pleopod pair.

Nine volts is sufficient to power the LED in this system, using a Monsanto No. MV-50 LED emitting in the red, 600–700 nm, and a No. CL903 photoresistor made by Clairex Electronics, New York. LED's emitting infra-red wavelengths of light have also been used successfully. However, malfunctions are difficult to detect since an infrared LED's output cannot be detected by sight. Caution is needed when varying the voltage to the LED to keep the current input to less than 40 mA for the MV-50. This is done by putting the appropriate resistance on the positive lead of the LED. In this case a 500 Ω resistor was used, which means a 20 V power source was the maximum power source that could have been used. The photoresistor acts on one arm of a Wheatstone bridge (Fig. 3). Each pleopod beat acts to unbalance the bridge and create an electrical pulse which can be time-averaged on a cardiotachometer or recorded directly on either analog or digital recording systems.

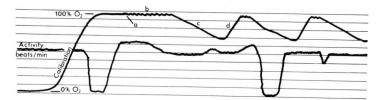

Fig. 4. Sequence of events during a respirometry run of *Gnathophausia ingens* in the intermittent flowing water respirometer. Activity is recorded continuously when the animal is in the respirometer. The oxygen sensor is calibrated, *a* the respirometer is sealed, *b* the mV controller controls the mixing pump to maintain a high level of oxygen in the respirometer until the animal's initial, high activity decreases, *c* the mV controller and mixing pump are turned off by the timed switch and the animal's oxygen consumption rate recorded, *d* the timed switch turns on the mV controller and mixing pump and the level of oxygen in the respirometer is increased

4 Experimental Procedure

After an animal is attached to the activity monitor it is placed in the respirometer and the top is sealed with the exception of the port for the POS which is removed for calibration. The POS is placed in a beaker of seawater at the experimental temperature and calibrated at saturation conditions while bubbling air via an airstone through the seawater. The corresponding concentration of dissolved oxygen is read from standard tables (App. A). Nitrogen is bubbled through the seawater for zero calibration. After the animal is allowed to recover for 4 to 6 h as water is continually circulated through the respirometer, the POS is inserted into the respirometer and the respirometer sealed with a rubber stopper (Fig. 4a).

With the mV controller on, the oxygen level in the respirometer is maintained within 0.20 kPa of a preset p_{O_2} until the activity of the animal stabilizes (Fig. 4b). The timed switch of the mV controller is set to (1) allow enough time to measure oxygen consumption (Fig. 4c), (2) keep the oxygen level in the respirometer within 15% of the p_{O_2} as regulated by the mV controller, and (3) allow adequate time to replenish the oxygen in the respirometer (Fig. 4d). In experiments with *Gnathophausia ingens* at temperatures between 3° and 7.5°C, the controller is turned on between 15 and 30 min every hour. The POS is recalibrated every 10–12 h by removing it through the port at the top of the respirometer. Disturbance to the animal is minimal. The preceding measurements of the animal's oxygen consumption rate were discarded if the sensor calibration had changed by more than 3%. At the end of an experiment the animal is removed from the respirometer and weighed. The respirometer is resealed without an animal and oxygen consumption recorded for at least 5 h with the same timing intervals used in the previous experimental run. This served as a control and was always less than 1% of the total decline in oxygen during each 30–45 min recording period at temperatures between 3° and 7.5°C. The lack of bacterial respiration was probably due to the low exprimental temperatures (Chap. II.9). In addition, the seawater reservoir and respirometer were washed with a mild acid and rinsed thoroughly with distilled water before each run and all seawater used was filtered through a 47-μm Millipore filter.

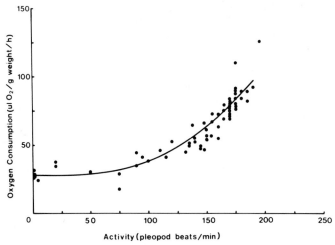

Fig. 5. An example of the relationship between the oxygen consumption rate and activity of *Gnathophausia ingens* recorded at 5.5°C and 8.0–12 kPa p_{O_2} using the method described in this section

The experimental results provide an accurate estimation of the relationship between oxygen consumption and pleopod beat rate of the animal (Fig. 5) at different temperatures and oxygen concentrations. This respirometry system would also be valuable in situations where an investigator requires the measurement of oxygen consumption rates while the animal is maintained under certain environmental conditions for long periods of time.

References

1. Belman BW (1978) Respiration and the effects of pressure on the mesopelagic vertically migrating squid *Histioteuthis heteropsis*. Limnol Oceonogr 23:735–739
2. Belman BW, Childress JJ (1973) Oxygen consumption of the larvae of the lobster *Panulirus interruptus* and the crab *Cancer productus*. Comp Biochem Physiol 44A:821–828
3. Belman BW, Gordon MS (1979) Comparative studies on the metabolism of shallow-water and deep-sea marine fishes. V. Effects of temperature and hydrostatic pressure on oxygen consumption in the mesopelagic zoarcid *Melanostigma pammelas*. Mar Biol 50:275–282
4. Childress JJ (1975) The respiratory rates of midwater crustaceans as a function of depth of occurrence and relation to the oxygen minimum layer off southern California. Comp Biochem Physiol 50A:787–799
5. Childress JJ (1977) Effects of pressure, temperature and oxygen on the oxygen consumption rate of the midwater copepod *Gaussia princeps*. Mar Biol 39:19–24
6. Mickel TJ, Childress JJ (1978) The effect of pH on oxygen consumption and activity in the bathypelagic mysid *Gnathophausia ingens*. Biol Bull 154:138–147
7. Quetin LB (1979) Metabolic and behavioral aspects of mid-water crustaceans. PhD Dissertation, Univ California, Santa Barbara, p 122
8. Quetin LB, Childress JJ (1976) Respiratory adaptations of *Pleuroncodes planipes* to its environment off Baja California. Mar Biol 38:327–334

9. Quetin LB, Mickel TJ, Childress JJ (1978) A method for simultaneously measuring the oxygen consumption and activity of pelagic crustaceans. Comp Biochem Physiol 59A:263–266
10. Sutcliffe DW, Carrick TR, Moore WH (1975) An automatic respirometer for determining oxygen uptake in crayfish [*Austropotamobius pallipes* (Lereboullet)] over priods of 3–4 days. J Exp Biol 63:673–688
11. Torres JJ, Belman BW, Childress JJ (1979) Oxygen consumption rates of midwater fishes as a function of depth of occurrence. Deep-Sea Res 26A:185–197

Chapter II.6 Sealed Respirometers for Small Invertebrates

L.B. Quetin and T.J. Mickel[1]

1 Introduction

The purpose of this section is to describe the sealed respirometers that we and others working at the Marine Science Institute, University of California, Santa Barbara have used in the past to obtain respiration measurements on a wide variety of crustaceans (for references see Chap. II.5).

2 Design Criteria

Respirometers that use POS to measure the oxygen consumption rate of small invertebrates may be of any shape or size as long as they satisfy two criteria: (1) adequate stirring to maintain a homogeneous solution, especially near the tip of the oxygen sensor, and (2) a volume large enough to provide at least an 8-h interval to measure the respiration rate, and small enough so that the decrease in oxygen over time can be adequately recorded by the POS. All of the above criteria involve compromises in the design of small sealed respirometers.

2.1 Stirring

It is essential that the POS be stirred adequately and also that the water in the respirometer be well mixed, otherwise the output of the sensor will be erratic, making it impossible to compute an oxygen consumption rate. On the other hand, increased turbulence will have a disturbing influence on the behavior of the animal in the respirometer. Thus we have two opposing demands, stirring should be minimized with regard to the organism, but maximized for a stable output of the POS. Large animals can usually tolerate more vigorous agitation than small animals and a robust, benthic, subtidal crab is more resistant in this respect than a fragile, deep-sea, pelagic shrimp. If

1 Marine Science Institute, University of California at Santa Barbara, Santa Barbara, CA 93106, USA

Polarographic Oxygen Sensors (ed. by Gnaiger/Forstner)
© Springer-Verlag Berlin Heidelberg 1983

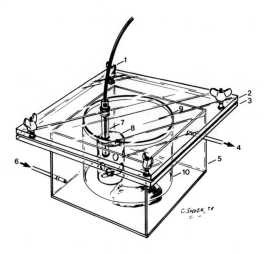

Fig. 1. A sealed respirometer with integrated water-jacket. *1* brass wing nut and bolt, *2* top of the animal chamber, *3* top of the water-jacket, *4* water outlet, *5* water-jacket, *6* water inlet, *7* POS, *8* stirring chamber, *9* O-ring, *10* glass animal chamber

the POS is partially isolated from the animal by a stirring chamber (Fig. 1, 8) within the respirometer, the immediate surrounding of the POS can be stirred more vigorously than the water around the animal. An alternative is to use a POS with a very small cathode that operates properly at a low level of stirring. In many instances the necessary mixing is provided by the activity of the animal.

2.2 Respirometer Volume

Ideally one would like a relatively small respirometer which does not restrict the movement of the animal to get a large change in p_{O_2} for a given oxygen consumption rate within a short period of time. Conversely, the respirometers must be large enough to allow the animal ample time to recover from handling and to return to a "normal" oxygen consumption rate before the oxygen in the respirometer is depleted to the point where the animal is severely stressed. The length of a useful respirometry "run" is the time from sealing the respirometer until the animal becomes stressed by lack of oxygen. We usually try to ensure that the length of a respirometry "run" of an animal lasts for at least 8–12 h. At the end of this time the oxygen level in the respirometer should have decreased to only 35% of air saturation. However, this is only a generalization and these limits may change with temperature, salinity, animal species, etc.

As an example of the above, consider a 3-g (wet weight) animal with an estimated routine metabolic rate of 2.23 μmol O_2 h^{-1} g^{-1} and an active metabolic rate of 11 μmol O_2 h^{-1} g^{-1}. Allowing for an initial 2-h period of high activity and an 8-h period at routine activity we can estimate that a total of 120 μmol O_2 will be consumed during the respirometry run. Therefore the total oxygen capacity of the respirometer should be 190 μmol O_2. The effects of temperature and salinity on the solubility of oxygen in water (App. A) need to be considered when calculating the volume of the respirometer. Air saturated water at 10°C and a salinity of 35‰ contains 284 μmol O_2 dm^{-3}. Thus in the above example the respirometer volume should be 0.67 dm^3. It is preferable to over- rather than underestimate the water volume required for a respirometer.

3 Jacketed Respirometer

The following is a detailed description of a respirometer that we have used to measure the oxygen consumption rates of crustaceans at oxygen levels from 100% to 0% saturation [4, 8]. Basically the respirometer consists of a glass vessel (animal chamber) surrounded by a Plexiglas water-jacket. A Plexiglas top and O-ring seal the animal chamber. A POS is inserted into the animal chamber from above (Fig. 1).

3.1 Construction

The respirometer is constructed by first making a square box from 5-mm Plexiglas that is 15 cm on a side and 8 cm high. The sides and bottom of the box are glued together with Plexiglas cement. Alternatively, a Plexiglas tube of appropriate diameter and a circular bottom can be used, but sheet material is generally more readily available.

Midway on one side of the Plexiglas box approximately 2 cm from the bottom a 8-mm hole is drilled for the inlet port and midway on the opposite side, 2 cm from the top, a 10-mm hole for the outlet port. A 3.5-cm length of appropriate size Plexiglas tubing is cemented into each of these ports using Plexiglas cement.

The top of the water-jacket is made of a 20 cm × 20 cm and 5-mm thick Plexiglas square with an O-ring groove cut into the surface. The inside diameter of the O-ring groove exceeds the inside diameter of the glass animal chamber by approximately 5 mm. A hole of equal size to the inside diameter of the glass animal chamber is cut through the center of the water-jacket top. Then 1 cm inward from each corner a 5-mm hole is drilled which is just large enough to fit a 1.5 inch (3.8 cm), 0.25 inch (6.35 mm) brass bolt. The glass animal chamber is then glued with silicone glue to the top of the water-jacket on the side opposite the O-ring groove. It is important to make a smooth joint between the top and the inside of the glass animal chamber to prevent bubbles from being trapped in the animal chamber when trying to seal it. The joint should be made wedge-shaped, wider at the outer side, by bevelling the rim of the glass chamber. This allows only a thin silicone layer inside, and provides sufficient surface area for a firm and tight joint. The larger cross-section enhances polymerization. A brass bolt is inserted through each corner hole so its threads protrude from the top on the same side as the O-ring groove, and is fixed to the water-jacket top with epoxy cement.

Next, the top with the attached animal chamber is fixed to the box forming the outer part of the water-jacket. This is done again with silicone glue so that a scalpel can be inserted between the silicone joints of the appropriate pieces to separate them and replace the glass animal chamber if it breaks. Just before gluing the top and bottom of the water-jacket together, a lump of silicone glue is squeezed onto the inside bottom of the water-jacket. The animal chamber is forced down on this mass of silicone glue when the top and bottom of the water-jacket are joined. When the silicone glue hardens it lends support to the glass animal chamber.

As a final step the top for the glass animal chamber (Fig. 1., 2) is made from a Plexiglas piece of the same dimensions as the top of the water-jacket: 20 cm × 20 cm × 5 mm. Holes are drilled in this top at each corner. These holes have the same center point as those drilled in the top of the water-jacket but are approximately 4 mm larger. Once these holes are drilled, the four bolt ends protruding from the top of the water-jacket will easily pass through the top of the animal chamber. When brass washers and wing nuts are then fastened to the bolts, the top of the animal chamber can be tightened onto the O-ring and the animal chamber sealed.

A port for the POS is cut into the top of the animal chamber. To form a tight seal the POS is inserted into a one-hole rubber stropper or a tight-fitting section of vinyl tubing. Tubing requires a more precise fit and the diameter of the port must be approximately 0.5 mm less than the diameter of the tubing when it is stretched over the POS. Tubing has the advantage that there is little surface area to trap bubbles when inserting the POS into the animal chamber. Much more care is needed to exclude bubbles with a rubber stopper.

The simplest method of stirring the animal chamber and oxygen sensor is to put a magnetic stirring bar in the bottom of the glass animal chamber with a magnetic stirrer underneath the respirometer. The animal can be isolated by putting a nylon mesh screen over the stirring bar creating a "false" bottom. This works well with POS that need little stirring. However, it may be necessary to build a separate stirring chamber for the POS. In this respirometer the stirring chamber (Fig. 1, 8) is 2.5 cm in diameter, 3 cm high and made of 2-mm thick Plexiglas. The bottom is solid and there is a POS port in the top. Small holes (approximately 3 mm) are drilled in the sides. The number and size of these holes can be varied to achieve the proper exchange rate between the glass animal chamber and the stirring chamber. The stirring chamber should be fixed with silicone glue to the bottom of the animal chamber to prevent it from rotating with the stirring bar.

3.2 Experimental Procedure

During an experiment, water from a constant temperature bath is circulated through the water-jacket. Care must be taken that the pressure inside the water-jacket will not burst the joints held together with silicone glue. Air-saturated water at experimental temperature is then poured into the animal chamber. The animal is placed in the chamber with as little disturbance as possible. After removing any gas bubbles, more water is added until a meniscus is formed above the O-ring. The top of the animal chamber is then lowered carefully onto the meniscus so as to exclude all bubbles. After tightening the wing nuts at the corners of the respirometer the animal chamber is sealed except for the port for the POS. A calibrated POS is then inserted into the respirometer with the tip passing through the top of the stirring chamber (Fig. 2a).

At the beginning there is a rapid decrease in oxygen due to the initial high activity of the animal (Fig. 2b). After this period (usually no longer than 2 h) the rate of oxygen consumption is relatively constant, as shown by the uniform decrease in oxygen partial pressure (Fig. 2c). Metabolism is hardly affected by the level of oxygen in the respirometer, until an oxygen level is reached where the animal is no longer able to regulate its oxygen uptake rate (Fig. 2d). Unless one is especially interested in performance in this low range, it is preferable to end the experiment before the animal succumbs from lack of oxygen.

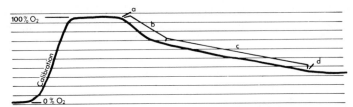

Fig. 2. The sequence of events during a sealed respirometry run. The POS is calibrated to 0% and 100% air saturation, *a* the respirometer is sealed, *b* the initial, rapid decrease in oxygen due to the high activity of the animal, *c* long-term, relatively uniform decrease in oxygen recorded as the normal oxygen consumption rate of the animal, *d* the level of oxygen below which the animal can no longer sustain itself

At the end of the experiment the animal is removed from the animal chamber, the POS recalibrated at air and nitrogen saturation, water added to replace the volume of the animal, and the animal chamber is resealed. Oxygen consumption in the control run is usually less than 5% of the total measured rate (temperature $< 10°C$) and is subtracted from the previous recording to obtain the oxygen consumption rate of the animal. The control run usually lasts 3–5 h. Then the POS is again recalibrated. If any calibration deviates by more than 3% from the previous one, the experimental results are discarded.

4 Other Respirometers

Other types of sealed respirometers have been used in our laboratory. One was used for a large crustacean 30–50 mm long maintained at high hydrostatic pressures (Chap. II.5) and the other was designed to measure the rates of oxygen consumption of small zooplankton between 0.5 and 5 mm long [2, 5]. A POS with a 0.0152 mm diameter cathode was used (Chap. I.8). For this micro-cathode POS, adequate stirring is produced by the movement of the animal in the respirometer.

5 A Respirometer for Small Zooplankton

The most versatile respirometer for small zooplankton is constructed as follows. The end of a 2-ml glass syringe (TOMAC Hospital Supply Corporation) is cut square and a 2-mm diameter hole drilled into the syringe body 10 mm from the cut end. This small hole is covered with a clear piece of 11-mm I.D. vinyl tubing lightly lubricated with silicon grease (Fig. 3).

This design has several advantages. Once the POS is inserted into the tip of the syringe it does not need to be removed. To calibrate the POS, water is circulated into the syringe from the opposite end so the sensor can remain in position. After calibration, the animal is deposited into the syringe and the plunger slowly inserted to ex-

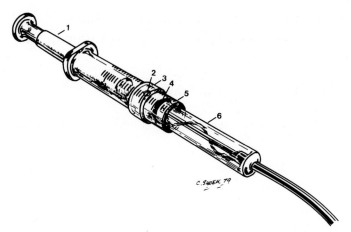

Fig. 3. Respirometer for small planktonic crustaceans. *1* syringe plunger, *2* 2-mm hole, *3* 11-mm I.D. vinyl tubing, *4* syringe barrel, *5* POS, *6* O-ring seal

clude any air and not harm the animal. The plunger is pushed gently downward while allowing water to leak out of the small 2-mm hole at the other end until the water in the syringe is reduced to the desired volume. If the animals are smaller than the small hole, it is left partially covered. The volume divisions make it very easy to use the same volume for a series of experiments. This respirometer would also work quite well at high hydrostatic pressures since the plunger would respond to the slight changes in water volume or pressure leaving the O-ring seal around the POS at the other end of the respirometer undisturbed. Pressure is applied to the respirometer by sealing it in a pressure vessel and using a high pressure hand pump.

6 Summary

In summary, there are many possible designs for sealed respirometers using a POS. With every design, however, some compromises will undoubtedly be necessary. Conditions for reliably measuring the physical parameter have to be fulfilled, but any type of respirometer must foremost be designed to minimize behavioral and physiological stress to the animal unless such responses are specifically desired. The most accurate measurement is meaningless unless due consideration is given to the reactions of the organism investigated.

For references see Chapter II.5.

Chapter II.7   A Method for the Simultaneous Long-Term Recording of Oxygen Evolution and Chloroplast Migration in an Individual Cell of *Acetabularia*

H.G. Schweiger, H. Broda, D. Wolff, and G. Schweiger[1]

A number of problems require the measurement of functional parameters in single cell systems. A major difficulty in such measurements is the low level of signal output that can be achieved from single cells. This problem can be avoided by performing measurements on populations of single cells. In cell populations, however, an interaction between cells cannot be excluded. Under these circumstances, it may be necessary to accept the low signal level of an individual cell and to develop appropriate methods for evaluation.

The methodological difficulties in terms of stability of measurement are further increased when long-term measurements are necessary, as is the case in studies of diurnal oscillations.

Some of the difficulties may be overcome when the size of the cell is such that it is large enough to produce a sufficiently strong signal. Such a cell is represented by the large unicellular and uninuclear green alga *Acetabularia,* which is found in tropical and subtropical seas. It can be cultured relatively easily in the laboratory and some of its special features, like the large size, with a length of up to 200 mm, the predictable location of the nucleus and the high survival rate after enucleation, make *Acetabularia* a suitable object for use in various biochemical and cell biology experiments [7].

In recent years, it has been shown that in *Acetabularia* the rate of photosynthesis is subjected to diurnal oscillations [5, 6, 9]. Furthermore, it was demonstrated that these oscillations represent a circadian rhythm. The oxygen evolution was continuously measured in a flow-through system by means of a polarographic oxygen sensor (POS). This method is sufficiently sensitive to allow recordings of the oxygen evolution of an individual cell of *Acetabularia,* which produces 0.2 to 0.4 μmol O_2 h^{-1} [4].

On the basis of a series of experiments, a model was constructed which attempts to explain this rhythmic behavior in biochemical terms [8]. Further justification of the model would be provided by experiments in which a second oscillating parameter could be recorded simultaneously with O_2 evolution. Such a parameter could be the intracellular chloroplast migration, which is subjected to diurnal oscillations [1, 3].

Therefore, it was desirable to develop a method by which oxygen evolution and chloroplast migration could be recorded simultaneously in an individual cell over a long period of time. Moreover, the method should allow the influence of various sub-

1 Max-Planck-Institut für Zellbiologie, 6802 Ladenburg (bei Heidelberg), FRG

Polarographic Oxygen Sensors (ed. by Gnaiger/Forstner)
© Springer-Verlag Berlin Heidelberg 1983

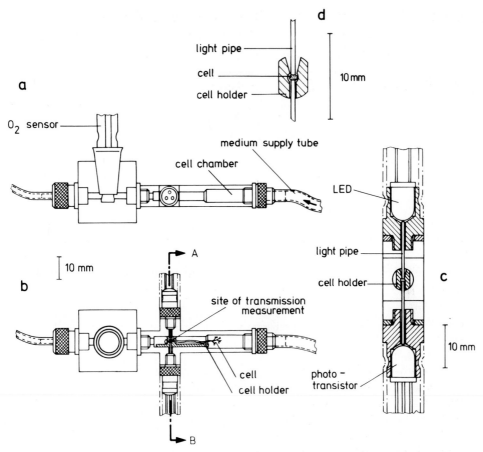

Fig. 1a–d. Measuring apparatus: **a** side view, **b** top view, **c** transverse section at $A-B$ (LED, light-emitting diode), **d** detail of cell holder

stances, particularly inhibitors of gene expression, upon the rhythm to be studied. The method of measurement and its characteristics are described and discussed in this paper.

Measurements were performed on cells of *Acetabularia mediterranea* with a length of about 35 mm. At this stage of development, the cells had not yet formed caps. Such a cell was placed inside a Plexiglas tube with an inner diameter of 5 mm (Fig. 1a). The tube was continuously perfused with medium passing the cell in an apical-basal direction. The flow rate was 0.4 cm^3 h^{-1}. Either supplemented natural seawater (Erd-Schreiber-Medium) or Müller-Medium, a defined artificial seawater, were used. Two millimeters above the rhizoid the cell was immobilized by slight pressure between two light guides which ended in a light source and a detector respectively (Fig. 1b). The pressure locally flattened the otherwise circular cross-section of the stalk.

The light source was a light-emitting diode (LED Typ ASBR 55-1, maximum light intensity 160 mcd), which was supplied with alternating current (ac) with adjustable amplitude and bias and a constant frequency of 1 kHz. A phototransistor (BPY62II)

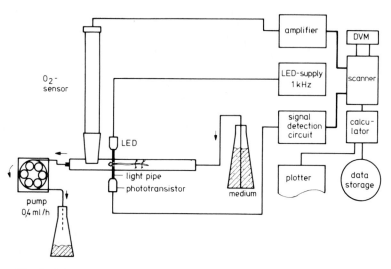

Fig. 2. Schematic drawing of experimental setup (LED, light-emitting diode; DVM, digital volt meter)

served as the signal detector. To separate the modulated measuring light from the ambient illumination, the signal was filtered and the 1 kHz band of the signal was then amplified and rectified. The resulting dc signal is proportional to the light transmission which is a measure of the local chloroplast concentration. The oxygen produced by the cell was measured electrochemically by directing the medium to a Clark-type POS. The platinum measuring electrode (1.5 mm diameter) and the silver/silver chloride reference electrode were mounted inside a glass tube filled with 3 mol dm^{-3} KCl and separated from the medium by a Teflon membrane (25 μm thick, area 25 mm^2). The platinum wire of the measuring electrode was insulated with glass such that only the tip was exposed to the electrolyte. The distance between this tip and the Teflon membrane was 12 μm and was kept constant by means of a cellophane membrane [4]. The electrode polarization voltage was 620 mV. The electrode current was converted to voltage and amplified. High frequency noise was eliminated by use of a low pass filter with τ = 22 s. The distance between the rhizoid and the middle of the POS corresponded to a volume of 0.2 cm^3, with a volume of 0.02 cm^3 at the sensor membrane. This small volume made stirring unnecessary. The whole apparatus was submerged in a water bath. The temperature of the water bath was regulated to $\pm$ 0.1 K by a kryostat.

The amplified electrode signal, which is proportional to the oxygen concentration in the medium, and the signal from the light transmission were fed each minute into a scanner and a digital voltmeter and then into a calculator. These values were summated for 30 min and then stored for further processing and evaluation (Fig. 2).

Under a 12/12 h light/dark regimen and otherwise constant conditions, oxygen evolution and chloroplast migration toward the apical end of the cell were induced almost simultaneously (Fig. 3). The slight delay of about 0.5 h was explained by the fact that the medium flowed with a speed of 0.4 cm^3 h^{-1} and thus needed about 30 min to cover the distance between the cell and the POS. The delay can be reduced

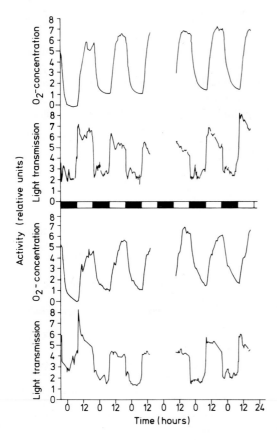

Fig. 3. Unprocessed data from two individual cells over 6 days. The *upper* and *lower sets of curves* represent data from one single cell each. The *bars* in the middle represent the light/dark (white/black) regimen (light 0800 to 2000 h). Between the 4th and the 5th day data were not recorded because an interruption of the measurements

by increasing the flow rate which, on the other hand, reduces the changes in concentration of oxygen. The temporal resolution of both the oxygen evolution and the light transmission measurement depends to a certain extent on the frequency of readings.

Oxygen evolution is considered an excellent marker for the switching on and off the light, since photosynthesis depends strictly upon light. This is in contrast to chloroplast migration, which clearly starts before the onset of illumination, indicating that it is not a response to the light itself, but rather is independent of the exogenous light/dark cycle.

Of course, in addition to oxygen evolution, oxygen consumption can be measured with this method also. The combination with light transmission measurements of different types opens a whole series of possibilities. Only slight changes in the geometry of the measuring chamber would permit fluorometric measurements also.

The method facilitates experimental manipulation of the culture medium. Another special advantage of the method is the independence of the measuring and the illuminating light which is achieved by frequency filtering. Moreover, the method allows the dependence or independence of two parameters in a single cell to be studied. If there is a dependence, one may analyze the temporal correlation between the two parameters. An additional possibility would be to measure light transmission at more than one site in the cell to gain greater insight into the dynamics of intracellular chloroplast migration.

The application of this method is by no means restricted to the unicellular organism *Acetabularia*. It should be possible to use it in studies of other unicellular and multicellular organisms with similar geometrical features. Additionally, this method should be applicable to even smaller unicellular organisms. In this case, however, the expenditure for more sensitive electronics has to be significantly increased. Finally, in combination with suitable test organisms, one might envision the use of this method for examining the biological value of different fluids. This might be important in pollution and environmental studies.

References

1. Broda H, Schweiger G, Koop H-U, Schmid R, Schweiger H-G (1979) Chloroplast migration: a method for continuously monitoring a circadian rhythm in a single cell of *Acetabularia*. In: Bonotto S, Kefeli V, Puiseux-Dao S (eds) Developmental biology of *Acetabularia*. Elsevier/North-Holland Biomedical Press, Amsterdam New York, pp 163–167
2. Karakashian MW, Schweiger H-G (1979) Evidence for a cycloheximide-sensitive component in the biological clock of *Acetabularia*. Exp Cell Res 98:303–312
3. Koop H-U, Schmid R, Heunert H-H, Milthaler B (1978) Chloroplast migration: a new circadian rhythm in *Acetabularia*. Protoplasma 97:301–310
4. Mergenhagen D, Schweiger H-G (1973) Recording the oxygen production of a single *Acetabularia* cell for a prolonged period. Exp Cell Res 81:360–364
5. Mergenhagen D, Schweiger H-G (1975) Circadian rhythm of oxygen evolution in cell fragments of *Acetabularia mediterranea*. Exp Cell Res 92:127–130
6. Schweiger E, Wallraff HG, Schweiger H-G (1964) Über tagesperiodische Schwankungen der Sauerstoffbilanz kernhaltiger und kernloser *Acetabularia mediterranea*. Z Naturforsch 19:499–505
7. Schweiger H-G, Berger S (1979) Nucleocytoplasmic interrelationships in *Acetabularia* and some other Dasycladaceae. Int Rev Cytol Suppl 9:11–14
8. Schweiger H-G, Schweiger M (1977) Circadian rhythms in unicellular organisms, an endeavor to explain the molecular mechanism. Int Rev Cytol 51:315–392
9. Sweeney BM, Haxo FT (1961) Persistence of a photosynthetic rhythm in enucleated *Acetabularia*. Science 134:1361–1363

Chapter II.8   A Respirometer for Monitoring Homogenate and Mitochondrial Respiration

W. Knapp[1]

1 Introduction

Over the past 20 years, the polarographic oxygen sensor (POS) has become an increasingly important instrument for studying respiration. Investigations in this field often rely upon information about the oxygen requirements of tissues, cell homogenates and cell organelle suspensions (e.g., mitochondria). Not until the introduction of the POS was it possible to obtain detailed information about oxidative cell metabolism.

Widely used alternatives to the POS are the manometric measurement of the oxygen absorbed [6, 16] and the volumetric method [7]. Although both methods are extremely well suited for registering the oxygen consumption of tissues, homogenates and cell organelle suspensions, the addition of substances during the investigation is relatively difficult. Furthermore, these methods generally do not provide a good resolution of small respiratory fluctuations.

The POS is becoming increasingly attractive for oxidative cell respiration studies. Although mostly employed by physiologists and biochemists [5, 11, 12], experience shows that the POS is ideal for investigating the pharmacological and toxic actions on oxidative cell metabolism of pharmaceutically active entities and should be included in the basic equipment of every toxicological and pharmacological laboratory. Unfortunately, its advantages are at present not widely enough appreciated, although pharmaceutical research offers sufficient problems with respect to increasing therapeutic safety.

2 Principle of the Measuring Apparatus

In principle a typical measuring system consists of a reaction vessel in which the POS is inserted. The vessel can be sealed, and a fine opening on the upper end allows the

---

1 Gebro G. Broschek KG, Chemisch-pharmazeutische Fabrik, Entwicklungs- u. Forschungsabteilung, A–6391 Fieberbrunn, Austria
and Institut für Zoologie, Abt. Zoophysiologie der Universität Innsbruck, A–6020 Innsbruck, Austria

Polarographic Oxygen Sensors (ed. by Gnaiger/Forstner)
©Springer-Verlag Berlin Heidelberg 1983

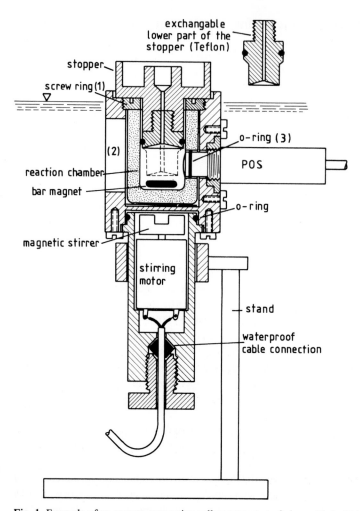

Fig. 1. Example of an oxygen-measuring cell constructed of glass with built-in magnetic stirrer. The glass chamber is built into a block of Plexiglas and fixed with the screw-ring (*1*). On three sides there are large openings (*2*), enabling the measuring cell to be surrounded with water circulated from a thermostatically controlled water bath. The POS is held horizontally in the Plexiglas block and is sealed with an O-ring where it enters the glass chamber (*3*). A further O-ring seals the Teflon stopper, the lower surface of which is slightly hollowed out in the form of a cone. This construction ensures the total expulsion of trapped air through the capillary during closure and offers a simple means of closing the chamber without the retention of air bubbles. The magnetic stirrer is mounted onto the Plexiglas block and sealed against the entry of water. The entire system is fixed in a thermostatically controlled water bath. Changing the size of the lower part of the stopper permits a variation of the reaction chamber volume

introduction of homogenate, cell suspensions, substrates, cofactors or pharmaceuticals. The medium under investigation is stirred using a magnetic bar driven by an externally applied rotating magnet. To ensure constant temperature, the reaction chamber can be fitted with a double wall with thermostated water circulating between the inner and outer walls. Alternatively, the entire vessel, including a waterproof motor, is suspended in a thermostated water bath (Fig. 1).

Commercially available reaction chambers are seldom ideally suited for a particular experiment. Self-construction of the reaction vessel is therefore often necessary [13, 14] according to the following design criteria:

1. The volume of the reaction chamber has to be adapted for the specific requirements (usually about $1-10$ cm^3).
2. The POS forms an integral part of the reaction chamber and thus need not be removed for cleaning purposes or during a change of the experimental medium. The position of the sensor should prevent attachment of air bubbles. These precautions protect the membrane from damage and minimize the time required for cleaning the chamber and changing the buffer.
3. The size of the electrode should be appropriate for the chamber volume. The high oxygen consumption of large electrodes (Chap. I.1) comprises a substantial source of error in small reaction chambers.
4. During sealing of the chamber the total exclusion of air bubbles must be ensured.
5. The medium must be well stirred to prevent the formation of a zone of reduced oxygen content directly on the membrane.
6. The entire reaction chamber, including the sensor tip, must be kept at constant temperature.
7. It must be possible to introduce substrate or test substances into the reaction chamber at any time during the measurement, without the infusion of foreign oxygen.
8. The material used for constructing the reaction chamber must possess good heat-exchange properties to permit measurements at defined temperatures.
9. Provision must be made for the insertion of a thermistor into the chamber during the experiment. It is sometimes advantageous to equip the chamber with a built-in thermistor.
10. Disassembly of the oxygen sensor must be easy, permitting simple replacement of the membrane.
11. The material chosen for the construction of the chamber must be easy to clean. It must not permit dissolution of oxygen or interfering substances.

3 Preparation of Buffer Solutions with a Defined Oxygen Partial Pressure

The measurement of oxygen consumption and the calibration of the sensor [15] should be conducted using thermostated solutions either saturated with oxygen or with a gas mixture at a defined partial pressure of oxygen. The apparatus shown in Fig. 2 proved useful for the preparation of such media.

Air, or a gas mixture, is blown continuously through a heat exchanger immersed in the water bath, then through a sterile filter (25 mm in diameter, pore size 0.45 μm) and through an aeration stone into the water contained in bottle B. The air or gas mixture thus becomes saturated with water vapor, and passes into bottle A which contains

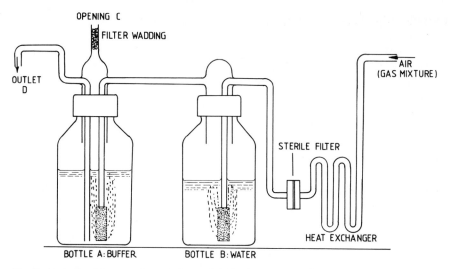

Fig. 2. Apparatus for preparing solutions of defined oxygen partial pressure. Bottles *A* and *B* contain the buffer (test medium) to be equilibrated, and water, respectively. Air (gas mixture) is thermostated in a heat exchanger, sterile filtered, saturated with water vapor in bottle *B* and finally flows through the buffer in bottle *A*. The entire apparatus stands in a temperature-controlled water bath

the buffer solution (test medium). For calculating the p_{O_2} at equilibrium see Appendix A and B. The use of a gas mixture saturated with water vapor prevents the vaporization of water in bottle A which would lead to an increased buffer concentration. The sterile filter protects the buffer from contamination by microorganisms. The excess air (gas mixture) escapes through the exit C. Buffer solution can be removed by closing the opening C with a finger: the increased pressure causes the expulsion of buffer through tube D.

Bottles and tubes should be sterilized before use.

4 Introduction of Substances During Measurements

The stopper of the reaction chamber is equipped with a capillary through which substances such as substrate, inhibitors, ADP etc. can be introduced. To prevent oxygen from entering the chamber, the capillary should have a length of at least 15–20 mm. Microliter syringes (e.g., Hamilton) are best suited for introducing substances; tuberculin syringes suffice for larger quantities up to 1 cm^3.

It is recommended that large volumes be brought to the experimental temperature before being injected. Ideal for this purpose are commercially available syringes with a temperature jacket. By the eddition of small volumes (relative to that of the reaction vessel), the temperature of the solutions introduced can be ignored in most cases.

5 Cleaning the Reaction Chamber

After each measurement, the reaction chamber is emptied with a water pump and thoroughly cleaned with water. An intensive treatment with suitable cleaning agents is, however, sometimes necessary as the measuring system can become contaminated with interfering substances.

5.1 Contamination with Water-Insoluble Chemicals

When water-insoluble substances (e.g., dissolved in ethanol) are added during the measuring process, traces of them remain on the chamber walls after emptying. For this reason, after being cleaned with water, the reaction vessel must be rinsed with solvent. Water-soluble solvents are finally removed by rinsing with water. The use of hydrophobic solvents necessitates cleaning with suitable detergents. Water is always employed for final rinsing.

Of particular importance is the thorough cleaning of awkward areas, for example those around the sensor. Possible incompatibility with the chamber material must always be considered in the choice of cleaning solvents.

5.2 Contamination by the Diffusion of Substances into Plastics

A large number of substances diffuse rapidly into many polymeric materials. A plastic chamber (e.g., of Plexiglas) may be incompatible with the substances to be used during the investigation (mostly inhibitors or pharmaceuticals). It should be noted that diffusion can lead to changes in the initial concentrations [1–4, 8–10]. With clomethiazole in plastic chambers we observed more than 20% material loss. Furthermore, substances which have diffused into the chamber material cannot be removed totally and traces will leak into the test medium during subsequent investigations. Glass reaction vessels offer the only alternative in such cases.

A further problem is the possibility of diffused oxygen in the plastic material of the reaction chamber. This oxygen diffusion may lead to false results particularly in very small chambers.

5.3 Contamination of the POS Electrolyte

Although polythene, polycarbonate, and Teflon are thought to be impermeable except for gases, it is possible for various substances to pass through the sensor membrane into the electrolyte solution [1, 8, 9]. This can lead to changes in the electrode characteristics and contamination of the samples subsequently examined, if the absorbed substances diffuse back through the membrane into the reaction chamber.

An effective solution is not possible here. One should, however, be aware of the situation and its implications. When necessary, the electrolyte should be renewed and exposure to contamination reduced to a minimum. Sometimes, improvement can be achieved by using thicker sensor membranes, even if this causes a decline in the sensitivity and a longer response time of the POS.

5.4 Contamination Due to Strong Basic Solutions and Detergents

Basic solutions and detergents are difficult to remove. Cleaning is considerably easier if the chamber is treated for a short time with 0.1 mol dm^{-3} hydrochloric acid and then thoroughly rinsed with water.

5.5 Contamination of the Reaction Chamber with Microorganisms

The chief source of error in the oxygen reaction chamber is due to contamination with algae and bacteria. Plastic vessels are problematic, as microorganisms grow particularly well on this material (Chap. II.9).

Membranes of POS which have been exposed to contaminated solutions for some hours or days are covered with bacteria. Such membranes must be replaced. The chamber must be cleaned with water after each measurement. Ideally, it should be rinsed hourly with 50% methanol or isopropanol.

During larger pauses between measurements (e.g., overnight or at weekends), the reaction vessel should be filled with 30% methanol. Before use, the chamber must be carefully washed with water. Instead of methanol, sodium azide or 0.06% phenylmercuricborate solution can be used.

Acknowledgment. I wish to express my thanks to Mr. John R. Owen for his help in preparing this paper.

References

1. Aspinall JA, Duffy TD, Saunders MB, Taylor CG (1980) The effect of low density polyethylene containers on some hospital-manufactured eyedrop formulations. I. Sorption of phenylmercuric acetate. J Clin Hosp Pharm 5:21–29
2. Baaske DM, Amann AH, Wagenknecht DM, Mooers M, Carter JE, Hoyt HJ, Stoll RG (1980) Nitroglycerin compatibility with intravenous fluid filters, containers, and administration sets. Am J Hosp Pharm 37:201–205
3. Cloyd JC, Vezeau C, Kenneth WM (1980) Availability of diazepam from plastic containers. Am J Hosp Pharm 37:492–502
4. Chiou WL, Moorhatch PA (1973) Interaction between vitamin A and plastic intravenous fluid bags. J Am Med Assoc 223:328
5. Estabrook RW (1967) Mitochondrial respiratory control and the polarographic measurement of ADP:O rates. In: Colowick SP, Kaplan NO (eds) Methods in enzymology, vol X. Academic Press, London New York, pp 41–47
6. Malek J (1965) Manometrische Apparatur nach Warburg. Theorie und Anwendungstechnik. In: Kleinzeller A et al (eds) Manometrische Methoden. G. Fischer, Jena

 7. Malek J (1965) Volumetrische Methoden. In: Kleinzeller A et al (eds) Manometrische Metho-
    den. G. Fischer, Jena
 8. Marcus E, Kim HK, Autian J (1959) Binding of drugs by plastics. J Pharm Sci 48:457—462
 9. McCarthy TJ (1970) Interaction between aqueous preservative solutions and their plastic con-
    tainers. Pharm Weekbl 105:557—563
10. Moorhatch P, Chiou WL (1974) Interactions between drugs and plastic intravenous fluid bags.
    Part I. Sorption studies on 17 drugs. Am J Hosp Pharm 31:72—78
11. Slater EC (1966) Oxidative phosphorylation. In: Florkin M, Stotz EH (eds) Comprehensive
    biochemistry, vol 14. Elsevier, Amsterdam, pp 327—396
12. Slater EC (1967) Application of inhibitors and uncouplers for a study of oxidative phosphory-
    lation. In: Colowick SP, Kaplan NO (eds) Methods in enzymology, vol X. Academic Press,
    London New York, pp 48—57
13. Starlinger H, Lübbers DW (1972) Methodische Untersuchungen zur polarographischen Messung
    der Atmung und des "kritischen Sauerstoffdrucks" bei Mitochondrien und isolierten Zellen mit
    der membranbedeckten Platinelektrode. Pfluegers Arch 337:19—28
14. Starlinger H, Lübbers DW (1973) Polarographic measurements of the oxygen pressure perform-
    ed simultaneously with optical measurements of the redox state of the respiratory chain in sus-
    pension of mitochondria under steady-state conditions at low oxygen tensions. Pfluegers Arch
    341:15—22
15. Turner AC, Hutchinson WF, Don Turner M (1978) Importance of medium solutes in polaro-
    graphic measurement of low oxygen consumption rates. Comp Biochem Physiol 60A:131—132
16. Umbreit WW, Burris RH, Stauffer JF (1957) Manometric techniques. Burgess, Minneapolis

Chapter II.9   Bacterial Growth and Antibiotics in Animal Respirometry

G.J. Dalla Via[1]

1 Introduction

The uncontrolled contribution of bacterial oxygen consumption in animal respirometry respresents a substantial problem and results in an ambiguous reading of the animal's metabolic rate. This problem became especially important when the application of polarographic oxygen sensors (POS) made possible long-term measurements of the dynamics of an animal's energy metabolism (which may be superimposed by bacterial growth). Bacterial growth is most rapid on free surfaces [42, 60] such as the inner walls of the animal chamber, but also in the stirring chamber of the POS, valves, connecting tubings, etc.

Some researchers resort to chemical bactericides to eliminate or inhibit the growth of bacteria in the system [1, 4, 9, 32], whereas others determine the bacterial oxygen consumption in a blank experiment and subtract it from the observed total oxygen uptake ([21, 24, 51], Chap. II.3). However, an inspection of the literature suggests that many workers have ignored the part played by bacterial action in oxygen consumption measurements.

2 Antibiotics

The range of application of antibiotics is very wide, and during the past decades new types have been continuously developed, while at the same time significant progress has been made in clarifying their mechanisms of action [23, 25]. Antibiotics have been used in respirometry in the belief that they exert no influence on the oxygen consumption of the experimental animals [18, 32]. It also should be noted that considerable differences exist in the chemical properties of various antibiotics. Their stability and antimicrobial activity depends to a large extent on the pH and temperature of the medium. Hence the physical and chemical properties of the experimental media largely determine the choice of antibiotic. The properties of some antibiotics have been com-

---

1 Institut für Zoologie, Abteilung Zoophysiologie, Universität Innsbruck, Peter-Mayr-Str. 1A, A–6020 Innsbruck, Austria

Polarographic Oxygen Sensors (ed. by Gnaiger/Forstner)
© Springer-Verlag Berlin Heidelberg 1983

piled from the literature in Tables 1 and 2 [12, 20, 23, 26, 29, 41, 46, 54]. Not every antibiotic is suitable for a particular experiment; for instance under anaerobic conditions kanamycin, neomycin and streptomycin are inactive.

The mechanism of action also determines the applicability of a specific antibiotic in respirometry. The following classification is given [41]:

a) Antibiotics interfering with nucleic acid or protein biosynthesis: actinomycins, adriamycin, amikacin, aminoglycosides, bleomycins, capreomycin, chloramphenicol, clindamycin, fusidic acid, lincomycin, macrolide antibiotics, mitomycins, puromycin, rifamycins, sisomicin, streptonigrin, streptozotocin, thiamphenicol, viomycin.
b) Antibiotics affecting the function of the cytoplasmatic membrane: amphotericin B, candicidin, gramicidins, nystatin, pimaricin, polymyxins, streptomycin, trichomycin, tyrocidins.
c) Antibiotics interfering with cell wall biosynthesis: bacitracin, D-cycloserine, cephalosporins, fosfomycin, novobiocin, penicillins, ristocetins, vancomycin.

Due to their unspecific attack on cell metabolism, the first two classes not only damage bacteria, but also disturb the metabolism of the experimental animals. The third class of antibiotics specifically inhibits cell wall synthesis in the bacteria and so hardly influences animal cells.

In selecting an antibiotic it is also important to know the antimicrobial spectra of the antibiotics (Table 2). They can be distinguished as follows [54]:

a) Wide-spectrum antibiotics: ampicillin, cephalosporins, chloramphenicol, rifampicin, tetracyclines.
b) Antibiotics preferentially effective against gram-negative bacteria: gentamicin, kanamycins, neomycins, paromomycin, polymyxin B and E, streptomycin.
c) Antibiotics preferentially effective against gram-positive bacteria: bacitracin, erythromycines, fusidic acid, lincomycin, novobiocin, penicillins, peptolid-antibiotics, rifomycin, ristocetin, vancomycin.
d) Fungicids: amphotericin B, griseofulvin, nystatin, trichomycin.
e) Antibiotics inhibiting protozoa: fumagillin, trichomycin.

For use in seawater, antibiotics effective against gram-negative bacteria are to be preferred. Penicillin is inapplicable, since it inhibits mucopeptide synthesis and is thus detrimental to gram-positive bacteria only.

Antibiotics can be classified according to their undesirable toxicity to animals [41]:

a) Antibiotics with low toxicity: cephalosporins, erythromycin, fusidic acid, lincomycins, penicillins, thiamphenicol.
b) Toxic antibiotics causing reversible and irreversible damage: aminoglycoside antibiotics (e.g. neomycin), amphotericin B, chloramphenicol, novobiocin, polymyxins, viomycin, paromomycin, tyrothricin, nystatin.

Table 1. Properties of some antibiotics: solubility in water, optimal pH range for maximum antimicrobial activity, inactivation rate (%) and inactivating agents in aqueous solutions, and cross-resistance (referring to antibiotic No.)

Antibiotic	Solubility mg cm^{-3}	pH	Inactivation rate pH	°C	days	%	Inactivating agents	Cross-resistance
1. Amphomycin	Soluble	4.5–7.5	7	24–25	30	Stable		
2. Ampicillin	Acid: 100	5.5	7	24	7	20		[32]
3. Bacitracin	1000	6.5 (6–7.5)	6	5	14 2 h	Stable 100	pH $\geqslant$ 9; high saline concentrations cause precipitation; sodium thiosulfate; ions of heavy metals Cd^{2+}, Mn^{2+}, Zn^{2+}; H$_2$O$_2$;	
4. Candicidin			7	4	7	Stable		
5. Carbenicillin	Good		2	21	2	50		[2]
6. Carbomycin		5–7		25	11	Stable		[16]
7. Cephaloridin	220–250	7.5–8.5		4 25	4 1–4	Stable Stable	pH > 8; light	[8]
8. Cephalothin	200–300	6.5 (6–7.3)		4 37 37 37	2 12 h 24 h 48 h	Stable 11–23 36–39 47–55		[7, 18]
9. Chloramphenicol	2.5–4.4	7.4–8.0	2.5–9 7 7	24 30 5	1 60 700	Stable Stable Stable	pH > 9.5; phenylalanine; cyclic amino-acids; metal ions do not deactivate chloramphenicol	[40]
10. Chlortetracycline	Base: 0.55 8.6	6.1–6.6	7 7 8 8 8.5	37 37 37 37	10 h 24 h 10 h 24 h 4–5 h	65 95 92 99.7 50	Bi- and trivalent metallic cations	[30, 39]
11. Cloxacillin	Good			4 25	7 4	5 15		

No.	Antibiotic	Solubility	pH	pH	Temp	Time	Stability	Remarks	References
12.	Cycloserine	100	6.4–7.4	7, 7	37, 37	7, 14	25, 38	D-alanine; Mycobyctin;	
13.	Colistin	> 1	6.5–7.5	6, 7, 7, 9	30, 30, 30	14, 3, 16, 1	Stable, 10, 57, 50	pH > 6; Ca^{2+}	[35]
14.	Dicloxacillin	150			24, 24, 4	2, 9, 14	Stable, 70, 10		[11, 24, 29, 32–34]
15.	Dihydro-streptomycin	250–500	7.5–8	4–7, 4–7	4, 28	90, 60	Stable, Stable	More stable than streptomycin; not deactivated by cysteine	[39]
16.	Erythromycin	Base: 2.1 10–20–200	8–8.5	6, 7, 8.5	25, –25–+4	1–3 h, 1, 56	100, 14, Stable	Acids; pH < 6; anaerobiosis does not modify activity; not affected by metallic ions	[6, 28]
17.	Framycetin	Soluble		2–9	25	30	Stable		[25]
18.	Fusidic acid	Alkali-salt	6			36 h	15	pH > 7	
19.	Gabromycin	High	7.4–8.0					pH < 6; proteins and anaerobiosis do not inhibit the activity	
20.	Gentamycin	Sulphate: > 20	7.8	2.2–10			Stable	Na-, K-, Ca-salts; CO_3^{2-}, SO_4^{2-}, Cl^-, PO_4^{3-}, NO_3^-	[15, 22, 25, 31]
21.	Griseofulvin	0.01		3–8.8			Stable	Purine	
22.	Kanamycin	Sulphate: 360	7.6–8.0	2.6–7.9, 2.6–7.9, 2.6–7.9	25, 37, 45	180, 180, 120	1.57, 1.9, 2.8	Phosphate, citrate, chloride, Mg^{2+}, anaerobiosis, CO_2^-, K^+, Ni^{2+}, Fe^{2+}, PO_4^-, Mg^{2+}	[15, 25, 31, 39]
23.	Lincomycin	Base: 500	8–8.5		70	180	Stable	Erythromycine	
24.	Methicillin	Good		7–7.2, 7–7.2, 7–7.2	4, 25, 37	5, 5, 2.5	20, 50, 50	pH 4–5; kanamycin; streptomycin	[11, 14, 29, 32–34]

Table 1 (continued)

Antibiotic	Solubility mg cm⁻³	pH	Inactivation pH	Inactivation °C	Inactivation days	Inactivation %	Inactivating agents	Cross-resistance
25. Neomycin B	6.3	7.4–8.0	8	25	360	Stable	Anaerobiosis, RNA, glucose, Cl⁻, Na⁺, K⁺, PO₄³⁻, Fe²,³⁺, Mg²⁻, Aluminium ions	[14, 15, 32–34, 39]
26. Nitrofurantion	0.21	5.5–6.0	7	37	1	Stable	pH > 9, < 4	
27. Novobiocin	100	5.5	7–10	24	60	50	Bivalent metallic cations	
28. Oleandomycin	Base: 5 Phosphate: >1000	8.0	2.2–9 / 5–7 / 9	37 / 24 / 25	1 / 1 / 21	Stable / Stable / Stable		[16]
29. Oxacillin	Good	6–6.6		4 / 24 / 24	15 / 7 / 15	6 / 9 / 38		[11, 14, 24, 32–34]
30. Oxytetracyclin	Base: 0.6 / 6.9	6.1–6.6	7 / 7 / 8 / 8	37 / 37 / 37 / 37	10 h / 1 / 10 h / 1	34 / 66 / 25 / 75	Bi- and trivalent metallic cations	[10, 40]
31. Paromomycin	Good	7.6–8.0						
32. Penicillin G	>20	6–6.5	6.5 / 6.5	4 / 24–25	3–7 / 2	Stable / 50	pH < 4, pH > 8, Pb, Cu, Hg, Zn, Sn, Cd, some amino acids, H₂O₂, rubber, glycerine, ethyl alcohol	[33, 34]
33. Penicillin V	>750		5	37	33 h	50		[32, 34]
34. Pheneticillin	Good		6.5	4–35	Stable	Stable		[32, 33]
35. Polymyxin B	25	6.5–7.5	6–7	38	360	Stable	pH 2, 8; bivalent cations;	[13]
36. Propicillin	Good		6	24	7	12		[34]
37. Rifomycin	5%	6.0–7.4	7	25	700	Stable		

38. Spectinomycin	225	> 8.0	8	30	50	50		
39. Streptomycin	250–500	7.5–8.0	4–7 4–7 7	28 4 15	60 90 30	Stable Stable Stable	pH < 2; pH > 8.5; t > 28°C; $KMnO_4$; KJO_4; H_2O_2, HNO_3, ascorbic acid, glucose, NO_3^-, OCN^-, Mg, Ca, DNA, urea, anaerobiosis	[15]
40. Tetracycline	Base: 1.7 10.9	6.1–6.6	7 7 8 8	37 37 37 37	10 h 24 h 10 h 24 h	2 42 36 82	Bi- and trivalent cations	[10, 30]
41. Vancomycin	> 100	8.0	3–7	37	6	10		
42. Viomycin	5.6–7.8	7.4–8.4	5–6	24	7	Stable	pH > 9	[22, 39]

Table 2. Antimicrobial spectra of some antibiotics (+ efficient, ± poorly efficient, − inefficient against gram-positive or gram-negative bacteria), minimal inhibitory concentrations and development of resistance (for bacteria)

Antibiotic	Bacteria				Fungi	Protozoa	Resistance
	G^+	G^-	G^+ $\mu g\ cm^{-3}$	G^-	$\mu g\ cm^{-3}$	$\mu g\ cm^{-3}$	
Adicillin	±	±	5–10	1.5			
Amphomycin	+	−	2.5–6.2		Resistant	Resistant	
Amphotericin B	−	−			0.04–3.7	Nearly all res.	None
Ampicillin	+	±	0.01–1.5	>8			
Bacitracin	+	−	0.07–27	−10000	Resistant	Nearly all res.	Slow
Boromycin	+	−	0.05		Active		
Candicidin	−	−			0.5–50		
Capreomycin	+	−	0.5–20				Rapid
Carbenicillin	+	±	0.2–1.56	250			
Carbomycin	+	−	2–5		Resistant	30–250	
Cephalexin	+	±	50	50			
Cephaloridin	+	±	0.01–5	0.1–100	Resistant	Resistant	Slow
Cephalothin	+	±	0.02–8	>100		Resistant	Slow
Chloramphenicol	+	+	0.5–20	0.4–30	Resistant	125–2000	Slow
Chlortetracycline	+	+	0.3–20	0.1–50	Resistant	25–1000	Moderate-fast
Cloxacillin	+	−					
Colistin	−	+		0.1–50	20	125	Slow
Cycloserin	±	+	6.5–50	<125			Slow
Demethyl- chlortetracycline	+	+	0.3–20	0.1–50	Resistant	Nearly all res.	Moderate-fast
Dicloxacillin	+	−					
Dihydrostrepto- mycin	±	+	1–25	0.5–25	Resistant	Resistant	Rapid
Erythromycin	+	−	0.01–2		Resistant	Nearly all res.	Rapid
Framycetin	+	+					Slow
Fumagillin	+	−			Resistant	Effective	
Fusidic acid	+	−	0.2–10		Resistant	Resistant	Rapid
Gabromycin	+	+	2–7	1–5			None-slow
Gentamycin	+	+	0.8–40	1–50	>100– >1000	Resistant	Moderate-fast
Gramicidin	+	−	1–10				
Griseofulvin					0.2–15		None
Hamycin	−	−			0.01–4		
Kanamycin	±	+	4–500	0.5–15	Resistant	Nearly all res.	Moderate-fast
Kitasamycin	+	−	0.2–3		Resistant		
Lincomycin	+	−	0.04–32	>200	Resistant		Rapid
Myxin	+	+	0.6–2.5	0.4–3.2	1–10		
Nactins	+	−	1.0		10		
Neomycin B	±	+	0.5–150	0.5–15	Resistant	40–3000	Moderate-fast
Nitrofurantoin	+	+	2–3	0.002–20			Slow
Novobiocin	+	+	0.1–6	2–25	10–1000	125	Rapid
Nystatin	−	−			0.6–5– 10–30	250	None
Oleandomycin	+	−	0.5–3				Rapid
Oxacillin	+	−	0.06–1.5				
Oxytetracycline	+	−	0.3–20	0.1–50	Resistant	30–250	Moderate-fast

Table 2 (continued)

Antibiotic	Bacteria			Fungi	Protozoa	Resistance
	G⁺ G⁻	G⁺ $\mu g\ cm^{-3}$	G⁻	$\mu g\ cm^{-3}$	$\mu g\ cm^{-3}$	
Paramomycin	− +		0.8−25	Resistant	Nearly all res.	Moderate-fast
Penicillin G	+ ±	0.006−3	>100	Resistant	Resistant	Slow
Penicillin V	+ −	0.01−8				Slow
Peptolid-antibiotics	+ −	0.1−4		Resistant	Resistant	Moderate-fast
Phenethicillin	+ ±	0.03−4	>100			
Pimaricin				1−12	Nearly all res.	
Polymyxin B	− +		0.05−5	125−250	125	Slow
Propicillin	+ ±	0.03−33				
Rifamid	+ ±	0.03−10	25−250			Rapid
Rifampicin	+ ±	0.01−0.1	1−50	>100		Rapid
Rifamycin	+ ±	0.03−0.1 (10)	25−250			Rapid
Ristocetin	+ −	0.5−4		Resistant	Resistant	None
Rolitetracycline	+ +	0.3−20	0.1−50	Resistant	Nearly all res.	Moderate fast
Spectinomycin	− +		4−500	Resistant	Resistant	Rapid
Spiramycin	+ −	0.1−5				Moderate-fast
Streptomycin	± −	1−25	0.5−25	Resistant	Resistant	Rapid
Tetracycline	+ +	0.3−20	0.1−50	Resistant	60−250	Moderate-fast
Trichomycin				0.6−10	250	None
Tyrothricin	+ −	1−10				None-slow
Vancomycin	+ −	0.5−3		Resistant	Resistant	None
Viomycin	− +		2−12			Rapid

Table 3 gives examples of the effect of individual antibiotics and their combinations on various aquatic animals. Penicillin has the lowest toxicity, which makes possible its use even in high doses. Kanamycin, neomycin, polymyxins and streptomycin may lead to a neuromuscular block. Kanamycin is less toxic than neomycin and paromomycin, but should not be used in combination with either of these since this leads to a considerable increase in toxicity. A combination of streptomycin and dihydrostreptomycin reduces the toxicity of the individual components. Of the tetracyclines, tetracycline, followed by oxytetracycline, combine optimal antibiotic properties, i.e., low toxicity and a high bactericidal effect. Streptomycin, chloramphenicol, kanamycin, and neomycin are antibiotics with long half-lives in seawater, but due to their high toxicity to invertebrates, their dosage may be critical [12]. Auromycin, terramycin, and polymyxin B are effective against marine bacteria but are toxic to phytoplankton, while penicillin, kanamycin, neomycin, and streptomycin are suitable for phytoplankton cultures [3, 47]. Chlortetracycline, oxytetracycline, bacitracin, carbomycin, Oleandomycin, tetracycline, erythromycin and penicillins have short half-lives in solution, and are therefore unsuitable for long-term experiments [12]. Antibiotics may not only disturb oxygen uptake in animals but also induce anoxic contributions to the total metabolic rate [19].

Table 3. Effect of antibiotics on physiological parameters in animals

Genera	Stage[a]	Antibiotics[b]	μg cm^{-3}	Days[c]	Parameter	Effect[d]	Ref.
Coelenterata							
Tubularia	a	PEN + STR	100 + 100	1	Regeneration	+	[17]
Tubularia	a	PEN + STR	400 + 400	1	Regeneration	+	[17]
Nemertina							
Amphiporus	a	STR + DISTR	750–5000	10	Regeneration	0	[52]
Cerebratulus	a	STR + DISTR	750–5000	10	Regeneration	0	[52]
Lineus	a	STR + DISTR	750–5000	10	Regeneration	0	[52]
Annelida							
Cognettia	a	FRA	250	4	Growth	+	[28]
Cognettia	a	PEN	250	4	Growth	+	[28]
Cognettia	a	STR	250	4	Growth	+	[28]
Lumbriculus	a	STR + NEO	20–200		Heat dissipation	+	[19]
Mollusca							
Adula	l	PENG + STR	50 + 100	5	Survival	0	[39]
Australorbis	l	PENG + STR	60*+ 100	3–4	Growth	–	[8]
Australorbis	l	STR	10	3–4	Growth	0	[8]
Australorbis	l	PENG	60*	3–4	Growth	0	[8]
Mya	l	PEN + STR	30*+ 100		Growth	0	[49]
Ostrea	l	PENG	30*	7	Yield of spat	+	[53]
Ostrea	l	PENG	30*	7	Growth	0	[53]
Ostrea	l	PENG + STR	30*+ 50	7	Growth	–	[53]
Tritona	l	STR + PENG	50 + 60	24	Growth	–	[27]
Tritona	l	STR + PENG	50 + 60	24	Survival	+	[27]
Crustacea							
Artemia	l + a	PENG	150	37	Growth	0	[12]
Artemia	l + a	PENG	200	7	Growth	–	[12]
Artemia	l	STR	200	37	Growth	0	[12]
Artemia	l + a	STR	300	7	Growth	–	[12]
Artemia	a	STR	250	37	Growth	0	[12]
Artemia	l	POL	30	37	Growth	0	[12]
Artemia	l + a	POL	50	7	Growth	–	[12]
Artemia	a	POL	40	37	Growth	0	[12]
Artemia	l + a	CHL	25	37	Growth	0	[12]
Artemia	l + a	CHL	50	7	Growth	–	[12]
Artemia	l + a	NAR	10	37	Growth	0	[12]
Artemia	l	NAR	15	7	Growth	–	[12]
Artemia	a	NAR	20	7	Growth	–	[12]
Artemia	l + a	CAN	2.5	37	Growth	0	[12]
Artemia	l	CAN	3	7	Growth	–	[12]
Artemia	a	CAN	5	7	Growth	–	[12]
Artemia	l + a	NYS	10	37	Growth	0	[12]
Artemia	l	NYS	15	7	Growth	–	[12]
Artemia	a	NYS	30	7	Growth	–	[12]
Artemia	l + a	KAN	250	37	Growth	0	[12]
Artemia	l	KAN	500	7	Growth	–	[12]
Artemia	a	KAN	750	7	Growth	–	[12]
Artemia	l + a	NEO	5	37	Growth	0	[12]
Artemia	l	NEO	10	7	Growth	–	[12]
Artemia	a	NEO	15	7	Growth	–	[12]

Table 3 (continued)

Genera	Stage[a]	Antibiotics[b]	$\mu g\ cm^{-3}$	Days[c]	Parameter	Effect[d]	Ref.
Artemia	1 + a	NAL	50	37	Growth	0	[12]
Artemia	1	NAL	100	7	Growth	−	[12]
Artemia	a	NAL	150	7	Growth	−	[12]
Artemia	1 + a	TET	5	37	Growth	0	[12]
Artemia	1	TET	7.5	7	Growth	−	[12]
Artemia	a	TET	9.5	7	Growth	−	[12]
Artemia	1 + a	TRI	25	37	Growth	0	[12]
Artemia	1	TRI	50	7	Growth	−	[12]
Artemia	a	TRI	75	7	Growth	−	[12]
Artemia	1 + a	MET	500	37	Growth	0	[12]
Artemia	1 + a	MET	1000	7	Growth	−	[12]
Calanus	a	CHL + STR	50 + 50		Respiration	0	[32]
Calanus	a	PEN	50	1	Feeding	+	[32]
Calanus	a	CHL	50	1	Feeding	−	[32]
Calanus	a	CHL + STR	50 + 50	1	Feeding	−	[32]
Calanus	a	STR	50	1	Feeding	0	[32]
Cancer	1	STR + PEN	100 + 100	40	Survival	+	[15]
Cancer	1	STR + PEN	50 + 50	40	Survival	+	[15]
Cancer	1	STR + PEN	10 + 10	40	Survival	+	[15]
Cancer	1	STR + PEN	100 + 100	100	Survival	+	[16]
Cancer	1	CHL	1	38	Survival	+	[16]
Cancer	1	CHL	5	38	Survival	+	[16]
Cancer	1	CHL	10	38	Survival	+	[16]
Cancer	1	STR + PEN	100 + 100	38	Survival	+	[16]
Cancer	1	CHL + KAN	1 + 100	30	Survival	−	[16]
Cancer	1	KAN	100	30	Survival	−	[16]
Cancer	1	CHL + NEO	1 + 50	27	Survival	−	[16]
Cancer	1	NEO	50	27	Survival	−	[16]
Cancer	1	PEN + STR	8 + 130		Survival	+	[39]
Daphnia	a	DISTR + CHL	25 + 25	1	Respiration	0	[59]
Daphnia	a	DISTR + TET	25 + 25	1	Respiration	0	[59]
Daphnia	a	DISTR + TET	50 + 50	1	Respiration	−	[59]
Eurytemora	a	CHL	20		Respiration	0	[22]
Palaemon	1	PENG + STR	30* + 50	30−36	Survival	0	[57]
Pleuroncodes	1	PPEN + PEN	75 + 25	74	Growth	−	[5]
Temora	a	CHL	50		Growth	−	[22]

[a] Abbreviations: a = adult; l = larval
[b] Abbreviations: BAC = bacitracin; CAN = candicidin; CHL = chloramphenicol; DISTR = dihydro-streptomycin; FRA = framycetin; KAN = kanamycin; MET = methenamine-mandalate; NAL = nalidixin; NAR = naramycin; NEO = neomycin; NYS = nystatin; PEN = penicillin; PENG = penicillin G; POL = polymixin; PPEN = procaine penicillin; STR = streptomycin; TET = tetracyclin; TRI = trichomycin
[c] Days of treatment
[d] The effect is related to the parameter against the untreated control (+ positive, − negative, 0 no effect)
* Calculated from international units

3 Other Chemotherapeutics and Bactericidal Agents

Before the discovery of antibiotics chemotherapeutics were of considerable importance. However, their use in respirometry is not recommended due to their high toxicity.

Chlorine is unsuitable in respirometry and its toxicity is dependent on temperature and species [34, 44]. Significant reductions in standard respiration rates were measured in larval lobsters [7]. Increasing concentrations of residual chlorine inhibit larval development and cause morphologic deformations in some fish [36]. These effects have to be borne in mind when using chlorinated tap water, in which chlorine is more persistent than in seawater [14].

Ozone is a potent germicide [20, 37, 55]. Its efficiency is affected by temperature and pH, and it is unstable in water where the contact time is more critical than the quantity of ozone used [48]. Its toxicity is still under debate [13, 50, 56]; but due to the liberation of oxygen it interferes with the polarographic measurement.

UV-irradiation is an effective disinfectant [6], but since its penetration is limited, bacteria which adhere to the inner surfaces of the respirometer are barely affected.

4 Bacterial Growth and Respiration

Inoculation of the respirometer with bacteria occurs via the medium and the experimental animals. Bacteria are seldom freely suspended in natural waters but are generally attached to solid suspended particles [11, 35, 45]. They can be eliminated by filtering the test water (0.45 μm). Bacteria growing on the body surfaces of the test animals, however, cannot be eliminated. Rinsing the animals in a bactericide solution is not advisable, because handling and the toxicity of the solution are likely to induce stress in the animals. Bacterial respiratory inhibition by antibiotics is dependent on the growth phase and the position within a phase of growth. Therefore uniform suppression of bacterial respiration cannot be assumed [58]. If the bacteria are inhibited, fungal growth increases [12, 38]. On the other hand, antibiotics may interfere with the metabolism of the experimental animal. In some cases the total metabolism of the animal in the experiment may be considerably increased, as in *Lumbriculus variegatus* and *Salvelinus fontinalis* ([19]; Chap. II.3).

The estimation of microbial respiration in a control chamber without experimental animals is not to be recommended, because the conditions for the growth of bacteria are not identical in test and control (bacteria inoculation by experimental animals, availability of soluble excreta). The subtraction of the mean background O_2-consumption rate, i.e. (initial + final background): 2, from the oxygen uptake rate is unsuitable if the respirometer has been previously sterilized or cleaned, since this does not take into consideration the form of the bacterial growth curve during the experiment (Figs. 1 and 2).

5 Experimental Observations

A possible solution is suggested by considering the growth curve of bacteria. After an initial log-phase and following the exponential log-phase bacterial growth in a chemostats culture remains constant (stationary phase). This microbiological principle proved to be valid for bacterial growth in a respirometer.

Oxygen consumption measurements were made with an automated polarographic respirometer (Chap. II.1). This is an intermittently closed system in which any of the four chambers can be automatically closed in sequence, while flow is maintained through the other three chambers. The oxygen uptake by eight prawns (*Palaemonetes antennarius*) was recorded as the decrease in dissolved oxygen during "closed" periods of 12 min. The interval between two closed periods of the same chamber was 1 h. The prawns were repeatedly put into the respirometer for 1 h and subsequently removed for 1 h to permit measurement of the bacterial respiration in the respirometer. In this way the bacterial contamination by the experimental animals and the presence of the excreta of the prawns were guaranteed. Animals were put into the compartments of

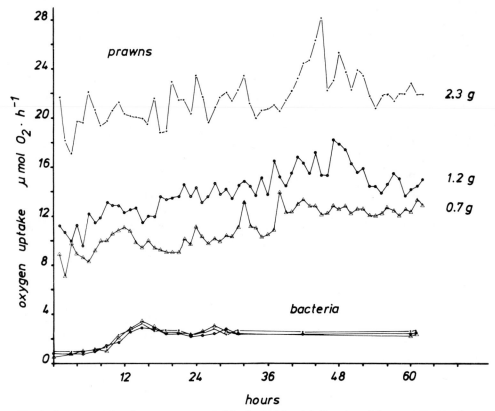

Fig. 1. Oxygen uptake of prawns uncorrected for bacterial metabolic rate and the respective simultaneous oxygen consumption rate of the bacteria in the system. The total biomass (wet weight) is given for each size group

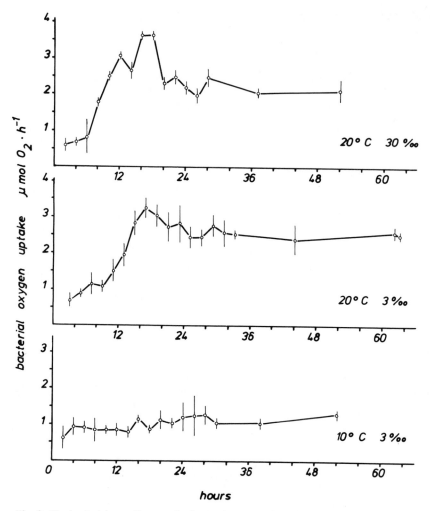

Fig. 2. The bacterial growth curve in the respirometer discontinuously inoculated with experimental animals during different conditions of salinity and temperature. Means ± S.D. are given

cages made of stainless steel netting to permit their rapid removal from the respirometer.

The bacterial growth in the respirometer increases exponentially after a lag-phase of 5 to 6 h, overshoots, and stabilizes after 12 to 20 h irrespective of animal size (Fig. 1). Hence, true animal respiration rates are obtained by determining the final bacterial background rate and substracting it from the total oxygen uptake in the experiment discharging the initial period. To estimate the initial rates, the changing bacterial respiration rate should be subtracted from the total oxygen consumption rate.

The curve of bacterial growth in the respirometer should be determined under experimental conditions. Salinity in contrast to temperature affects bacterial growth only slightly (Fig. 2). The microbial growth curve must also be determined for each system and respirometer because the relative surface area and the materials of the respirometer exert a considerable influence [31, 33, 43]. The results also confirm that over the first 2–5 h no significant bacterial growth is observable [18] although in long-term measurements bacterial respiration can attain high levels. The bacterial oxygen consumption may be as high as 30% of that of the specimens under observation (Fig. 1). The population densities of bacteria occupying the surfaces are relatively constant and independent of the quantity of the available energy source [40, 48].

The bacterial respiratory effect is decreased by reducing the surfaces in the respirometer, avoiding plastic materials in the construction, maintaining a flow-through and increasing the biomass in the animal chamber.

6 Bacterial Interference with Stability of POS

A new membrane mounted on the POS constitutes a surface which may be gradually overgrown by bacteria. There was a correlation between deviation of the POS signal from the original calibration point and bacterial oxygen uptake (Fig. 3). This may be

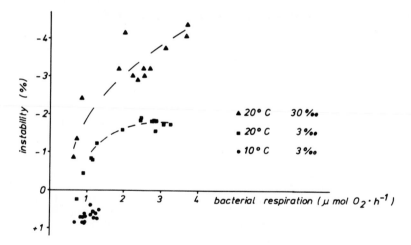

Fig. 3. Deviation of the POS stability from the beginning of the experiments as a function of bacterial respiration. Without intermittent calibrations the expected concentration of oxygen, c'_t, at time t would be

$$c'_t = I_t \times \frac{c_o}{I_o} \, ,$$

where c_o and I_o are the oxygen concentration and signal of the POS at the calibration point. The relative instability of the POS is then calculated as

$$\frac{c'_t - c_t}{c_t} \times 100 \, .$$

The *broken lines* were fitted by eye to indicate the trends considered significant due (1) to O_2-depletion of the membrane boundary layer by bacterial respiration, and (2) to increased thickness of the diffusion layer (membrane + bacteria) which reduces the membrane permeability. This source of error should be considered in long term in situ measurements (Part III) where uncontrolled bacterial Aufwuchs on the membrane will occur.

7 Conclusions

The bacterial problem in the respirometry of aquatic animals arises from excessive bacterial growth on the inner walls of the respirometer, especially in long-term measurements. The use of antibiotics may be considered in combatting this problem, but is limited by their physicochemical properties and because they are often more toxic to the animals than to the bacteria. Chemotherapeutics are also unsuitable in respirometry because of their high toxicity.

The time course of bacterial growth shows that the bacterial oxygen consumption rises to constant levels within 12 to 20 h and has to be subtracted from the total oxygen consumption rate. By this method no chemicals need be used to inhibit bacterial growth, thus avoiding any possible influence on the experimental animals.

Acknowledgments. I wish to express my gratitude to E. Gnaiger for proposing critical comments and suggestions on the manuscript. The study was supported by the Oesterreichische Nationalbank, project no. 1208 and by the Österreichischer Forschungsfond, project no. 3307.

References

1. Alcaraz M (1974) Respiración en crustáceos: Influencia de la concentración de oxígeno en el medio. Invest Pesq 38:397–411
2. Bell GR, Hoskins GE, Hodgkiss W (1971) Aspects of the characterization, identification, and ecology of the bacterial flora associated with the surface of stream-incubating Pacific Salmon (*Oncorhynchus*) eggs. J Fish Res Board Can 28:1511–1525
3. Berland BR, Maestrini SY (1969) Study of bacteria associated with marine algae in culture. II. Action of antibiotic substances. Mar Biol 3:334–335
4. Booth CE, Mangum CP (1978) Oxygen uptake and transport in the lamellibranch mollusc *Modiolus demissus.* Physiol Zool 51:17–32
5. Boyd CM, Johnson MW (1963) Variation in the larval stages of a decapod crustacean, *Pleuroncodes planipes* Stimpson (Galatheidae). Biol Bull 124:141–152
6. Brown C, Russo DJ (1979) Ultraviolet light disinfection of shellfish batchery sea water. I. Elimination of five pathogenic bacteria. Aquaculture 17:17–23
7. Capuzzo JM (1977) The effects of free chlorine and chloramine on growth and respiration rates of larval lobsters (*Homarus americanus*). Water Res 11:1021–1024
8. Chernin E (1959) Cultivation of the snail *Australorbis glabratus* under axenic conditions. Ann NY Acad Sci 77:237–245
9. Childress JJ (1975) The respiratory rates of midwater crustaceans as a function of depth of occurrence and relation to the oxygen minimum layer of Southern California. Comp Biochem Physiol 50A:787–799

10. Colberg PJ, Lingg AJ (1978) Effect of ozonation on microbial fish pathogens, ammonia, nitrate, nitrite and BOD in simulated reuse hatchery water. J Fish Res Board Can 35:1290–1296

11. Cooke WB (1956) Colonization of artificial bare areas by microorganisms. Bot Rev 22:613–638

12. D'Agostino A (1975) Antibiotics in culture of invertebrates. In: Smith WL, Chanley MH (eds) Culture of marine invertebrate animals. Plenum Press, New York London, pp 109–133

13. Dungworth DL, Cross CE, Gillespie JR, Plopper CG (1975) The effects of ozone on animals. In: Murphy JS, Orr JR (eds) Ozone chemistry and technology. Franklin Inst Press, Philadelphia, pp 29–54

14. Duursma EK, Parsi P (1976) Persistence of total and combined chlorine in sea water. Neth J Sea Res 10:192–214

15. Fisher WS, Nelson RT (1977) Therapeutic treatment for epibiotic fouling on Dungeness crab (*Cancer magister*) larvae reared in the laboratory. J Fish Res Board Can 34:432–436

16. Fisher WS, Nelson RT (1978) Application of antibiotics in the cultivation of Dungeness crab, *Cancer magister*. J Fish Res Board Can 35:1343–1349

17. Fulton C (1959) Re-examination of an inhibitor of regeneration in *Tubularia*. Biol Bull 116:232–238

18. Giese AC, Farmanfarmaian A, Hilden S, Doezema P (1966) Respiration during the reproductive cycle in the sea urchin *Strongylocentrotus purpuratus*. Biol Bull 130:192–201

19. Gnaiger E (1980) Energetics of invertebrate anoxibiosis: Direct calorimetry in aquatic Oligochaetes. FEBS Lett 112(2):239–242; Pharmacological application of animal calorimetry. Thermochim Acta 49:75–85

20. Goldberg HS (1959) Antibiotics, their chemistry and nonmedical uses. Van Nostrand Co, New Jersey

21. Green JD, Chapman MA (1977) Temperature effects on oxygen consumption by the copepod *Boeckella dilatata*. N Z J Mar Freshwater Res 11:375–382

22. Gyllenberg G (1973) Comparison of the Cartesian diver technique and the polarographic method, an open system, for measuring the respiratory rates in three marine copepodes. Commentat Biol 60:1–13

23. Hahn FE (1979) Antibiotics, vol V. Mechanism of action of antibacterial agents. Springer, Berlin Heidelberg New York

24. Harding GCH (1977) Surface area of the Euphansiid *Thysanöessa raschii* and its relation to body length, weight, and respiration. J Fish Res Board Can 34:225–231

25. Hash JH (1972) Antibiotic mechanisms. Annu Rev Pharmacol 12:35–56

26. Helwig H (1973) Antibiotika-Chemotherapeutika. Thieme, Stuttgart

27. Kempf SC, Dennis Willows AO (1977) Laboratory culture of the nudibranch *Tritonia diomedea* Bergh (Tritoniidae: Opisthobranchia) and some aspects of its behavioral development. J Exp Mar Biol Ecol 30:261–276

28. Latter PM (1977) Axenic cultivation of an enchytraeid worm, *Cognettia sphagnetorum*. Oecologia 31:251–254

29. Lorian V (1966) Antibiotics and chemotherapeutic agents in clinical and laboratory practice. CC Thomas, Springfield, Illinois, USA

30. Lough RG, Gonor JJ (1973) A response-surface approach to the combined effects of temperature and salinity on the larval development of *Adula californiensis* (Pelecypoda: Mytilidae). I. Survival and growth of three and fifteen-day old larvae. Mar Biol 22:241–250

31. Marshall KC (1972) Mechanism of adhesion of marine bacteria to surfaces. Gaitherburg, Maryland, USA, Proc 3rd Int Congr Mar Corrosion Fouling, October 1972, pp 625–632

32. Marshall SM, Orr AP (1958) Some uses of antibiotics in physiological experiments in sea water. J Mar Res 17:341–346

33. Meadows PS (1964) Experiments on substrate selection by *Corophium* species: Films and bacteria on sand particles. J Exp Biol 41:499–511

34. Middaugh DP, Crane AM, Couch JA (1977) Toxicity of chlorine to juvenile spot, *Leiostomus xanthurus*. Water Res 11:1089–1096

35. Moebus K (1972) Factors affecting survival of test bacteria in sea water: marine bacteria, test bacteria and solid surfaces. Helgol Wiss Meeresunters 23:271–285

36. Morgan RP, Prince RD (1978) Chlorine effects on larval development of striped bass (*Morone saxatilis*), white perch (*M. americana*) and blueback herring (*Alosa aestivalis*). Trans Am Fish Soc 107(4):636–641

37. Murphy JS, Orr JR (1975) Ozone chemistry and technolog: A review of the literature 1961–1974. Franklin Inst Press, Philadelphia

38. Nilson EH, Fisher WS, Shleser RA (1976) A new mycosis of larval lobster (*Homarus americarus*). J Invertebr Pathol 27:177–183

39. Poole RL (1966) A description of laboratory-reared zoeae of *Cancer magister* Dana, and megalopae taken under natural conditions (Decapoda Brachyura). Crustaceana 11:83–97

40. Quastel JH, Scholefield PG (1951) Biochemistry of nitrification in soil. Bacteriol Rev 15:1–23

41. Reiner R (1977) Antibiotics. In: Korte F, Goto M (eds) Natural compounds. Part 2: Antibiotics, vitamins and hormones. Thieme, Stuttgart, pp 1–68

42. Relini G (1974) Colonization patterns of hard marine substrata. Mem Biol Mar Oceanogr 4 (4-5-6):201–261

43. Sechler GE, Gundersen K (1972) Role of surface chemical composition on the microbial contribution to primary films. Gaithersburg, Maryland, Proc 3rd Int Congr Mar Corrosion Fouling, 1972, pp 610–616

44. Seegert GL, Brooks AS (1978) The effects of intermittent chlorination on *Coho salmon,* Alewife, Spottail Shiner and Rainbow Smelt. Trans Am Fish Soc 107:346–353

45. Sheldon RW, Evelyn TPT, Parson TR (1967) On the occurence and formation of small particles in sea water. Limnol Oceanogr 12:367–375

46. Spector WS (1957) Handbook of toxicology, vol II. Antibiotics. W.B. Saunders, Philadelphia London

47. Spencer CP (1952) On the use of antibiotics for isolating bacteria-free cultures of marine phytoplankton organisms. J Mar Biol Assoc UK 31:97–106

48. Spotte S (1979) Fish and invertebrate culture. Wiley, New York

49. Stickney AP (1964) Salinity, temperature, and food requirements of soft-shell clam larvae in laboratory culture. Ecology 45:283–291

50. Stokinger HE (1965) Ozone Toxicology: A review of research and industrial experience, 1954–1964. Arch Environ Health 10:719–731

51. Swiss JJ, Johnson MG (1976) Energy dynamics of two benthic crustaceans in relation to diet. J Fish Res Board Can 33:2544–2550

52. Tuker M (1959) Inhibitory control of regeneration in nemertean worms. J Morphol 105:569–600

53. Walne PR (1958) The importance of bacteria in laboratory experiments on rearing the larvae of *Ostrea edulis* (L.). J Mar Biol Assoc UK 37:415–425

54. Walter AM, Heilmeyer L (1969) Antibiotika-Fibel. Antibiotika und Chemotherapie. Thieme, Stuttgart

55. Wedemeyer GA, Nelson NC (1977) Survival of two bacterial fish pathogens (*Aeromonas salmonicida* and the enteric redmouth bacterium) in ozonated, chlorinated, and untreated waters. J Fish Res Board Can 34:429–432

56. Wedemeyer GA, Nelson NC, Yasutake WT (1979) Physiological and biochemical aspects of ozone toxicity to Rainbow trout (*Salmo gairdneri*). J Fish Res Board Can 36:605–614

57. Wickins JF (1972) Developments in the laboratory culture of the common prawn, *Palaemon serratus* Pennant. Fish Invest London Ser II 27(4):1–24

58. Yetka JE, Wiebe WJ (1974) Ecological application of antibiotics as respiratory inhibitors of bacterial populations. Appl Microbiol 28:1033–1039

59. Zeiss FR (1963) Effects of population densities on zooplankton respiration rates. Limnol Oceanogr 8:110–115

60. Zobell CE, Anderson DQu (1936) Observations on the multiplication of bacteria in different volumes of stored sea water and the influence of oxygen tension and solid surfaces. Biol Bull 71:324–342

Chapter II.10 pO_2 and Oxygen Content Measurement in Blood Samples Using Polarographic Oxygen Sensors

C.R. Bridges[1]

1 Introduction

A major function of blood in most organisms is to supply the tissues with oxygen. The measurement of oxygen partial pressure (p_{O_2}) and oxygen content (c_{O_2}) in blood are therefore important criteria through which the state of the gas exchange system of the organism can be assessed.

Baumberger [2] initially used a polarographic technique to measure p_{O_2} in blood. Further improvements, notably the construction of the membrane-covered oxygen sensor [7], led to rapid advances, and today the measurement of p_{O_2} in blood is routinely performed with a polarographic oxygen sensor (POS) in most medical laboratories. The increasing importance of accurate blood p_{O_2} measurements in research laboratories either for the construction of dissociation curves [9, 29] or the measurement of oxygen content [6, 19, 22] has given more emphasis to the critical assessment of p_{O_2} measurement in blood. Several reviews are available on methodology and techniques used in human blood p_{O_2} measurements at 37°C [4, 12, 15, 19, 30–32]. However, the ecophysiologist working with nonmammalian vertebrates or invertebrates may encounter problems not experienced with human blood, e.g., nucleated red blood cells, different respiratory pigments, lower temperatures, limited availability of blood. The present chapter therefore attempts to review available information and provide some guide lines for p_{O_2} measurement for those who have ecophysiological interests.

2 p_{O_2} Measurement in Blood Using a POS

2.1 Materials and Apparatus Design

2.1.1 Polarographic Oxygen Sensor

A number of reviews are available concerning the performance of different commercial sensors in the laboratory [11, 15]. When measuring p_{O_2} in blood, the sample size is

---

1 Max-Planck-Institut für experimentelle Medizin, Abteilung Physiologie, Hermann-Rein-Straße 3, D–3400 Göttingen, FRG

Polarographic Oxygen Sensors (ed. by Gnaiger/Forstner)
© Springer-Verlag Berlin Heidelberg 1983

usually small and therefore a small cathode diameter (0.012–0.020 mm) is preferable. This is partly due to the reduction of the stirring effect and also due to the lower oxygen consumption of the small cathode. A 2 mm cathode has an oxygen consumption of 90 Pa min^{-1} in 0.1 cm^3 H_2O compared with 0.2 Pa min^{-1} for a 0.020 mm cathode [30].

The membrane is important in controlling the sensitivity and response time of the POS (Chap. I.1). A Teflon membrane is approximately 60 times as permeable to oxygen as a polypropylene membrane of similar thickness and therefore Teflon has a shorter response time τ_{99} = 10 s at 25 μm and 37°C compared with 25 s for a polypropylene membrane, [30]). Teflon membranes may exhibit greater gas/blood differences in p_{O_2} readings [21] and may undergo sensitivity changes with time due to their flexibility [15]. Polypropylene has a higher temperature coefficient for the response time. When working at temperatures of 15°C and lower, Teflon gives a much better response time. At higher temperatures, around 37°C, polypropylene may provide an adequate response time. Membranes should ideally be fitted to the sensor 24 h in advance of measurements and then they usually remain stable for at least an 8-day period, depending upon the amount of use [15].

Hahn et al. [13] have shown that the use of alkaline electrolytes (pH = 11.2) and a polarizing voltage of 0.9 V gives a marked improvement in both stability and response time. Details for an electrolyte giving a pH of 11.2 at 37°C are Na_2HPO_4 = 0.0587 g; KCL = 0.0745 g; and NaOH = 0.170 g in 100 cm^3 bi-distilled water.

Oeseburg et al. [24] have confirmed Hahn's findings and recommend the use of bicarbonate instead of phosphate buffer to add to stability (electrolyte = $NaHCO_3$ 0.08 mol cm^{-3}, KCL 0.1 mol dm^{-3}; NaOH added until pH = 11.2). They also suggest that the membrane holder and electrolyte chamber should be degassed after filling to improve stability. For further information see Chapter I.5.

2.1.2 Experimental Setup

Figure 1 shows an experimental setup (after N. Heisler and W. Nüsse) for replicate determination of p_{O_2} in blood samples. It consists of two sensor cuvettes (Radiometer D616) made of stainless steel and glass which enclose two POS (Radiometer E5047). The cuvettes are mounted in a perspex chamber (volume 1.25 dm^3). A large volume is used to damp down any temperature fluctuations and the chamber is thermostatted by an external source. Calibration gases are fed into the system via needle valves and PVC tubing (4 mm inner diameter, 1.5 mm outer wall thickness). From here the gases are passed through the humidifier at a flow rate of 25 cm^3 min^{-1}, which ensures that the gases are fully saturated with water vapor at experimental temperature. The gas calibration line (polypropylene, 1 mm inner diameter, 0.5 mm wall thickness) can be connected to the sensors via the Luer-Lock injection port (3 mm inner diameter) through which the blood sample is introduced. The POS are mounted such that sequential filling requires 0.2 cm^3 of blood which can be withdrawn and used for other measurements.

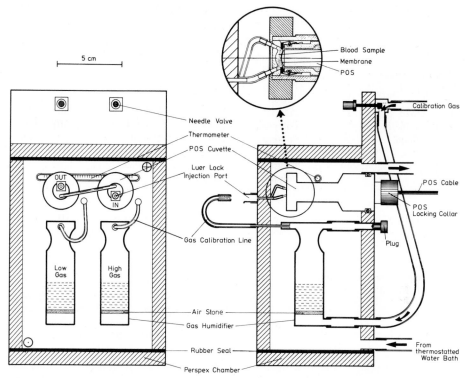

Fig. 1. p_{O_2} measurement system designed by N. Heisler and W. Nüsse, consisting of two POS mounted in a Perspex chamber, together with a calibrating gas humidification system

2.2 Calibration

Due to the fact that all commercially available sensors show deviation from linearity above 40 kPa it is suggested that, when working at p_{O_2} ranges above this value, a bracketing calibration is used [15]. When working in the normal physiological range for blood p_{O_2}, i.e., 1–13 kPa, calibration bracketing gives added accuracy but is not essential. Since the POS may show hysteresis or memory effects equilibration of the POS at the approximate p_{O_2} of the sample before injection is recommended. Moran et al. [21] have also suggested exposing the sensor to nitrogen before the calibration gas or blood sample. Alkaline electrolytes reduce the hysteresis effect [13].

Since the micro-cathode POS is designed to operate in unstirred solutions it has been found that liquid samples give lower p_{O_2} readings than their corresponding equilibrating gas [12, 30, 34]. Table 1 illustrates some gas/liquid factors measured at different temperatures with different membranes. Doubly distilled deionized water gives the closest approximation to the gas value and there is considerable variation between blood and water. Therefore when calibrating with water it is important to determine the water/blood factor by equilibrating the blood with the same gas as in the water calibration and measuring the p_{O_2} in both liquids. If gas calibration is used, the gas/blood factor must be ascertained. A number of parameters may influence the gas/blood

Table 1. Gas/blood factors

Sample	Temp. (°C)	Membrane a	p_{O_2} Range [kPa]	$\dfrac{p_{O_2} \text{ gas}}{p_{O_2} \text{ sample}}$	Ref.
Human blood	37	Polypro.	0–93	1.100	[21]
Human blood	37	Teflon	0–93	1.230	[21]
Human blood	37	Polypro.	0–93	1.044	[30]
50% Glycerine/H_2O	37	Polypro.	0–93	1.126	[30]
Chicken blood	40	Polypro.	0–87	1.145	[23]
Dog blood	40	Polypro.	0–87	1.057	[23]
1.0 M NaCl	40	Polypro.	0–87	1.035	[23]
H_2O^b	40	Polypro.	0–87	1.003	[23]
Dogfish blood	15	Teflon	–	1.150	N. Heisler (pers.comm.)
Trout blood	15	Teflon	0–13	1.101	This study
Eel blood	15	Teflon	0–13	1.141	This study

Polypro. = Polypropylene

[a] All membranes approximately 25 μm thick
[b] Doubly distilled deionized water

factor, such as the type of membrane and its thickness, Teflon giving larger gas/blood factors than polypropylene [21]. Nightingale et al. [23] found that the gas/blood factor for bird blood increased with increasing hematocrit and therefore increasing viscosity. It is clear that each experimental setup with different blood will give different gas/blood factors. The gas/blood factor must therefore be measured for each experimental series on the same blood as used for the p_{O_2} measurements. This can be done by pooling the used blood samples and then tonometering the blood at known p_{O_2} values and measuring both tonometer gas and blood p_{O_2}.

2.3 Measurement

The following procedure is used when making measurements with the system as shown in Fig. 1. Whenever possible the instruments are allowed to warm up and the POS to adjust to the measuring temperature overnight. A calibration for the zero point is made by flowing oxygen-free nitrogen through the humidifier system and past the sensors in their cuvettes for 10 min. A high calibration gas at around 12 kPa p_{O_2} is then provided by a gas mixing pump (301 a-f Wösthoff, Bochum, West Germany) and the p_{O_2} reading adjusted. The linearity of the response is then checked with a low calibration gas supplied by the mixing pump at around 2.7 kPa p_{O_2}. Prior to the injection of the sample, the cuvettes are rinsed with physiological saline, to prevent any hemolysis of blood by water contact, and then either the high gas or low gas calibration is applied, depending on whether arterial or venous samples are to be measured. When the POS has come into equilibrium, the gas line is disconnected and the blood sample slowly injected, being careful not to generate pressure gradients or trap gas bubbles in the cuvettes. The response of the sensors can be followed on a calibrated chart recorder

(BD 9, Kipp & Zonen, Holland). When no recorder is available, a stopwatch can be used to make readings after 1, 1.5, 2, and 3 min. Depending on the response of a given blood (Fig. 2B–D) an extrapolation back to time zero can be made to determine the p_{O_2}. After a reading the sample is pushed out of the system and pooled for gas/blood factor measurement. The measured p_{O_2} is then multiplied by this factor to give the actual p_{O_2}. The cuvettes are flushed slowly until clean with a saline solution which has been equilibrated at the same temperature and p_{O_2} as the calibration gas. The measuring procedure is repeated for the next sample. After completion of all measurements the POS cuvette is filled with a bio-detergent (Radiometer enzyme electrode detergent S4160) and left overnight.

2.4 Problems and Sources of Error in p_{O_2} Measurements

2.4.1 Reproducibility

If measurements are made repeatedly on the same sample, rinsing with saline and calibrating with air in between, a coefficient of variation of around 2% is expected [32]. Gleichmann and Lübbers [12] reported a reproducibility of 0.5% for double determinations on blood up to 40 kPa and Moran et al. [21] found that repeated determinations on blood, equilibrated with gas tensions within the physiological range, gave a standard deviation of 1.7% and 1.2% for Teflon and polypropylene membranes respectively. Using the system shown in Fig. 1 Scheid and Meyer [29] found in experiments on human, rabbit and duck blood that the difference in p_{O_2} reading of the first and second POS, filled sequentially from a syringe, for duplicate determinations, averaged 0.01 ± 0.05 (SD) kPa for 165 measurements.

2.4.2 Temperature

Temperature control to within ± 0.1°C is of critical importance. Human blood, and probably blood of other animals as well, has a high temperature coefficient and a 1°C change in temperature will change the blood p_{O_2} by 6%–7% [31]. The temperature change will effect the sensor response itself giving a total error of 10%. Measurements should be carried out at the in vivo temperature of the blood for the greatest accuracy. Where this is not possible then a series of nomograms is available for human blood [14, 16] which, with some reservations, can be applied to temperature corrections for other bloods.

2.4.3 Blood Metabolism

Mammalian blood has non-nucleated red blood cells and a relatively low oxygen consumption. The fall in p_{O_2} of stored blood samples is dependent on the temperature and the initial p_{O_2} [10, 26]. At 37°C and normal physiological p_{O_2} values the fall in p_{O_2} is less than 0.07 kPa min^{-1}, compared with 0.3–0.4 kPa min^{-1} at p_{O_2} values above 40 kPa. Cooling of the sample to around 0°C reduced the p_{O_2} decline to 0.4 kPa

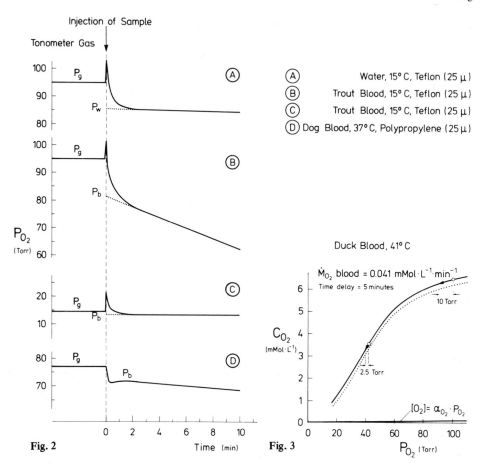

Fig. 2. Actual recordings of p_{O_2} measurements made with a POS for various liquids and different equilibrating gas levels. p_g = partial pressure of oxygen in tonometer gas; p_w = partial pressure of oxygen in water; p_b = partial pressure of oxygen in blood; p_w and p_b = "instrumental" p_{O_2}'s and have to be corrected by the gas/liquid factor (see text)

Fig. 3. Oxygen dissociation curve for duck blood (modified after [28]) shows the effect of blood metabolism on the measurement of p_{O_2}. *Continuous line* is the true dissociation curve. Time delay of 5 min results in a down shift in curve by the amount of oxygen which is metabolized (*broken curve*). *Arrows* indicate drop of c_{O_2} due to metabolism and thus p_{O_2} change during storage

min^{-1} even at high p_{O_2} values. The problem of blood metabolism is more severe when working with nucleated red blood cells as found in nonmammalian vertebrates and some annelids. Here metabolism may be several times higher than that of human blood even at lower temperatures. Figure 2B–D illustrates some case studies of p_{O_2} measurement in blood. When water is equilibrated with tonometer gas (p_g) at around 13 kPa (= 95 mm Hg) and injected into the cuvette system (as in Fig. 1), the sensor gives a relatively stable reading for the water p_{O_2} (p_w) after 2 min with negligible oxygen consumption thereafter (Fig. 2A). If trout blood is equilibrated at the same gas tension, which approximates to arterial values, and injected into the cuvette, then a rapid fall in p_{O_2} is seen (Fig. 2B) until after 2 min, a steady decline in p_{O_2} is given at approximately 0.27 kPa min^{-1} (= 2 mm Hg min^{-1}) which is mainly due to blood metabolism. Therefore to determine the accurate p_{O_2} of the blood (p_b), an extrapolation back to time zero is made. If the blood is equilibrated at low p_{O_2} values (2 kPa) which approximate to venous blood, then a similar response to that shown by water is seen (Fig. 2C) with a steady reading after 2 min followed by a very slow decline. Extrapolation back to zero in this case only gives a change in p_b of 0.07 kPa (= 0.5 mm Hg) compared with 0.5 kPa (= 4 mm Hg) at high p_b values. In comparison, when dog blood, which has no nucleated red cells, is measured at arterial values a plateau is given between the first and second minute after injection (Fig. 2D) and this value is generally taken as p_b.

These differences in the rate of change of p_{O_2} between high and low p_{O_2} values for blood in general can be explained by the shape of the oxygen dissociation curve for blood. As an example Fig. 3 shows the oxygen dissociation curve for duck blood [28]. Avian blood has a very high metabolism, up to ten times that of human blood, and is similar to fish blood at the same temperature. Scheid and Kawashiro [28] found that the oxygen content of a blood sample would decrease by 0.2 mmol dm^{-3} during a 5-min delay in measurement. At high p_{O_2} values around 13 kPa (= 100 mm Hg), where the dissociation curve flattens off as the hemoglobin approaches 100% saturation (Fig. 3), a fall in oxygen content of 0.2 mmol dm^{-3} gives a 1.3 kPa (= 10 mm Hg) change in p_{O_2}. For the same oxygen content change at around the P_{50} only a fall of 2.5 mm Hg is given due to the steepness of the dissociation curve. It is therefore evident that when making p_{O_2} measurements in metabolically active blood at p_{O_2} ranges where the respiratory pigment approaches 100% saturation, some knowledge of the rate of metabolism is necessary in order to avoid errors in the measurement. Scheid and Kawashiro [28] calculated that for duck blood at 20 kPa a drop of p_{O_2} of 4 kPa min^{-1} could be expected with virtually all of the oxygen consumption being satisfied by physically dissolved oxygen. To avoid metabolic effects in blood samples arterial blood should, whenever possible, be measured immediately. Failing this, storage on ice, together with venous samples, whose rate of p_{O_2} change with time is small, is recommended.

2.4.4 Calibration and Tonometry

If borate/sulfite solutions are used for zero calibration, they must be fully rinsed from the system as even small residual amounts can cause depletion of p_{O_2} in blood samples. Ideally for liquid calibration equilibrated blood should be used, but this is not practical when available blood volumes are small. Therefore when determining either water/blood factors (p_w/p_b) or gas/blood factors (p_g/p_b), as in Fig. 2B–C, with metabolically active blood care must be taken to ensure complete equilibration between gas and blood in the tonometer. Thin film tonometers should be used and the equilibrating gas should contain a carbon dioxide tension similar to that found in vivo to ensure that the blood remains within the range of the normal dissociation curve. When working at low temperatures, equilibration times should be adjusted accordingly to ensure complete equilibration. Sampling syringes should be carefully rinsed with tonometer gas before taking up the sample, and transfer to the POS accomplished in as short a time as possible, avoiding large temperature gradients. Gas/blood factors should be ascertained at both arterial and venous p_{O_2} ranges as these may differ [11, 26]. The use of solutions with similar physical properties to blood such as glycerine/water mixtures has been advocated [15, 30] to overcome tonometry problems. However, these provide only a partial answer since the gas/blood factor varies with viscosity and thus differs with hematocrit and different respiratory pigments. Therefore, blood should be used to determine the correction factors.

2.5 Summary

The following is a summary of guidelines for p_{O_2} measurement in blood:

1. Cathode diameter, membrane, electrolyte and polarizing voltage are selected to give a fast and stable response at the selected measuring temperature, which must be maintained to within $\pm 0.1°C$.
2. For liquid calibration borate/sulfite solutions and air-equilibrated water are used to provide zero p_{O_2} and the high calibration point respectively.
3. Calibration with humidified gases should be used whenever possible, checking linearity of response by calibration bracketing.
4. The POS must be equilibrated at a p_{O_2} similar to that of the sample before each measurement.
5. The response of the POS should be followed and a consistent measuring procedure determined for a given blood.
6. Water/blood or gas/blood correction factors should be determined for each measurement series.
7. To avoid errors due to metabolism, arterial samples should be measured immediately or else stored on ice.
8. The metabolic activity of the blood should be determined when working with nucleated red blood cells at high p_{O_2} ranges.

3 Oxygen Content Measurement Using a POS

3.1 Theory

Consider a volume, V [cm^3], of a solution which displaces chemically bound oxygen from a respiratory pigment. The initial partial pressure of oxygen in this solution, p_s [kPa], can be determined with a POS. A small volume of blood (v) is injected into the solution such that it displaces a corresponding amount of the solution from the containing vessel which is immediately closed. The oxygen content of the solution, c_s [μmol O_2], can be determined if the solubility of oxygen in the solution, S_s [μmol cm^{-3} kPa^{-1}] (App. A), is known.

$$c_s = p_s \times S_s \times (V-v).$$

(1)

When the blood and solution are mixed, chemically bound oxygen is released from the pigment and since the system is a closed one then the chemically bound oxygen is converted into physically dissolved oxygen and the oxygen partial pressure in the system rises to a new level (p_{s+b}). The oxygen content of the mixture will now consist of that of the solution (c_s) and that of the blood (c_b) such that

$$c_s + c_b = p_{s+b} \times S_s \times V. \tag{2}$$

Since the ratio V/v is usually large, then the added blood will not significantly affect the solubility of oxygen in the solution. The total oxygen content of the blood, c_b [μmol O_2 cm^{-3} blood] can be calculated from Eq. (2) such that

$$c_b = S_s \times \frac{V}{v} \times (p_{s+b} - \frac{V-v}{V} \times p_s). \tag{3}$$

S_s, V and v are known and p_s and p_{s+b} are measured using a POS, therefore total oxygen content of the blood can be determined.

3.2 Materials and Apparatus

3.2.1 Displacement of Chemically Bound Oxygen

Previous manometric methods for oxygen content measurement [27, 38] utilized chemical agents to release oxygen bound to respiratory pigments and these can be utilized for oxygen content measurement with the POS. Table 2 illustrates the use of different reagents for various respiratory pigments. Potassium ferricyanide solutions (6 g dm^{-3}) have generally been used for hemoglobin oxygen content determinations, with saponin added (3 g dm^{-3}) to aid cell lysis. Due to the instability of saponin, it is recommended to make up the solution freshly every day and store it in a dark bottle. The use of carbon monoxide-saturated saline has been preferred in some measurements where small bore stopcocks are used, as it does not lyse cells and thus prevents the accumulation of cell debris in the apparatus [17]. The addition of a small quantity of an anti-foaming agent such as caprylic alcohol to the solution is also recommended. Rawlinson [25] has suggested that potassium cyanomercurate be used instead of potassium cyanide to avoid oxygen reabsorption problems, but Truchot [36] found no significant difference in oxygen content determinations on hemocyanin using both cyanide and cyanomercurate.

Table 2. Chemical reagents for the release of bound oxygen from respiratory pigments

Respiratory pigment	Reagent	Ref.
Hemoglobin	Potassium ferricyanide K_3 [Fe(CN)$_6$]	[8, 18, 22, 37, 38]
	Carbon monoxide CO	[1, 3, 17]
Hemocyanin	Potassium cyanide KCN	[6, 8, 36]
	Potassium cyanomercurate K_2 [Hg(CN)$_4$]	[25]
Hemerythrin	Potassium ferricyanide	[8, 39]

3.2.2 Apparatus

Neville [22] employed a simple mixing vial for blood and ferricyanide and measured the increase in p_{O_2} with a mercury dropping electrode. A simple and accurate mixing system can be constructed out of graduated syringes and a three-way tap, utilizing a glass or mercury bead to facilitate mixing, and then measuring p_{O_2} with a separate POS [18, 35].

Mayers and Forster [20] utilized the bores of two stopcocks to give precise mixing volumes of blood and ferricyanide. The method was modified [17] using a five-port, double-bore stopcock, replacing ferricyanide with carbon monoxide equilibrated saline and using a volume limited syringe to assure reproducibility of dilution volumes.

A number of chamber designs in perspex, glass, and stainless steel are available which incorporate the POS in the mixing chamber and thus avoid excessive handling of the samples [1, 5, 6, 33, 37]. Since the change in oxygen tension (Δp_{O_2}), due to the release of chemically bound oxygen from the pigment, is proportional to $c_b \times v/V \times S_s^{-1}$, then the chamber volume (V) and injection volume (v) are selected such that Δp_{O_2} is large for a given blood oxygen content c_b. When working with hemoglobins which have a relatively high oxygen content even at low hemoglobin saturations compared with hemocyanins, chamber volumes of $2.2-2.55$ cm^3 and injection volumes of $22-72$ mm^3 have been used [1, 5]. Tucker [37] using a small chamber volume (0.5 cm^3) and a 7 mm^3 injection volume obtained a $\Delta p_{O_2} > 13$ kPa for fully saturated human hemoglobin. With hemocyanin even at full saturation, oxygen content is usually below 5 vol-% and therefore a small chamber volume is preferable especially if the available blood volume is limited.

Figure 4 illustrates a chamber constructed out of perspex [6] for the measurement of oxygen content in blood containing hemocyanin. It consists of a POS (Radiometer E5047) with the outer sleeve removed and a special perspex sleeve constructed which allows the O-ring, which normally secures the membrane at the tip of the POS, to also

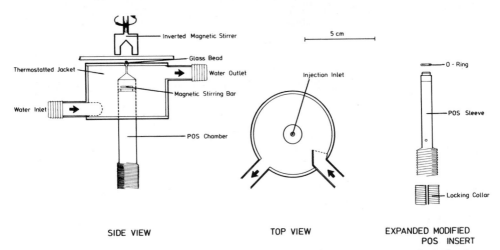

Fig. 4. Combined mixing and p_{O_2} measurement chamber for oxygen content determinations (after [6])

form a watertight seal with the floor of the sample chamber (0.5 cm^3). A magnetic stirring bar is enclosed within the chamber and is driven by an external inverted magnetic stirrer (Cenco, Neth.) placed on top of the chamber. The chamber can be sealed with a small glass bead which is dropped into a concave depression at the top of the injection inlet (1 mm diameter) after the injection of the blood sample. The roof of the sample chamber is cone-shaped to aid in the removal of air bubbles trapped in the chamber. Chamber and POS are thermostatted at $\pm 0.1°C$ of the selected temperature by a water jacket which is supplied by an external source.

3.3 Measurement

For measurements of oxygen content in blood containing hemocyanin, a temperature of 32°C was selected for the POS and the sample chamber as this temperature facilitates the displacement of chemically bound oxygen from the respiratory pigment. The sample chamber is first filled with a zero p_{O_2} solution (1 mg $Na_2 SO_3$ + 5 ml 0.01 M $Na_2 B_4 O_4$ solution) and the zero point set. The solution is then aspirated from the chamber using a shortened hypodermic needle with a flattened tip. A needle tip diameter of 0.8 mm (21G) is used which allows air to enter the chamber as the fluid is aspirated from the chamber. The sample chamber is rinsed with distilled water to remove all traces of sulfite and then filled with an air equilibrated solution of potassium cyanide (6 g dm^{-3}) which is maintained in a water bath at the same temperature as the POS and the sample chamber. A small air bubble is left in the sample chamber to maintain air equilibration and the magnetic stirrer switched on. This second calibration point is set and the calibration solution aspirated from the chamber. A 20-cm³ syringe fitted with a three-way tap is used to draw up 15 cm³ of the KCN stock solution from the water bath. The oxygen tension of the cyanide solution is lowered by closing the tap and pulling on the syringe plunger, causing a decrease in pressure in the syringe. Dissolved gases come out of solution and the gas bubbles are expelled via the three-way tap. This operation is repeated a number of times until the solution p_{O_2} is low ($p_{O_2} < 5$ kPa). Alternatively, the KCN solution can be bubbled with nitrogen in the water bath if this is convenient.

The sample chamber is now filled with the low p_{O_2} KCN, the stirrer bar switched on and after equilibration (usually 30 s) a steady recording is obtained (p_s). A 100-mm³ syringe (Hamilton 1710), fitted with a Chaney-adaptor for reproducibility, is used to inject a 65 mm³ sample of blood into the chamber. A long blunt-ended needle (21G) is used to introduce the sample near the bottom of the chamber displacing an equivalent amount of KCN solution from the chamber. The needle is withdrawn and the injection port closed with a glass bead and the stirrer switched on. The response of the POS is followed, as previously, with a recorder until a stable reading is obtained (p_{s+b}). After the measurement the blood/KCN mixture is aspirated from the chamber and either the aerated calibration solution or the low p_{O_2} solution used to prepare for the next measurement. Each determination takes approximately 5 min to complete. The volume of the chamber and the injection volume can be found by weighing before and after filling them with distilled water. Values for S_s can be found in Tucker [37] or

Laver et al. [19] for a given chamber temperature. Equation (3) is then used to calculate the total oxygen content of the blood.

3.4 Reproducibility, Accuracy and Problems

Previous manometric methods for oxygen content determinations have the disadvantage of being time-consuming and generally require skilful manipulation of the apparatus. The measurement of oxygen content using a POS is relatively rapid and simple, requiring no special skills. Table 3 illustrates values for reproducibility of the various POS methods and their accuracy compared with the Van Slyke determination. Reproducibility is relatively good, being within $\pm$ 0.3 vol-% for all methods, which compares favorably with similar Van Slyke determinations [17, 18, 20]. Accuracy varies with the different methods but can be high [37]. Both accuracy and reproducibility will depend upon oxygen tension measurements and are therefore open to similar errors as mentioned previously for blood p_{O_2} measurements. Ideally Δp_{O_2} should be kept large to avoid errors and also to ensure that the injection volume is known accurately. A weighing procedure similar to that suggested by Scheid and Meyer [28] for O_2 equilibrium curves could be adopted to improve accuracy. Since calibration and measurements are made in the same liquid, gas/blood factors are not required. The use of the combined mixing chamber and POS, with a stirring bar, prevents the formation of diffusion gradients in the sample. However, problems may arise when working with metabolically active bloods, especially when there is a large difference between in vivo blood temperature and that of the mixing chamber and POS, as experienced with fish blood. Again careful assessment of the POS response is necessary. Errors due to the change in solubility of the solution after the addition of blood are very small and Laver et al. [18] could find no significant difference in solubility between ferricyanide solutions and blood/ferricyanide mixtures. Since the solutions themselves are very

Table 3. Reproducibility and accuracy of oxygen content measurements using a POS

Method	Blood volume [mm^3]	Reproducibility [vol-%][a]	Accuracy [vol-%][b]	Ref.
Mixing vial and separate POS	250	$\pm$ 1.0% R	$\pm$ 0.40	[22]
Syringe mixing and separate POS	50	0.22 $\pm$ 0.3 D	0.42 $\pm$ 0.54	[18]
Syringe mixing and separate POS	5	0.21 $\pm$ 0.13 D	0.16 $\pm$ 0.14	[35]
Syringe mixing and separate POS	10	0.16 $\pm$ 0.13 D	0.16 $\pm$ 0.11	[35]
Stopcock bore and separate POS	120	2% $\pm$ 0.29 D	0.37 $\pm$ 0.50	[20]
Stopcock bore and separate POS	100–200	$\pm$ 0.07 D	0.54 $\pm$ 0.49	[17]
Mixing chamber and POS combined	72	$\pm$ 0.18 R	–	[1]
Mixing chamber and POS combined	7	$\pm$ 0.14 R	0.03 $\pm$ 0.17	[37]
Mixing chamber and POS combined	150–1000	0.14 $\pm$ 0.11 R	0.20 $\pm$ 0.16	[33]
Mixing chamber and POS combined	25	$\pm$ 0.25 R	–	[5]
Mixing chamber and POS combined	65	$\pm$ 0.064 R	–	[6]

[a] Reproducibility expressed as the mean difference between duplicates (D) or replicates (R) $\pm$ S.D.
[b] Accuracy expressed as the mean difference between a Van Slyke determination and POS method $\pm$ S.D.

dilute, then there is little difference in their solubility compared with water and for a 2 g dm^{-3} solution solubility changes by less than 0.5% [18]. Care must be taken to maintain a large V/v ratio, for if this becomes too small then the blood will affect the solubility of the solution and sufficient reagent may not be present to release all the chemically bound oxygen from the pigment. The addition of 0.01 g of ferricyanide to 2 cm^3 of blood is sufficient to liberate all the oxygen from oxyhemoglobin [38]. Therefore for a chamber volume of 0.5 cm^3 and a 6 g dm^{-3} solution of ferricyanide sufficient reagent is present for up to 0.6 cm^3 of blood.

3.5 Summary

1. Select reagent for displacement of chemically bound oxygen, according to respiratory pigment and ease of use.
2. Select chamber or syringe volume and injection volume to give a large Δp_{O_2} for a given oxygen content range.
3. Calibrate POS with (a) zero p_{O_2} solution, (b) aerated reagent solution.
4. Lower initial p_{O_2} of reagent, such that Δp_{O_2} is within the calibrated range of the POS.
5. Measure initial p_{O_2} of the reagent, inject known blood volume and measure p_{O_2} of the mixture, monitoring the response of the POS.
6. Total oxygen content of the blood is then determined from the known values of chamber or syringe volume, injection volume and the solubility of oxygen in the reagent, together with the measured p_{O_2} values.

References

1. Awad O, Winzler RT (1961) Electrochemical determination of the oxygen content of blood. J Lab Clin Med 58:489–494
2. Baumberger JP (1938) Determination of the oxygen dissociation curve of oxyhaemoglobin by a new method. Am J Physiol 123:10
3. Baumberger JP (1940) The accurate determination of haemoglobin oxyhaemoglobin and carbon monoxide haemoglobin (or myoglobin) by means of a dropping mercury electrode. Am J Physiol 129:308
4. Bishop JM, Pincock AC, Hollyhock A, Raine J, Cole RB (1966) Factors affecting the measurement of the partial pressure of oxygen in blood using a covered electrode system. Respir Physiol 1:225–237
5. Borgström L, Hägerdal M, Lewis L, Pontén U (1974) Polarographic determination of total oxygen content in small blood samples. Scand J Clin Lab Invest 34:375–380
6. Bridges CR, Bicudo JEPW, Lykkeboe G (1979) Oxygen content measurement in blood containing haemocyanin. Comp Biochem Physiol 62A:457–462
7. Clark LC Jr (1956) Monitor and control of blood and tissue oxygen tensions. Trans Am Soc Artif Int Organs 2:41–48
8. Cook SF (1927) The action of potassium cyanide and potassium ferricyanide on certain respiratory pigments. J Gen Physiol 11:339–348
9. Duvelleroy MA, Buckles RG, Rosenkaimer S, Tung C, Laver MB (1970) An oxyhemoglobin dissociation analyser. J Appl Physiol 28:227–233

10. Eldridge F, Fretwell LK (1965) Change in oxygen tension of shed blood at various temperatures. J Appl Physiol 20:790–792

11. Flenley DC, Millar JS, Rees HA (1967) Accuracy of oxygen and carbon dioxide electrodes. Br Med J 2:349–352

12. Gleichmann U, Lübbers DW (1960) Die Messung des Sauerstoffdruckes in Gasen und Flüssigkeiten mit der Pt-Elektrode unter besonderer Berücksichtigung der Messung im Blut. Pfluegers Arch Ges Physiol 271:431–455

13. Hahn CEW, Davis AH, Albery WJ (1975) Electrochemical improvement of the performance of p_{O_2} electrodes. Respir Physiol 25:109–133

14. Hedley-Whyte J, Radford EP Jr, Laver MB (1965) Nomogram for temperature correction or electrode calibration during p_{O_2} measurements. J Appl Physiol 20:785–786

15. Heitmann H, Buckles RG, Laver MB (1967) Blood p_{O_2} measurements: Performance of microelectrodes. Respir Physiol 3:380–395

16. Kelman G, Nunn JF (1966) Nomograms for correction of blood p_{O_2}, p_{CO_2}, pH and base excess for time and temperature. J Appl Physiol 21:1484–1490

17. Klingenmaier CH, Behar MG, Smith TH (1969) Blood oxygen content measured by oxygen tension after release by carbon monoxide. J Appl Physiol 26:653–655

18. Laver MB, Murphy AJ, Seifen A, Radford ER Jr (1965) Blood O_2 content measurements using the oxygen electrode. J Appl Physiol 20:1063–1069

19. Laver MB, Seifen A (1965) Measurement of blood oxygen tension in anesthesia. Anesthesiology 26:73–101

20. Mayers LB, Forster RE (1966) A rapid method for measuring blood oxygen content utilizing the oxygen electrode. J Appl Physiol 21:1393–1396

21. Moran F, Kettel LJ, Cugell DW (1966) Measurement of blood p_{O_2} with the macrocathode electrode. J Appl Physiol 21:725–728

22. Neville JR (1960) A simple, rapid polarographic method for blood oxygen content determination. J Appl Physiol 15:717–722

23. Nightingale TE, Boster RA, Fedde MR (1968) Use of the oxygen electrode in recording p_{O_2} in avian blood. J Appl Physiol 25:371–375

24. Oeseburg B, Kwant G, Schut JK, Zijlstra WG (1979) Measuring oxygen tension in biological systems by means of Clark-type polarographic electrodes. Proc K Ned Akad Wet Ser C 82:83–90

25. Rawlinson WA (1940) Cristalline hemocyanins: some physical and chemical constants. Austral J Exp Biol Med Sci 18:131–140

26. Rhodes PG, Moser KM (1966) Sources of error in oxygen tension measurement. J Appl Physiol 21:729–734

27. Roughton FJW, Scholander PF (1945) Micro-gasometric estimation of blood gases I. Oxygen. J Biol Chem 148:541–550

28. Scheid P, Kawashiro T (1975) Metabolic changes in avian blood and their effects on determination of blood gases and pH. Respir Physiol 23:291–300

29. Scheid P, Meyer M (1978) Mixing technique for study of oxygenhemoglobin equilibrium: A critical evaluation. J Appl Physiol 45:818–822

30. Severinghaus JW (1968) Measurement of blood gases, p_{O_2} and p_{CO_2}. Ann NY Acad Sci 148:115–132

31. Severinghaus JW, Bradley BA (1971) Blood gas electrodes or what the instructions didn't say. Radiometer, Copenhagen, ST 59, pp 1–64

32. Sigaard-Andersen O (1974) The acid-base status of the blood. Munksgaard, Copenhagen, pp 175–181

33. Solymar M, Rucklidge MA, Prys-Roberts C (1971) A modified approach to the polarographic measurement of blood O_2 content. J Appl Physiol 30:272–275

34. Sproule BJ, Miller WF, Cushing IE, Chapman CB (1957) An improved polarographic method of measuring oxygen tension in whole blood. J Appl Physiol 11:365–370

35. Tazawa H (1970) Measurement of O_2 content in microliter blood samples. J Appl Physiol 29:414–416

36. Truchot JP (1978) Variations de la concentration sanguine d'hémocyanine fonctionnelle au cours du cycle d'intermue chez le crabe *Carcinus maenas* (L.). Arch Zool Exp Gen 119:265–282

37. Tucker VA (1967) Method for oxygen content and dissociation curves on microliter blood samples. J Appl Physiol 23:410–414

38. Slyke Van DD, Neill JM (1924) The determination of gases in blood and other solutions by vacuum extraction and manometric measurements. I. J Biol Chem 61:523–573

39. Weber RE, Fänge R, Rasmussen KK (1979) Respiratory significance of priapulid hemerythrin. Mar Biol Lett 1:87–97

Chapter II.11   Determination of the In Vivo Oxygen Flux into the Eye

I. Fatt[1]

The cornea of the mammalian eye is a unique tissue in two respects. It is the only transparent connective tissue and it is the only tissue that obtains some of the oxygen needed for its metabolic processes directly from the air. Although not obvious, it is nevertheless true that the normal transparency of the cornea can be maintained only if there is an adequate oxygen supply to its front surface. The open eye receives this oxygen from the air, the closed eye receives oxygen from the capillary bed on the underside of the eyelid.

The relationship between corneal transparency and oxygen supply to this tissue is a subtle one. It is worth explaining here because this relationship forms the basis for the application of polarographic oxygen sensors (POS) to studies of the eye.

The transparency of the cornea could be a result of equal refractive index of the protein fibers and the gelatinous ground substance surrounding each fiber. Maurice [8] separated the fibers from the ground substance and showed that their refractive indices were not equal, and therefore the fibers should scatter light. Light scattering in the cornea would lead to a milky appearance of the cornea rather than the normal transparency. He further showed that the absence of light scattering in the cornea was a result of a combination of uniform size fibers, regular spacing, and a specific spacing dimension. Each fiber was a scattering center, but the scattered wave was destructively cancelled by the scattered waves from its nearest neighbors. This destructive scattering was critically dependent upon water content of the tissue because water content controlled fiber spacing. In an independent and earlier study Smelser [11] and Smelser and Ozanic [12] showed that reduction of the supply of oxygen to the front surface of the cornea caused a loss of transparency. Combining the observations of Maurice with those of Smelser and Ozanic points to the oxygen supply as the key to normal water content of the cornea and therefore its transparency. It now appears that in some way not yet clearly understood, the aerobic metabolism in the epithelial cells on the front surface of the cornea is essential to the process that keeps excess water out of the cornea in the presence of an imbibition pressure in the gelatinous ground substance that is always trying to draw in water.

Quantification of the oxygen flux into the front surface of the in vivo cornea became possible only after the development of the Clark POS. At first, quantitative in-

---

1 School of Optometry, University of California, Berkely, CA, USA

Polarographic Oxygen Sensors (ed. by Gnaiger/Forstner)
© Springer-Verlag Berlin Heidelberg 1983

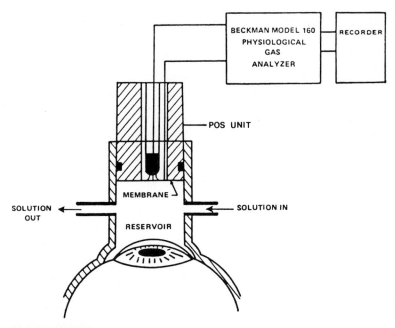

Fig. 1. POS fitted into a tube cemented to a scleral contact lens as used by Hill and Fatt [5] to measure oxygen flux into the cornea

formation on oxygen flux into the cornea was only of academic interest. It was soon realized, however, that the tolerance of an eye to a contact lens might be a function of the ability of oxygen to get around or through the lens to the eye. The designers and fitters of hard contact lenses, that are impermeable to oxygen, need to know the oxygen flux into the cornea to decide whether or not a given contact lens will allow enough air-saturated tears to bathe the cornea. Knowledge of this oxygen flux is even more critical in the design of the new soft contact lenses. These lenses are fitted tightly to the eye, not allowing significant tear flow under the lens. All of the oxygen needed at the front surface of the cornea must come by diffusion through the contact lens.

The availability of the POS and the need for data on oxygen flux into the cornea started a research effort in 1963 that is still being carried out in many laboratories throughout the world.

The first attempt to measure oxygen flux into the front surface of the in vivo human cornea was made by Hill and Fatt [5] using the apparatus shown in Fig. 1. The POS was the modified Clark sensor described by Fatt [2]. The reservoir over the cornea had a volume that could be adjusted between 0.2 cm^3 and 0.8 cm^3. Air-saturated saline at 37°C was flooded through the reservoir until both oxygen tension and temperature were constant. This flow was then stopped and the decline in oxygen tension in the reservoir recorded. The oxygen flux from the reservoir to the cornea was calculated from the known volume of the reservoir and the Henry's law solubility coefficient for oxygen in isotonic saline. The reservoir was assumed to be well mixed. The results of this first attempt at measuring oxygen flux into the front surface of the cornea was a value of 0.21 nmol O_2 cm^{-2} h^{-1} [= 4.8 mm^3 O_2 (STP) cm^{-2} h^{-1}].

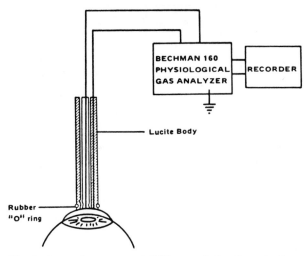

Fig. 2. A membrane-covered POS pressed directly onto the cornea in the method of Hill and Fatt [6]

Although the apparatus shown in Fig. 1 was relatively simple, it was not easy to use. The scleral contact lens had to fit tightly to the cornea to prevent loss of solution from the reservoir. Such tight fit is uncomfortable for the subject. The sensor assembly was heavy enough to pull the scleral lens from the eye unless the subject was supine and the sensor weight was counterbalanced by springs or weights and pulleys. Another serious problem was the long time, about 30 min, that the subject had to wear the lens and sensor assembly to get an accurately measurable decline in oxygen tension of the reservoir. Clearly the method was not suitable for studies on large numbers of subjects or on a population of clinic patients.

Hill and Fatt [6] achieved a solution to the problems posed by the scleral lens and sensor of Fig. 1 by limiting the oxygen reservoir to the polyethylene membrane covering the sensor face. The membrane-covered POS is pressed against the cornea as shown in Fig. 2. When the data are replotted, with recorder readings normalized to the reading at time zero, on a logarithmic scale and time on the arithmetic scale the graph of Fig. 3 is obtained. The flux J_{O_2}, into the cornea is related to the clope, $d\,(\log p_{O_2})/dt$, of the line in Fig. 3 by the equation [3],

$$J_{O_2} = 2.3\, p_{O_2}\, S_m\, z_m\, d\,(\log p_{O_2})/dt, \tag{1}$$

where $S_m\, z_m$ is the product of oxygen solubility and membrane thicknesses, and p_{O_2} is the oxygen tension at which the flux is to be calculated.

There are at least two important advantages of the method of flux measurement that uses only the sensor pressed onto the cornea. The scleral lens is not needed, so the procedure can be carried out on any human or lower animal without first fitting a scleral contact lens. Second, the sensor is held on the eye for only 30–60 s. Problems still remain, however. Most human subjects and lower animals cannot tolerate the sensor pressed against the cornea for even 30 s unless a topical anesthetic is applied first.

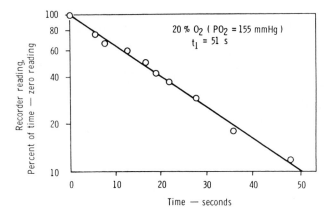

Fig. 3. Semilogarithmic plot of data obtained from a membrane-covered POS when pressed onto the cornea

Unfortunately the topical anesthetic often affects the metabolic rate of the corneal tissue, thereby confusing the data obtained by this procedure. Also, the sensor can, in the hands of an inexperienced investigator, abrade the cornea and create sites for development of infection.

The most serious difficulty with the procedure that uses a sensor pressed directly onto the cornea lies in the analysis of the data. The analysis that leads to Eq. (1) is only an approximation to the real system. As soon as the sensor is pressed to the cornea, the oxygen tension in both sensor membrane and cornea begins to decrease. The oxygen flux measured in this way is not for a constant oxygen tension at the corneal surface, or even for a small decline in this tension, but rather for a system that is rapidly changing in oxygen tension. Farris et al. [1] have suggested that the slope of the recorded oxygen tension near zero time is representative of the flux into the air-exposed cornea. Unfortunately the record is usually noisy near time zero. Hill and Fatt [7] have suggested that a standard procedure be adopted in which the slope of the record between 19 and 13 kPa (140 and 100 mm Hg) be taken as a measure of oxygen flux into the cornea.

Measurements of oxygen flux into the cornea made by pressing a membrane-covered POS onto the cornea are rapid and convenient but at best they can only be used for comparative purposes. Hill has made extensive use of this procedure to examine the effects of contact lens wear on oxygen flux into the cornea seconds after the lens has been removed. He claims that this flux is representative of the steady-state oxygen tension under the lens while it is worn [7].

In a preliminary report Peterson and Fatt [9] showed that by pressing the membrane-covered sensor onto a soft contact lens on the rabbit eye they could obtain an indication that there was an oxygen flux through the contact lens to the cornea. Zantos and Holden [13] applied this procedure to human eyes, but they too did not use the data to calculate oxygen flux into the cornea.

The quantitative interpretation of the data obtained from a POS pressed onto a soft contact lens on the human eye was given by Fatt [3]. He showed that by proper

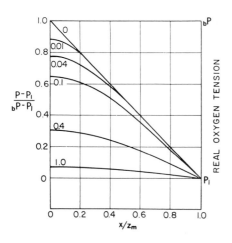

Fig. 4. Dimensionless oxygen tension, $(p-p_1)/(_ap-p_1)$, and real oxygen tension, p, as a function of position in a soft contact lens worn on the eye. p_1 is the oxygen tension at the lens-cornea interface, $_bp$ is atmospheric oxygen tension. z_m is the thickness of the lens and x is the thickness variable. *Numbers on the curves* are the values of the dimensionless group $D_m t/z_m^2$ (see text)

choice of contact lens material and lens thickness, a relatively simple solution of the transient diffusion equation could be applied to the sensor data. The mathematical solution he used assumed that during the first 30 s after placing the POS on the lens, the decrease in oxygen tension observed by the sensor at the front surface of the lens had not yet reached the back surface. Furthermore, the slope of the oxygen tension versus position curve at the cornea–lens interface is assumed to remain close to the steady state initial value, as shown in Fig. 4, where only dimensionless times of less than 0.1 are used in the solution. Dimensionless time is the group $D_m t/z_m^2$ appearing on each curve in Fig. 4. It allows the graph to be used for any contact lens when the diffusion coefficient D_m and the thickness z_m are known.

The oxygen sensor pressed against the contact lens gives an oxygen tension that is decreasing with time. The record shows the oxygen tension at $x/z_m = 0$, the vertical axis of Fig. 4, as a function of time. The data from the record are plotted on semilogarithmic paper with the ratio of recorded oxygen tension to starting oxygen tension on the vertical logarithmic scale and time on the horizontal arithmetic scale as shown for two sample lens in Fig. 5. From the straight line on this plot we get,

$$d\,(p_{x=0})/dt = 2.3\,\Delta\,(\log p_{x=0})/\Delta t. \tag{2}$$

Then the oxygen flux, J_{O_2}, is calculated from,

$$J_{O_2} = -\frac{S_m \times z_m \left[\dfrac{d\,(p_{x=0})}{dt}\right]}{2\sum\limits_{n=0}^{n=\infty} \exp\left[-\dfrac{D_m\,(2n+1)^2\,\pi^2\,t}{4\,z_m^2}\right]}, \tag{3}$$

where D_m is the diffusion coefficient for oxygen in the contact lens material, S_m is the solubility of oxygen in this material, and z_m is the lens thickness. An average value of thickness can be used [4] or if the lens thickness does not vary by more than 10% from center to edge the center thickness may be used for z_m. Typical values for D_m and S_m (or α) are 2×10^{-4} cm^2 s^{-1} and 1.8×10^{-8} mol O$_2$ cm^{-3} kPa^{-1} [or 4×10^{-2} cm^3 O$_2$ (STP) cm^{-3} atm^{-1}] respectively. Membrane thickness is in the range 0.15 to 0.50 mm.

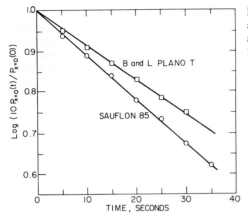

Fig. 5. Semilogarithmic plot of the data from a membrane-covered POS when pressed onto a soft contact lens worn on the eye of a human subject

When oxygen flux is calculated by means of Eq. (3), the POS membrane can be ignored because it contributes very little oxygen to the flux into the cornea. A comparison of $S_m z_m$ for the contact lens and a typical POS membrane of polyethylene shows that the contact lens holds fivefold more oxygen than the POS membrane.

The oxygen flux into the in vivo human cornea was found to be 0.18 nmol cm^{-2} h^{-1} (= 4.0 mm^3 cm^{-1} h^{-1}) for a very thin soft lens (0.10 mm) under which the oxygen tension would be expected to be about 1.5 kPa (= 11 mm Hg) and the cornea swollen about 2% over its normal thickness. A thicker soft lens (0.50 mm) of the same material gave a flux of 0.09 nmol cm^{-2} h^{-1} (= 2 mm^3 cm^{-2} h^{-1}). The oxygen tension under this lens is about 0.3 kPa (2 mm Hg) and the cornea has swollen 8% above its normal thickness.

These data demonstrate that the POS can measure oxygen flux into a cornea under a soft contact lens. The thicker lens leads to less flux and this in turn leads to corneal swelling. The direct cause of the swelling may be the osmotic attraction of water into the cornea because of the production of excess lactate by the hypoxic epithelium.

Rasson [10] has extended the above treatment to soft contact lens materials of any material and thickness. In Rasson's mathematical treatment, the oxygen tension at the contact lens-cornea interface need not remain constant after the sensor is applied and

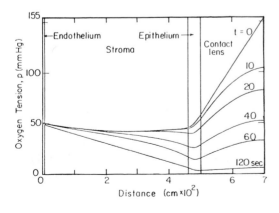

Fig. 6. Oxygen tension profiles in the cornea and soft contact lens before the sensor is pressed onto the lens ($t=0$) and for various times after the sensor is on the lens. For the lens acting as a membrane $D_m \times S_m/z_m$ = 3.3 $\times$ 10^{-12} cm s^{-1} mol cm^{-3} kPa^{-1} or $D_m \times \alpha/z_m$ = 1.0 $\times$ 10^{-8} cm s^{-1} cm^3 O_2 cm^{-3} mm Hg^{-1} [10]

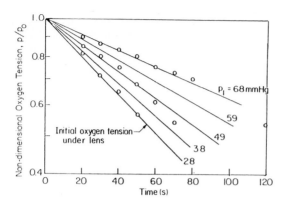

Fig. 7. Oxygen tension data from a sensor pressed onto a soft contact lens worn on the eye (logarithmic vertical scale) as a function of time. Each *line* represents initial oxygen tension at the cornea-contact lens interface (see Fig. 6)

the record can be analyzed for all times. Figure 6 (from Rasson) shows the oxygen tension at the cornea-contact lens interface falling from 6.7 to 0.67 kPa (50 to 5 mm Hg) during the 120 s the sensor is on the contact lens. The semilogarithmic plot of normalized oxygen tension versus time is shown in Fig. 7. From the slope of this line one can use Fig. 8 to read the initial oxygen tension under the lens. From this tension and the known oxygen tension in air (20.7 kPa or 155 mm Hg), one can easily calculate the oxygen flux into the cornea while it is covered by a contact lens.

Rasson's analysis of the data obtained from a POS pressed onto a soft contact lens on a cornea may permit the clinician to ascertain quickly if the lens allows delivery of the minimum amount of oxygen needed by the cornea.

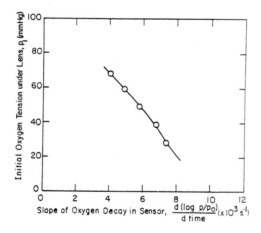

Fig. 8. Slope of the lines in Fig. 7 as a function of the initial oxygen tension at the cornea-contact lens interface (see Fig. 6)

References

1. Farris RL, Takahashi GH, Donn A (1965) Oxygen flux across the in vivo rabbit cornea. Arch Ophthal 74:679–682
2. Fatt I (1964) An ultramicro oxygen electrode. J Appl Physiol 19:326–329
3. Fatt I (1978) Measurement of oxygen flux into the cornea by pressing a sensor onto a soft contact lens on the eye. Am J Optom Physiol Optics 55:294–301

4. Fatt I (1979) The definition of thickness for a lens. Am J Optom Physiol Optics 56:324–337
5. Hill RM, Fatt I (1963) Oxygen uptake from a reservoir of limited volume by the human cornea in vivo. Science 142:1295–1297
6. Hill RM, Fatt I (1963) How dependent is the cornea on the atmosphere? J Am Optom Assoc 35:873–875
7. Hill RM, Fatt I (1964) Oxygen deprivation of the cornea by contact lenses and lid closure. Am J Optom 41:678–687
8. Maurice DM (1957) The structure and transparency of the cornea. J Physiol 136:263–286
9. Peterson JF, Fatt I (1973) Oxygen flow through a soft contact lens on a living eye. Am J Optom 50:91–93
10. Rasson JE (1979) Transient diffusion in multilayer biological tissue. PhD thesis, Univ California, Berkeley, California 1979. Available on microfilm in Europe from University Microfilm International, 18 Bedford Row, London WC1R 4EJ, England
11. Smelser GK (1952) Relation of factors involved in maintenance of optical properties of cornea to contact lens wear. Arch Ophthal 47:328–343
12. Smelser GK, Ozanics V (1953) Structural changes in corneas of guinea-pigs after wearing contact lenses. Arch Ophthal 49:335–340
13. Zantos SG, Holden BA (1977) Research techniques and materials for continuous wear of contact lenses. Aust J Optom 60:86–95

Part III Field Applications of Polarographic Oxygen Sensors

Chapter III.1 In Situ Measurement of Oxygen Profiles in Lakes: Microstratifications, Oscillations, and the Limits of Comparison with Chemical Methods

E. Gnaiger[1]

1 Introduction

Molecular oxygen is the substance most extensively monitored in scientific and routine investigations of aquatic ecosystems. Oxygen distributions in stratified lakes and departures from atmospheric equilibrium concentrations provide more information on lake characteristics and for water management than any other chemical parameter [8]. Accordingly, the most common application of POS in field ecology is the measurement of dissolved oxygen concentrations along vertical or horizontal transects of aquatic systems. The situation is reflected by the abundance of commercially available in situ oxygen probes (e.g., Delta Scientific, Electronic Instruments Limited, International Biophysics Corporation, Kahlsico International Corporation, Orbisphere Laboratories, Wissenschaftlich-Technische Werkstätten, Yellow Springs Instruments). These sensors incorporate macrocathodes and are commonly equipped with a simple battery-operated stirring device, and most instructions leave no doubt about the ease of the method.

The problem of measuring oxygen concentration in large bodies of water may appear simple, but, in fact, according to personal communications, difficulties with field applications of POS are experienced by many investigators although seldom mentioned in the literature ([2, 10, 21, 30], Chap. I.9). This justifies the discussion of some practical aspects which may be fundamental for reliable in situ oxygen measurements.

This contribution contains some guidelines for choosing between the chemical Winkler titration method involving discrete sampling and in situ measurement with polarographic oxygen sensors (POS). The unique advantages of the latter are demonstrated by studies of the oxygen dynamics in Kalbelesee, a shallow, productive subalpine lake. The high resolution of oxygen microstratification revealed the phenomenon of "spring stratification" which is probably widespread in this type of lake during the period after icebreak. Measurement of the small-scale patterns of oxygen in space are complemented by continuous monitoring of dissolved oxygen during daily cycles. The use of POS provides the most economic method for the study of physicochemical

1 Institut für Zoologie, Abteilung Zoophysiologie, Universität Innsbruck, Peter-Mayr-Str. 1A, A–6020 Innsbruck, Austria

Polarographic Oxygen Sensors (ed. by Gnaiger/Forstner)

processes in aquatic systems, especially in combination with the automatic measurement of additional ecological parameters.

2 Instrumentation

2.1 In Situ Oxygen Measurement

A YSI 5700 dissolved oxygen probe was used with a 15 m electrode cable. The POS was mounted in a modified stirring chamber which was originally designed for large volume respirometers (Chaps. II.1, III.3). A perspex chamber (Fig. 1) with a rotating magnet acts as a centrifugal pump. The magnet is driven by a 6 V dc motor running on 4 V which represents a compromise between the danger of decoupling and high stirring and pumping rates. Water is drawn through the central inlet in the bottom and discharged through the peripheral slots in the upper part of the chamber. Thus stirring does not disturb the natural oxygen gradient at the depth of submersion, and, additionally, turbulences are prevented by the 20 × 30 cm mounting plate. Within the chamber the turbulence not only provides an optimum current across the membrane, but is intense enough to tear off gas bubbles that might, if they collected in the chamber or at the membrane, distort the sensitivity and response time of the POS. This function of the stirring device is generally important when the probe is immersed in water after contact with air, and is essential when measurements are made in supersaturated surface waters where spontaneous bubble formation might occur. The supraoptimally high current in the stirring chamber serves another important purpose in continuous in situ measurements by keeping the growth of bacteria and algae on the membrane within limits (Chaps. I.9, II.9). If only gently stirred, membranes start "fouling" and have to be replaced. In fast rivers and in aeration tanks of sewage works the water currents may be high enough to render such stirring unnecessary [20].

The YSI 5700 and many other POS designed for in situ measurements are pressure-compensated. This is important if the electrolyte is not entirely free of gas bubbles. A pressure-transducing membrane on the side of the sensor shaft offsets changes in hydrostatic pressure on the oxygen transducing membrane. However, the hydrodynamic pressure generated by natural currents or in the stirring chamber is not compensated. This imposes problems, especially if a centrifugal pump is used as the stirring device, and therefore POS without pressure compensation are preferable. The pressure-transducing membrane can also be replaced by a pressure-tight seal, in which case special care is required to exclude all air bubbles when applying the membrane (Chap. I.1). This is recommended in any case and a small magnifying lens should be used for inspecting the electrolyte and membrane before taking measurements in the field.

The signal of the POS was measured with a YSI model 51 or 54 dissolved oxygen meter or with a digital oxygen-temperature meter (Mountain and See Instruments) (Chap. I.10). For continuous registration of oxygen and other parameters, a six-channel strip chart recorder (Goerz, Miniscript 6D) was used. A 12 V rechargeable battery served as the power supply. At the lake the actual atmospheric pressure was measured with an altitude meter as necessary for calibrating the POS (Chap. I.2; App. A).

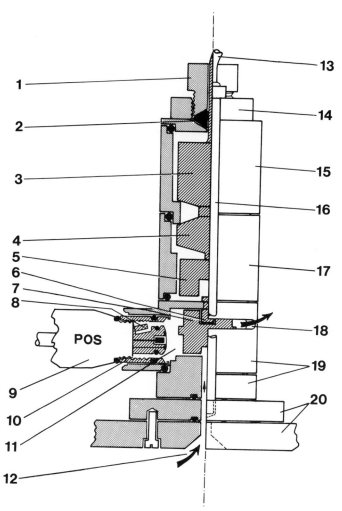

Fig. 1. The in situ stirring chamber for the POS; combined cross-section (*left*) and external view (*right*). *1* PVC screw, *2* conical rubber seal, *3* 6 V dc motor, *4* reduction gear, *5* drive magnet, *6* bearing pin, *7* magnetic follower, *8* thermistor, *9* POS, *10* stainless steel mounting sleeve, *11* stirring chamber, *12* inlet, *13* power supply cable for motor, *14* clamping plate, *15* perspex motor housing, *16* stainless steel mounting bolt, *17* perspex housing of drive magnet, *18* ejection slot, *19* perspex housing of stirring chamber, *20* lower mounting plate

2.2 Winkler Analysis

Water samples were collected with either a 2 dm^3 Ruttner sampler or a 5 dm^3, 0.5 m high Schindler sampler with built-in mercury thermometers calibrated at ± 0.05°C. Winkler reagents (App. D) were added immediately after filling 120 or 300 dm^3 Winkler brown glass bottles. These were completely submersed in cool lake water for storage. In a field laboratory the samples were titrated according to the amperometric

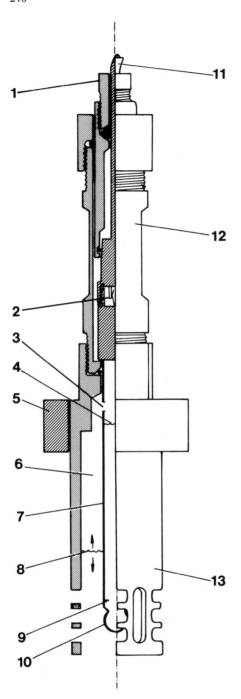

Fig. 2. Pressure compensation holder for the pH glass combination electrode; combined cross-section (*left*) and external view (*right*). *1* PVC screw, *2* Schott cable connection, *3* hole for electrolyte reservoir, *4* electrolyte level, *5* stabilizing weight, *6* pressure compensation air chamber, *7* pH electrode shaft, *8* water level rising with increasing depth, *9* diaphragm, *10* glass membrane, *11* electrode cable, *12* water- and pressure tight housing of cable connection, *13* electrode housing

method [12, 25] using the same voltmeter as used for the field pH-Eh measurements. Titrations were performed either with a 10 cm^3 automatic glass burette or with a 0.5 cm^3 microburette (Agla micrometer syringe, Burroughs Wellcome & Co., London) for near-zero oxygen concentrations.

2.3 In Situ Measurement of Temperature and pH

Temperature was measured with the thermistor of the YSI 5700 dissolved oxygen probe (YSI 400) and with separate thermistors (YSI 44006) linearized over 0° to 10°, 5° to 15°, or 10° to 20°C ranges (Chap. I.10).

While in situ determinations of redox potential and conductivity were performed only occasionally, pH was routinely measured to ± 0.01 pH units with a Radiometer specific ion meter (PHM 53) calibrated at 1 or 2 pH units full scale deflection. The Schott (N 62) pH combination glass electrode was mounted in a pressure compensation holder (Fig. 2). As the probe is lowered to greater depths, the water level within the holder rises and keeps the slight overpressure that is exerted by the electrolyte on the diaphragm constant. Two-point calibrations of the pH electrode at the temperature of the surface water rendered temperature corrections insignificant (< 0.01 pH units per °C). Special care is required in keeping the cable connections absolutely water tight. The mutual interference between the pH electrode and POS that was seen on a few occasions, was dependent on the distance between the two probes in the water. This might indicate defective insulation of the electrodes or cable connections.

3 Laboratory Tests

3.1 POS

Before commensing a field program, it is advisable to test the equipment under controlled laboratory conditions in order to gain some confidence in the method and to discover its limitations. Many commercial oxygen-temperature meters do not give an absolute accuracy of temperature measurements better than ± 0.5 to 1°C. This is definitely unsatisfactory for most applications and, due to a false account of the oxygen solubility and erroneous temperature compensation, will also adversely affect the oxygen measurements. With electronic linearization and digital display (Chap. I.10) improvement to ± 0.05°C was achieved by an additive correction term over the temperature range expected in the field (Table 1). Electronic temperature compensation of the oxygen solubility and membrane permeability should be tested for every combination of membrane, sensor, and electronic circuit (Chap. I.10). If these match, then the precision of oxygen determinations in the whole ecological temperature range is good (± 3 μmol dm^{-3} or 0.1 mg dm^{-3} at air saturation in pure water; Table 1), although conformity with absolute values may be less accurate (± 15 μmol dm^{-3} or 0.5 mg dm^{-3}) after simple calibration in air above the water surface (Chap. I.2).

The time course of the oxygen signal is determined by the water renewal time of the stirring chamber and by the time constant of the POS, as in respirometric systems

Table 1. Temperature compensation of a POS (YSI 5700) calibrated in air ($17°C$; $_b p = 95.2$ kPa) and linearity of the YSI thermistor 400 measured with a digital oxygen-temperature meter (Mountain and Sea Instruments). The measured values and the deviations (ΔT, Δc_s) from the expected values are listed. For the deviations, the means (calibration errors) ± standard deviations (linearization errors) are given below

Temperature [°C]			Oxygen [μmol dm^{-3}]		
θ	Signal	ΔT	c_s	Signal	Δc_s
2.50	2.27	0.23	400	388	12
4.05	3.80	0.25	384	–	–
4.21	3.96	0.25	382	372	10
6.26	6.03	0.23	346	334	12
9.94	9.77	0.17	331	319	12
12.08	11.89	0.19	315	303	12
12.10	11.90	0.20	315	300	15
14.14	14.02	0.12	301	288	13
15.92	15.78	0.14	290	275	15
17.85	17.69	0.16	278	263	15
19.82	19.57	0.25	267	253	14
		0.20 ± 0.045			13.4 ± 2.0

(Chap. II.3). Since the response time is the first function of the sensor to be affected by aging and poisoning, the manufacturer's specifications should not be uncritically relied upon when applying a sensor for some period of time in the field. With the present system (Fig. 1) and with a sensor that has been periodically used for more than one year, the 95% response was 60 s for oxygen and 40 s for temperature. The steady state values, however, were reached only after 8 and 5 min respectively. An equilibration time of 120 s is sufficient for applications in the field.

3.2 The Winkler Method

Since Winkler dissolved oxygen analysis was chosen as a reference for the field measurements with POS, the accuracy of the chemical method was also tested under standardized conditions. After equilibrating thermostated tap water either with air or nitrogen for 1 h, samples were siphoned into Winkler bottles exactly as in the field. To reduce volatilization of iodine [25], a large fraction of the expected volume of thiosulfate was added directly to the previously acidified water sample. Amperometric (Fig. 3) and redox titrations showed similar results (Table 2) with a high precision that cannot be achieved by the common starch indication method [12]. Practically, the amperometric method is superior to others, since only a few data points are required (in series determinations only one in the region of the linear slope, Fig. 3), and a step by step approach to the actual titration end point is unnecessary [25]. The samples equilibrated with nitrogen were taken after flushing the Winkler bottles with N_2. The non-zero values are partly due to the oxygen dissolved in the Winkler reagents.

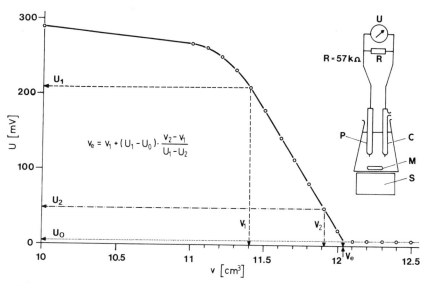

Fig. 3. Amperometric titration. Instrumental setup: C calomel reference electrode, M magnetic follower, P platinum electrode, R electrical resistance, S magnetic stirrer, U high impedance mV-meter. The titrated sample was 114.6 cm³ air-saturated water, 20.01°C; $_bp$ = 95.2 kPa (= 714 mm Hg); R = 57 kΩ; titrant 0.01 mol Na₂S₂O₃ dm⁻³. The titrant volume at the titration end point, v_e, is calculated according to the equation given in the figure. Since the background potential, U_0, and the slope, $x = (v_2 - v_1)/(U_1 - U_2)$, are sufficiently constant, they need only be determined once. In the following titrations only one titrant volume/mV reading, v and U, in the linear region is required, and the calculation of v_e simplifies to

$$v_e = v + (U - U_0) \times x.$$

The oxygen concentration in the sample, c_s, is then calculated as

$$c_s = \frac{v_e}{V_s} \times TF,$$

where V_s is the sample volume [cm³] corrected for the volume displaced by adding the Winkler reagents. For a 0.01 mol dm⁻³ thiosulfate solution the thiosulfate factor, TF, equals 2500 μmol O₂ dm⁻³ or 80 mg O₂ dm⁻³

Table 2. Comparison of the amperometric and redox titration methods in Winkler analysis. The means ± standard deviation of four replicates and the percentage of the theoretical air saturation value (at 20.0°C) are given. The sample volume was 116 cm³ minus 2 × 1 cm³ Winkler reagents (App. D). Other amperometric determinations in different years ranged from 99.3% to 100.0% of the expected air saturation value and the precision varied from date to date between ± 0.1% and ± 0.5%

	Amperometric		Redox	
	[μmol O₂ dm⁻³]	% air	[μmol O₂ dm⁻³]	% air
Air	261.6 ± 1.06	98.56	263.8 ± 0.75	99.36
Air	264.7 ± 0.31	99.06	256.3 ± 0.28	99.34
N₂ techn.	1.8 ± 0.09	0.67	1.9 ± 0.06	0.71
N₂ pure	1.5 ± 0.19	0.58	1.5 ± 0.16	0.57

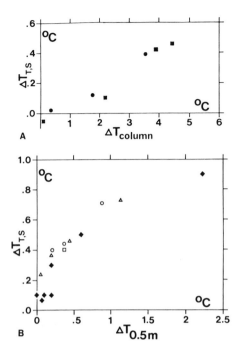

Fig. 4A,B. Error of temperature measurements in volume samplers, $\Delta T_{T,S}$, as a function of the vertical temperature gradient in the water column. **A** 2 dm³ Ruttner sampler; ΔT_{column} is the temperature difference in the water column from the surface to the sampling depth. **B** 5 dm³ Schindler sampler; $\Delta T_{0.5\ m}$ is the temperature difference in the half meter above the sampling depth. *Open symbols* are for vertical series during summer stratification, e.g., for 78-09-18 (Fig. 5). In this case the temperature in the water sampler was always higher than the in situ thermistor measurements serving as the reference for $\Delta T_{T,S}$. *Closed symbols* are for vertical series during inverse winter stratification. In this case the temperature in the water sampler was lower than in situ. The data points are the means of 3 to 10 measurements at every sampling depth (0.3 to 3.8 m) in Kalbelesee. Situations of anomalous temperature stratification (e.g., Fig. 6) were excluded

4 Comparison of Field Measurements in Situ and in Water Samples

4.1 Thermistor Versus Mercury Thermometer

A well-known problem with various volume samplers is the partial displacement of water from upper layers and its mixing with the water at the actual sampling depth. This explains the differences between temperature measurements in situ using the thermistor probe associated with the POS in the stirring chamber and measurements with mercury thermometers built into volume samplers (Fig. 4). The difference was a function of the temperature gradient above the sampling depth. During inverse stratification the temperature in the water sampler was too low, while during summer stratification it was too high with respect to in situ measurements. However, the simple relationship broke down during anomalous temperature stratification (Fig. 6). The two methods corresponded only in situations of temperature homogeneity throughout the water column. Otherwise differences of $0.5°C$ were fairly common. The speed of lowering the water sampler to the sampling depth had no influence on the result.

4.2 POS Versus Winkler Method

Whatever the method employed, practical experience is required, not only under favorable laboratory conditions, but also in the adverse environment often experienced by the field ecologist. An investigator with freezing, wet fingers, handles a pipette with reduced precision, and a heavy rain and cold wind may reduce his patience in following the standard procedures required with an electronic instrument. At low subzero temperatures Winkler bottles crack soon after sampling when the water freezes, and air calibration of the POS is unreliable due to ice formation on the membrane. On a hot, sunny day air calibrations may again be difficult since temperature stability of the POS is hard to achieve. Limitations in applying any field method arise in a variety of situations and have to be dealt with individually.

In my experience with different calibration chambers, the quickest and the most reliable procedure was one involving as little handling of the POS as possible. After setting the zero point in an approximately 5% solution of Na_2SO_3 (adding $CoCl_2$ as a catalyst), the sensor was thoroughly rinsed and mounted in the stirring chamber, which was then submersed in surface water for some time. For air calibration the chamber was quickly emptied, taking care to shake water droplets gently off the membrane. Without dismounting the POS, the stirring chamber was left in air and protected from the sun and wind until the temperature and p_{O_2} reading stabilized in the humid atmosphere of the chamber. The 100% air calibration point was then set according to standard equations relating oxygen concentration to temperature and the barometric pressure as corrected for water vapor saturation (App. A). The whole procedure was repeated to test for the expected precision of the calibration method ($\pm$ 1%, about $\pm$ 3 μmol dm^{-3}). Zero calibrations in the field agreed with previous calibrations in the laboratory, but this was frequently not the case for air calibrations.

The first measurement in the vertical profile was made after equilibration in water for 5 to 10 min, while the following readings at subsequent depths were taken at 2-min intervals. Most of the time during calibration can be efficiently used for preparing for other measurements, for sampling, etc. The 2-min exposure time required for the oxygen measurement is just sufficient for the registration of simultaneous measurements of temperature, pH and conductivity. Elucidation of the microstructure of vertical profiles was one of the aims in this investigation. For this purpose a stable 3.5 × 3.5 m float allowed positioning of the sensors in situ at precisely 0.05 m intervals in depth. Steps of 0.2 m depth were chosen as the standard for completing descending and ascending vertical series within a reasonable period of time. For plotting and data analysis all values were typed into a computer and interpolated at 0.05 m intervals with a third-degree polynomial spline fitting function. The following plots show the records on descent as dotted lines (one measurement per four dots), those on ascent as full lines.

Samples for Winkler determinations were collected by lowering the volume sampler until the middle of the column reached the sampling depth. After raising the sampler,

Table 3. Precision of the Winkler method in studies of oxygen depth profiles during homogeneous (78-09-18) and inhomogeneous (78-03-04) oxygen stratification in Kalbelesee. The means ± standard deviation of two replicates (5 dm³ Schindler sampler) are given

Depth [m]	Homogeneous [μmol O_2 dm⁻³]	Depth [m]	Inhomogeneous [μmol O_2 dm⁻³]
0.3	319.4 ± 1.75		
1.0	322.8 ± 1.28	1.3	212.5 ± 30.9
		1.6	179.4 ± 5.19
2.0	326.9 ± 0.97	2.1	80.0 ± 4.47
		2.6	39.4 ± 3.34
3.0	318.1 ± 0.75	3.1	18.75 ± 5.06
		3.6	1.53 ± 2.16

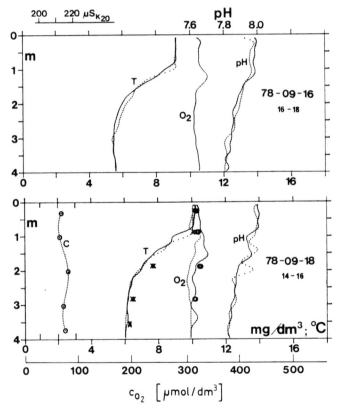

Fig. 5. Depth [m] profiles of environmental parameters during summer stratification in Kalbelesee. In situ measurements were taken at 0.2 m intervals moving downward (*dotted lines*) and upward (*full lines*). C conductivity [μS]; O_2 oxygen concentration given as c_{O_2} and ρ_{O_2}, *open circles* show the Winkler determinations; T temperature [°C]; *crosses* show the mercury thermometer readings in the Schindler sampler. The time of measurement is given below the dates

water was continuously siphoned into Winkler bottles until the sampler was half empty. If oxygen stratification was homogeneous, the reproducibility of the Winkler determinations was as good as in the laboratory (± 0.55%; cf. Tables 2 and 3). Even then, however, the in situ method revealed a characteristic microstructure of oxygen (Fig. 5). In the upper meter an interval of 2 h separated the POS (78-09-18) measurements made during descent and ascent. Samples for the Winkler analysis taken in between these registrations corresponded well with the mean values of the air-calibrated POS. However, the remarks on sampling errors in temperature measurements have to be kept in mind when comparing the remaining Winkler dissolved oxygen analyses and in situ POS measurements. Deviations between the two methods do not only originate from the oxygen analysis per se, but are largely due to the variability of oxygen concentration over short intervals of depth and time. Likewise, the high variability of Winkler determinations during clinograde oxygen stratification is due to the sampling method and microstratification, and not to the chemical analyses (Table 3). Therefore only the results for the mixed water layer (the top meter in Figs. 5 and 6) can be used for chemical in situ calibration of the POS.

Another apparent limitation to the comparability of measurements in situ and chemical analyses arises in the case of oxygen supersaturation (Fig. 6). This phenomenon is commonly known as the metalimnetic maximum in lakes with a positive heterograde oxygen distribution [8]. It occurs regularly in eutrophic waters [13, 15, 26], when high photosynthetic production leads to the accumulation of excessive amounts of oxygen, the concentrations of which may increase up to 300% to 500% air saturation. In depths of some meters, however, these supersaturation values do not

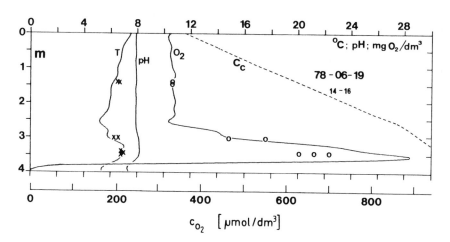

Fig. 6. Depth profiles of oxygen, temperature and pH during "spring stratification" in Kalbelesee. In the zone of homogeneously mixed melt water (78-06-19, 0–2.5 m) the O_2 measurements in situ by POS (*full line*) agreed with Winkler determinations (*circles*). Gassing-out reduced the O_2 content of the three samples from 2.5 m. The absolute saturation concentration, C_c, is shown by the *broken line* ($_bp$ = 81.9 kPa; see App. A, biogenic supersaturation). On the ordinates 0.0 m does not denote the water surface but the standard water mark. Since the actual water surface at 14–16 h on 19 June was 0.3 m above the water mark, C_c for 0.0 m standard depth applies to the actual hydrostatic pressure of a 0.3 m water column. For other explanations see Fig. 5

necessarily exceed the absolute saturation concentration which is a linear function of hydrostatic pressure and the volume fraction of other dissolved gases (App. A). When the hydrostatic pressure is released during hauling of the sample to the surface, gas bubbles form spontaneously and their oxygen content is lost in the Winkler analysis. Gassing out was actually observed in the samples taken from the depth of maximum oxygen concentration (Fig. 6; 3.5 m; 280% air saturation, about 87% absolute saturation), so that the large discrepancy between the two methods could be anticipated.

The limitations of the chemical method in the field resulting from short-range inhomogeneities of oxygen in natural waters and from formation of gas bubbles have to be taken into account in comparing the two methods and establishing a calibration point for the POS. When measurements were made on homogeneous bodies of water, agreement between the two methods was better than ± 5 μmol O_2 dm^{-3} (1.5%) in most cases, but occasional and unexplained deviations sometimes amounted to 5%. Such measurements with an air-calibrated POS are an order of magnitude less accurate than chemical determinations in unstratified waters. Part of the error of the air calibration value may be due to inadequate water vapor equilibration and to temperature differences between the membrane and compensating thermistor (see also Chap. II.10, gas/liquid factors). Proper chemical calibration, however, ensures high accuracy of the electrochemical in situ method which can then be combined with the unique advantage of the high spatio-temporal resolution of the POS.

5 Case Studies in a Shallow Mountain Lake (Kalbelesee)

The in situ method for measuring oxygen with POS was tested and refined in the course of a 4-year study of the oxygen dynamics in Kalbelesee (Hochtannberg, Vorarlberg, Austria). The calcareous catchment area of about 1.5 km^2 is predominantly characterized by alpine meadows. The lake is 300 m long and divided into a western and a smaller eastern basin. The results discussed here pertain to the western basin (Table 4).

Table 4. Morphometric parameters for the western basin of Kalbelesee (1650 m above sea level, latitude 47°16' N, longitude 10°08' E)

Maximum depth[a]	4.0 m
Mean depth	2.3 m
Relative depth	2.88%
Length	210 m
Maximum breadth	130 m
Area	17.5 × 10³ m²
Volume	35.3 × 10³ m³
Shore line	660 m
Development of shore line	1.41 m

[a] The water level may change from 3.7 to 4.8 m during the year. All values are for a standard depth of 4.0 m. For explanation of other expressions see [8]

5.1 Oxygen Dynamics During "Spring Stratification"

In winter up to 1 m of the water column above the bottom becomes anoxic, as in many shallow, snow-covered lakes and ponds [1, 3, 5, 7, 16, 18, 19, 23]. By the middle or end of March the winter cover reaches its maximum of 1.7 to 2.7 m. The oxygen

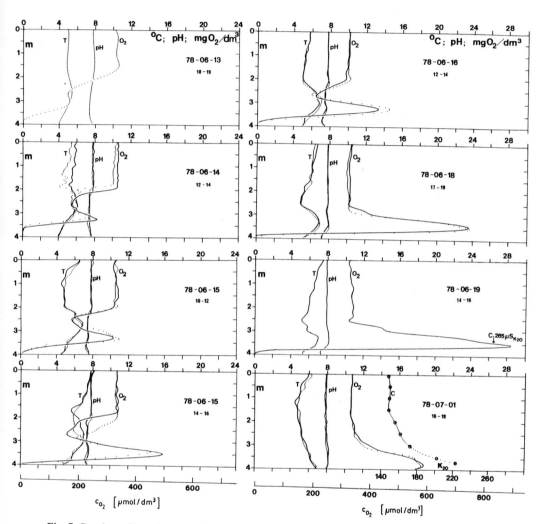

Fig. 7. Depth profiles of oxygen, temperature and pH during "spring stratification" after icebreak in Kalbelesee. Measurements were taken at depth intervals of 0.5 m (78-06-13), 0.1 m (78-06-14), and 0.2 m (all others) and plotted using a spline fit interpolation. Much detail of the actual depth profile remains elusive with 0.5 m depth intervals. However, a high confidence in the interpolations of 0.2 m intervals is justified on the basis of comparison with the profiles taking 0.1 m intervals. Note the close correspondence of downward (*dotted line*) and upward traces (*full line*), especially in the region of the O_2 peak and the first and second O_2 minimum. The fact that O_2, temperature and pH changed simultaneously in 2 to 3 m in the downward and upward profile in one case (78-06-15, 14–16 h) indicates a short term displacement of the whole water body in this region

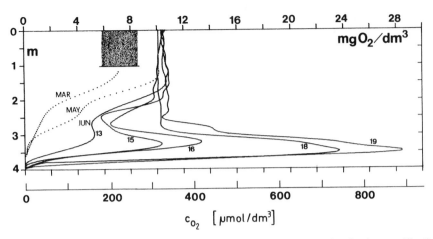

Fig. 8. Dynamics of dissolved oxygen before and after icebreak during "spring stratification" in Kalbelesee. In March and May (*dotted profiles*) the winter cover extended to about 1 m depth as indicated by the dark area. Icebreak occurred on the 11th of June. The *numbers* are the dates in June for the corresponding profiles (*full lines*, plots from Fig. 7)

conditions then become severe, occasionally causing winter fish kill [1]. With the onset of thaw, melt water passes the lake below the ice and forms a layer of water with low conductivity and increasing oxygen concentration. In 1978 icebreak occurred on the 11th of June. Two days later the surface temperature had risen above 4°C, but, due to chemical stratification, there was no immediate spring overturn in spite of homeothermy. After icebrak, a peak of oxygen concentration increased from day to day in 3.5 m depth (Fig. 7), where the conspicuous concentration of algae (mainly *Gymnodinium* sp.) set a sharp limit to the Secchi disk transparency. Isolated from atmospheric oxygen the stagnant, previously anoxic water became gradually supersaturated with oxygen. The photosynthetic source of oxygen and the mud, acting as an oxygen sink, maintained an extremely steep oxygen gradient in the free water column. The period of oxygen increase is summarized in Fig. 8. This illustrates the striking similarity of the development of the oxygen distribution in this shallow lake with that observed in the top millimeters of mud in the marine littoral (Chap. III.2). Surprisingly, however, oxygen concentrations decreased only negligibly during the night.

The same pattern of oxygen stratification evolved after icebreak in all years of observation. Even though this is usually restricted to 2 to 4 weeks, the occurrence of the "spring stratification" exerts a profound influence on the limnology of the lake. The high nutrient levels in the stagnant water layer in combination with high light intensities after icebreak support an efficient fixation of nutrients in algal biomass and secondary production. While the spring flood passes over the lake without mixing, the chemical stratification disappears gradually with the decreasing supply of melt water to the mixolimnion and with the photosynthetic reduction of bicarbonate concentration in the stagnant water layer. The chemical destabilization in the presence of anomalous temperature stratification (Fig. 7) leads to a delayed spring overturn. As a consequence of the previous spring stratification, the lake enters the summer stagnation period in a much more eutrophic state than in the case of an immediate spring over-

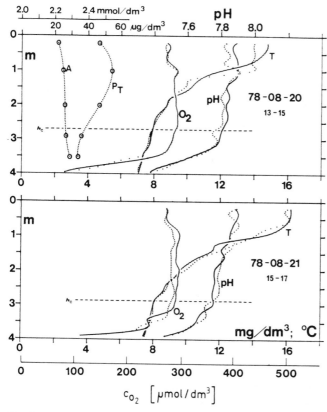

Fig. 9. Depth profiles of environmental parameters during summer stratification in Kalbelesee. *A* alkalinity (mmol dm⁻³), P_T total phosphorus (µg dm⁻³). The *horizontal line* indicates the critical depth, h_c, separating two ecotopes on the basis of different correlations between ecological parameters (cf. Fig. 11)

turn, which together with the flood of melt water would effect an efficient washout of nutrients. For similarly limited "meromictic behavior" of lakes compare [5, 22].

5.2 Continuous Measurement of Oxygen, Oscillations, and Correlations with pH and Temperature

Continuous recording of oxygen adds another dimension to the measurement of oxygen profiles in lakes. An example from the summer stagnation is presented here, showing the typical increase in oxygen at the level of the thermocline and a sharp decrease near the bottom, which is an exception for Kalbelesee during summer (Fig. 9). On calm, cloudless days the diurnal variation of surface temperature amounted to about 4°C with minimum temperatures at 0600 to 0700 h, and a maximum at 1800 h (Fig. 10). The temperature increase at 1 m depth was delayed until 1000 h and was connected with the onset of a slight breeze. At 1 m the oxygen concentration remained above the air-saturation value at surface temperature throughout the night. Even near

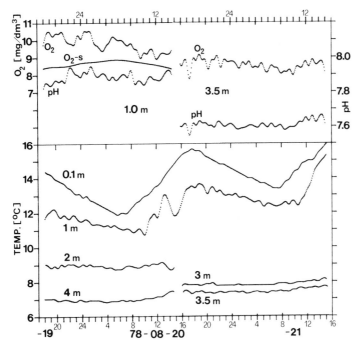

Fig. 10. Continuous records of oxygen, pH, and temperature at different depths in Kalbelesee. The corresponding vertical profiles are shown in Fig. 9. Two thermistors were continuously exposed at 0.1 and 1 m depth; two were changed from 1 and 4 m to 3 and 3.5 m respectively. During the first period (19/20 August) the POS and pH electrode were exposed at 1.0 m and then transferred to 3.5 m depth (20/21 August). O_2-s is the saturation concentration corresponding to equilibration with air at the surface temperature

the bottom, in the region of the steep oxygen gradient, the concentration fell only very slowly during the night (Fig. 10). This contrasts with the conspicuous diurnal oxygen minima characteristic for eutrophic ponds [13, 24, 26].

Short-term oscillations of the physicochemical parameters at fixed depths are superimposed on their daily fluctuations. The amplitude of these oscillations (Fig. 10) is related to the vertical gradient of the respective parameter at the depth of measurement (Fig. 9). Oxygen concentration, pH, and temperature were correlated both in the vertical profiles and in the continuous registration at 3.5 m (Fig. 11). This correlation at a depth where oscillations predominate over daily fluctuations supports the interpretation of these rapid changes in terms of internal seiches and excludes the possibility of simple noise in the registration. The amplitude of the seiches can be estimated by comparison of the amplitudes and vertical gradients of the parameters; values between 0.1 and 0.2 m were independently deduced from the oxygen, temperature, and pH measurements at the depth of 3.5 m (Figs. 9 and 10). For the interpretation of vertical profiles measured on descent and ascent, the rapid, oscillatory displacement of different bodies of water may be even more important than the relatively slow changes in dissolved oxygen within a body of water (for an extreme case of the former type see Fig. 7, 78-06-15, 14—16 h; note the simultaneous changes in oxygen, pH, and temperarature).

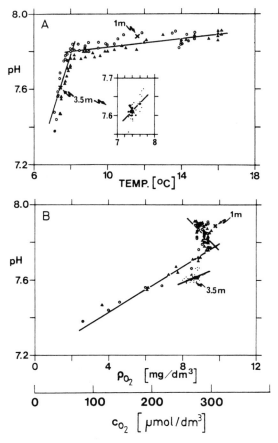

Fig. 11. Correlations of oxygen concentration, pH, and temperature in the vertical profiles shown in Fig. 9, and in the continuous registrations shown in Fig. 10. *Circles* and *triangles* are for the vertical profiles of August 20 and 21 respectively. The *crosses* show the mean values of the continuous registrations at 1 and 3.5 m depth as indicated by *arrows*. The inset shows the pH-temperature relationship in the continuous registration at 3.5 m with an enlarged temperature scale. The intercepts of the pH-temperature and pH-oxygen regressions are at 7.80 and 7.79 pH and at 8.08°C and 308 μmol O_2 dm^{-3} (9.87 mg O_2 dm^{-3}) respectively. Inserting these values in the vertical profiles (Fig. 9) gives the critical depth separating two ecotopes

Various ecological parameters in aquatic biotopes are correlated [6, 14, 29] and the study of these physicochemical relationships is substantially facilitated by the simultaneous in situ measurement of these parameters. For example, the temperature-pH relationship of a shaded and an unshaded sediment surface is not constant [6]. A comparable change in the temperature-pH and the oxygen-pH relationship was observed between the well-irradiated water column and the water column near the sediment in the lake (Figs. 9 and 11). The intercepts of the regression lines define a critical depth separating the water column into different ecotopes (Fig. 9). The point of inflection in the oxygen profile provides a definition of the critical depth in static terms (concentrations). However, changes in the correlations of ecological parameters indicate that different mechanisms prevail in each ecotope. This paves the way toward a better understanding of the dynamics and regulatory interactions in ecosystems and may have a bearing on the functional interpretation of algal distribution, critical light intensities, community respiration and other processes.

6 Discussion

In this study I have investigated some of the pitfalls and limitations of discrete sampling of the water column for the measurement of oxygen stratification by the Winkler method that result from high spatio-temporal resolution. In situ measurement by POS gives an undistorted picture of the oxygen dynamics in time and space. The need for continuous or semicontinuous profiling of physicochemical parameters was recognized long ago [17, 27, 28]. The determination of microstratifications of temperature and salinity revealed an unexpected complexity and heterogeneity of the water column in the oceans, and the application of in situ profilers equipped with additional physicochemical sensors, namely POS, stimulated a revolutionary development of various disciplines in oceanology.

With this in mind, it is hard to conceive why the in situ measurement with POS is not more generally and rigorously applied in limnological surveys and for routine monitoring of aquatic systems. The reasons may be economic. Most chemical and biological analyses require some kind of sampling procedure combined with determinations after transport to the laboratory. In a purely sampling-type study, relatively little time is saved by using POS, and the comparability of measurements in situ and in samples may possibly be restricted. Apart from methodological objections to discrete sampling, however, the economic rationale shifts if additional parameters such as depth, temperature, pH, conductivity or light intensity and transmission are transmitted by in situ sensors. In semicontinuous profiling simple registration is possible by reading from the electronic instruments. A continuous record of the water column requires automatic recording of the signals on an X-Y recorder, high currents on fast-reacting POS or mathematical correction for time delays [11, 27]. Continuous profiling is the method of choice for investigations in deep lakes and oceans, but imposes technical and economic difficulties in studies of shallow aquatic systems.

Apart from the problems associated with volume samplers, there are probably equally as many sources of error in chemical and electrochemical methods of dissolved oxygen analysis. Interference has to be expected for both methods, especially in natural low-oxygen environments and anoxic waters. In the presence of hydrogen sulfide a special modification of the Winkler method with the simultaneous measurement of H_2S should be applied [9]. An ordinary POS will also produce erroneous results, and should be replaced by an H_2S-insensitive sensor (Chap. I.6). Just as a short description of chemical oxygen analysis (App. D) cannot provide a general guideline for every kind of field situation, the instructions provided by manufacturers of POS merely constitute an aid to users, but are not a guarantee for precision in the frequently cited range of ± 1% (see Chap. I.2). A few unsatisfactory encounters with POS may elicit aversion to the apparently fallible electrochemical method and motivate a return to the reassuring safety of the traditional Winkler titration method. The chemically trained limnologist may find it easier to judge the reliability of his Winkler dissolved oxygen analysis than to estimate confidence limits of electrochemical in situ measurements. In fact, theoretical and practical guidelines for the Winkler analysis are exhaustively reviewed in many text books, while field applications of POS are treated with an apparent restriction to some theoretical aspects. Thus it seems that tradition and prejudice may be responsible for the continued preference shown by many laboratories for the chemical analysis of dissolved oxygen despite the various advantages of POS in routine measurements in situ.

Acknowledgments. This work was supported by the "Fonds zur Förderung der wissenschaftlichen Forschung in Österreich", Project Number 3917 and by the Forschungsförderungsbeitrag der Vorarlberger Landesregierung. I thank H. Forstner for placing the stirring chamber (Fig. 1) at my disposal.

References

1. Amann E, Gnaiger E (1979) Jahreszeitliche Abhängigkeit der Nahrungszusammensetzung von Regenbogenforellen (*Salmo gairdneri*) im Kalbelesee (Hochtannberg, Vorarlberg). Oesterr Fisch 32:32–39
2. Atwood DK, Kinard WF, Barcelona MJ, Johnson EC (1977) Comparison of polarographic electrode and Winkler titration determinations of dissolved oxygen in oceanographic samples. Deep-Sea Res 24:311–314
3. Barica J, Mathias JA (1979) Oxygen depletion and winterkill risk in small prairie lakes under extended ice cover. J Fish Res Board Can 36:980–986
4. Boyd CE, Romaire RP, Johnston E (1978) Prediciting early morning dissolved oxygen concentration in channel catfish ponds. Trans Am Fish Soc 107:484–492
5. Elgmork K (1959) Seasonal occurrence of *Cyclops strenuus strenuus.* Folia Limnol Scand 11: 1–196
6. Gnaiger E, Gluth G, Wieser W (1978) pH fluctuation in an intertidal beach in Bermuda. Limnol Oceanogr 23:851–857
7. Greenbank JT (1945) Limnological conditions in ice-covered lakes, especially as related to winter-kill of fish. Ecol Monogr 15:343–349
8. Hutchinson GE (1957) A treatise on limnology. I. Geography, physics, and chemistry. John Wiley and Sons, New York, pp 1015
9. Ingvorsen K, Jørgensen BB (1979) Combined measurement of oxygen and sulfide in water samples. Limnol Oceanogr 24:390–393
10. Kanwisher JW, Lawson KD, McCloskey LR (1974) An improved, self-contained polarographic dissolved oxygen probe. Limnol Oceanogr 19:700–704
11. Kersting K (1978) Automatic continuous oxygen- and temperature-profile measurements. Verh Int Ver Limnol 20:1216–1220
12. Knowles G, Lowden GF (1953) Methods for detecting the end-point in the titration of iodine with thiosulfate. Analyst 78:159–164
13. Kushland JA (1979) Temperature and oxygen in an Everglades alligator pond. Hydrobiologia 67:267–271
14. Lair N, Restituite F (1976) Projoect alpin O.C.D.E. pour la lutte contre l'eutrophisation. Lacs du massif central francais. II. Le Lac de Tazenat, interrelation entre paramètres. Ann Stn Biol Besse Chandesse 10:100–144
15. Lingeman R, Flik BJG, Ringelberg J (1975) Stability of the oxygen stratification in a eutrophic lake. Verh Int Ver Limnol 19:1193–1201

16. Mathias JA, Barica J (1980) Factors controlling oxygen depletion in ice-covered lakes. Can J Fish Aquat Sci 37:185–194

17. Mortimer CH (1974) Lake hydrodynamics. Mitt Int Ver Limnol 20:124–197

18. Nagell B, Brittain JE (1977) Winter anoxia. General feature of ponds in cold temperature regions. Int Rev Gesamten Hydrobiol 62:821–824

19. Pennak RW (1968) Field and experimental winter limnology of three Colorado mountain lakes. Ecology 49:505–520

20. Poole R, Morrow J (1977) Improved galvanic oxygen sensor for activated sludge. J Water Pollut Contrib Fed March 1977:422–428

21. Reynolds JF (1969) Comparison studies of Winkler vs. oxygen sensor. J Water Pollut Contrib Fed, Washington, Dec 1969, pp 2002–2009

22. Ruttner F (1955) Über die Entstehung meromiktischer Zustände in einem kaum drei Meter tiefen Quellsee. Mem Ist Ital Idrobiol 8:265–280

23. Schindler DW, Comita GW (1972) The dependence of primary production upon physical and chemical factors in a small, senescing lake, including the effects of complete winter oygen depletion. Arch Hydrobiol 69:413–451

24. Seki H, Takahashi M, Hara Y, Ichmura S (1980) Dynamics of dissolved oxygen during algal bloom in Lake Kasumigaura. Jpn Water Res 14:179–183

25. Talling JF (1973) The application of some electrochemical methods to the measurement of photosynthesis and respiration in fresh waters. Freshwater Biol 3:335–362

26. Uhlmann D (1966) Produktion und Atmung im hypertrophen Teich. Verh Int Ver Limnol 16:934–941

27. Landinham Van JW, Greene MW (1971) An in situ molecular oxygen profiler. A quantitative evaluation of performance. Mar Technol Soc J 4:11–23

28. Westerberg H (1972) A free falling polarographic oxygen sensor. Medd Havsfiskelab Lysekil 126:1–25

29. Webb KL, D'Elia CF (1980) Nutrient and oxygen redistribution during a spring neap tidal cycle in a temperature esturay. Science 207:983–985

30. Wilcock RJ, Stevenson CD, Roberts CA (1981) An interlaboratory study of dissolved oxygen in water. Water Res 15:321–325

Chapter III.2   In Situ Measurement of Oxygen Profiles of Sediments by Use of Oxygen Microelectrodes

N.P. Revsbech[1]

1 Introduction

Until recently, the measurement of oxygen was restricted to the water column (Chap. III.1), since no adequate methods were available for its measurement in sediments. The use of microsensors is essential for the resolution of the steep O_2 gradients often found along the sediment core. Measurements of oxygen in extracted pore water were made by several authors [2, 3, 8, 9, 12], but it is difficult to obtain reliable results this way. Teal and Kanwisher [22] attempted to measure the oxygen concentration of sediments in situ by insertion of a polarographic oxygen sensor (POS), but their sensor was large and was sensitive to stirring. Although unable to record an accurate oxygen profile, their measurements indicated that the sediment was anoxic below a few mm depth. In situ polarographic analysis of sediments was made by several authors [4, 11] using large platinum electrodes without membranes [15]. Because of their sensitivity to environmental conditions, e.g., interstitial water flow and the diffusion coefficient of oxygen in the sediment, it is difficult to calibrate these electrodes in terms of oxygen concentration. The readings are interpreted as oxygen availabilities, defined as the amount of oxygen reduced per unit time and per unit area of the active platinum surface. Standardized size and shape of the platinum electrode and a constant polarization voltage are necessary for reproducible results. This method is useful only in sandy sediments, and there are still difficulties in obtaining reliable results [5, 16]. Oxygen availability measurements made in coastal marine sediments generally indicate an oxygen penetration down to several cm depth in contrast to the results reported below.

The use of membrane-covered oxygen microelectrodes largely eliminates the experimental errors described above. Such electrodes, with tip diameters ranging from 2 to 10 μm, constructed according to Baumgärtl and Lübbers ([1]; Chap. I.4) have been succesfully used for the measurement of dissolved oxygen in marine sediments [17]. Oxygen microelectrodes offer three advantages: accurate measurement of oxygen tension in water without the need of stirring; excellent spatial resolution due to the very small oxygen-reducing platinum or gold surface [7]; extremely fast response to change in oxygen tension (Chap. I.4). The application of oxygen microelectrodes was previ-

1 Institute of Ecology and Genetics, University of Aarhus, Ny Munkegade, DK–8000 Aarhus C, Denmark

Polarographic Oxygen Sensors (ed. by Gnaiger/Forstner)
© Springer-Verlag Berlin Heidelberg 1983

ously restricted to a physiological context. The results presented in this section were obtained from marine sediments. The microelectrodes could, without doubt, also be used to measure oxygen in other media, such as soils, activated sludge, and bacterial colonies.

2 Microelectrode Design

The electrode design described by Baumgärtl and Lübbers has been modified for oxygen analysis in sediments. Experiments were conducted with electrodes having an active surface consisting of platinum or gold. The glass tubes used for the construction of the electrodes must be fairly thick to prevent water currents from causing vibration of the electrode tip. I used 5 mm (outer diameter) AR-glass tubes with a wall thickness of 0.9 mm (Glaswerk Wertheim, W. Germany). Risk of vibration was further reduced by limiting the length of the thin part of the electrode to 2–4 cm (Fig. 1). Vibration of the electrode tip was difficult to avoid when very long shafts were employed.

The best results are obtained with electrodes with tip diameters from 5 to 10 μm. With a tip diameter of less than 5 μm, the electrodes are generally too fragile and the current output too low. Electrodes with a tip diameter greater than approximately 10 μm must be covered by very thick membranes for accurate measurements in non-stirred media. For application in coarse-grained sediment the glass casing near the electrode tip should have a wall thickness of at least 1 μm. Electrodes with a reasonably thick glass casing can be pushed through sandy sediment, but any electrode will break if it hits a larger particle. The electrodes operate best if the membrane is protected by a recess of a few μm at the tip. The electrode tips were coated with polymer by immersion in polymer solution. Various polymers were tested and the best results were obtained with solutions of D.P.X., which is used for mounting microscope slides. Most problems during the use of oxygen microelectrodes are caused by poor membranes, and the procedure I use for the application of the membrane is therefore described in detail. The viscosity of the D.P.X. solution is adjusted with xylene so that the solution

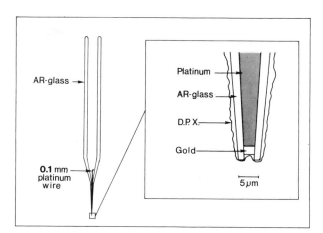

Fig. 1. Oxygen microelectrode used in sediment. The electrode tip is shown in detail on the *right*. The gold layer at the tip is not essential and also electrodes without gold were used

is just able to float down the wall of a test tube when this is put upside down. One cm^3 of the solution is then poured into a test tube (10 cm high, 1.3 cm inner diamter). Seven to eight mm of the microelectrode tip is immersed into the solution and fixed at this level with a slit stopper. The test tube with the microelectrode is then placed in a desiccator and vacuum is applied for 5 min, followed by 10 min at atmospheric pressure. The test tube is now placed upside down at an angle of 45°. When the solution slowly flows down along the wall of the test tube, the tip of the microelectrode leaves the solution before the shaft and this is very important. If the microelectrode were simply pulled out of the solution, large drops of polymer would aggregate near the tip and this might increase the diameter of the microelectrode considerably. The microelectrodes with the new membranes are baked at 60°C for 1 h after which the tips are immersed into distilled water. The performance of the new electrodes may be tested after hydration for 1 day. Baumgärtl and Lübbers coat the glass insulation of their microcoaxial needle sensors with a Ag/AgCl layer which forms a reference electrode (Chap. I.4). However, in environments where sulfide may occur, another type of reference electrode should be used (Chap. I.6).

3 Equipment for Field Use

The circuit for oxygen measurements with microelectrodes is very similar to those used for ordinary POS. The use of a highly sensitive ammeter and efficient electrical shielding is essential. All coaxial cables should have an additional graphite shielding underlying the copper shielding ("low noise" quality). If several electrodes are glued together (N.P. Revsbech, B.B. Jørgensen, and Y. Cohen, in preparation), cables with several individually shielded conductors contained within a common shield may be used.

The ammeter used by us in the field was a battery-operated Keithley 480 picoammeter. Since picoammeters do not operate satisfactorily at high air humidities, the meter was housed in a watertight Plexiglas box dried with a desiccant. A mercury cell connected to an adjustable voltage divider yielded the polarization voltage.

The reference electrode should have a large capacity. If free sulfide was present in the medium, a double junction reference electrode (Orion, model 90-02-00) was used to prevent contamination. Otherwise a normal calomel electrode (Radiometer, model K401) sufficed, but care should be taken to prevent clogging of the porous electrolyte bridge. Low-capacity reference electrodes caused erratic readings during field experiments when all electrical shielding must "float" at the potential of the reference electrode.

The holder for the p_{O_2} microelectrode is an extremely critical part of the equipment. It should provide electrical contact with the platinum wire of the microelectrode without adding electrical noise and must ensure absolute dryness of the inner surfaces of the electrode. Even if the inner surfaces of the electrode are only slightly humid, short circuits interfere with the readings and cause unstable zero currents. An electrode holder with an internal reservoir of desiccant (Fig. 2) solves all these problems. This electrode holder can be submersed into water and is adjustable to electrodes

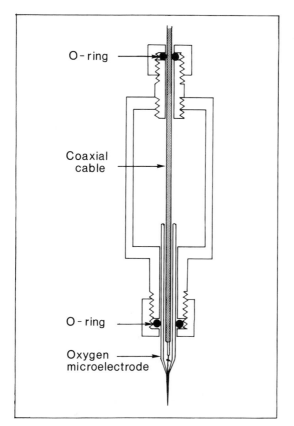

Fig. 2. Adjustable electrode holder made of Plexiglas. The shielding of the coaxial cable is removed from the lowest mm. The uninsulated wire thus protruding from the lower end of the cable is in electrical contact with the platinum wire of the microelectrode. The space inside the electrode holder can be filled with a desiccant, in which case a cotton plug should be used to protect the inner surfaces of the electrode from contamination

of different lengths. In less than 20 cm of water it is more convenient to use long electrodes to avoid submersion.

The microelectrodes were pushed into the sediment by means of a micromanipulator mounted on a sturdy rack, which was placed on a concrete slab during field measurements.

4 Microelectrode Operation

The electrodes were checked in a 2% NaCl solution for satisfactory current output at N_2 saturation and air saturation prior to the application of the membranes. Before each experiment the membrane-covered electrode should be tested in water from the sampling locality. The signal drift should be small and the stirring effect in water should be less than 5%. In a pure NaCl solution linear calibration curves are usually obtained, but in seawater there may be significant deviation from linearity. However, electrodes with thick membranes showed good linearity even in a hypersaline brine (tested up to 80‰ salinity). Measurements with such electrodes gave current ratios for air-saturated seater/oxygen saturated seawater (both currents corrected for zero current) of 4.6 to 4.8. The theoretical ratio is 4.77. Some electrodes show a transient increase in sensitivity to oxygen ("anoxic overshoot") after exposure to low oxygen tension [14]. Electrodes used in illuminated sediments often signalled oxygen saturations changing from 0% to 100% within a few seconds and the anoxic overshoot therefore represents a potential source of error. Large anoxic overshoots indicate poor membranes and only electrodes with negligible overshoots should be used. Electrodes with the characteristics described above normally had a current output (at $20°C$) of about 100 pA for air saturation and 3 to 12 pA for N_2 saturation (zero current). Electrodes with a current output as low as 20 pA for air saturation and 1 pA for N_2 saturation have also been successfully used.

Two different techniques were used for the measurement of oxygen in sediments. In the "fixed point" technique, a p_{O_2}-electrode was fixed in a certain depth of the sediment and the oxygen tension was continuously recorded. The sensitivity of the electrode must be constant during these experiments. Electrodes which show a drift in the sensitivity to oxygen can only be used with the "dip" technique, in which the electrode is pushed stepwise into the sediment and the oxygen tension recorded at 0.5 mm or 1.0 mm intervals. Only a few seconds are needed to record the oxygen tension at each depth and the electrode can be recalibrated between "dips", thus minimizing the error of drift. The reading in the deeper, anoxic part of the sediment was used as zero calibration. Winkler analysis of the water above the sediment ([20], App. D) yielded the second calibration point for a linear calibration curve.

The electrodes could be pushed through a sheet of latex rubber without damage so that a diffusion barrier of this kind can be introduced experimentally between the sediment and the overlying water [17]. The fact that the electrodes can be pushed through soft polymers may be advantageous in several other contexts.

5 Environmental Factors Interfering with the Readings

Mg^{2+} and Ca^{2+}, which are very aboundant in seawater, may both poison the electrodes (Chap. I.4). These ions and dissolved organic matter account for the observation that oxygen microelectrodes are often much less stable in seawater than in a pure solution of NaCl and that electrodes with thin membranes show non-linear calibration curves in seawater.

Hydrogen sulfide interferes with the measurement of oxygen in sediments [17]. A new electrode with an active surface consisting of platinum often showed a decline in oxygen sensitivity of up to 50% after the first exposure to sulfide, but once poisoned its sensitivity to sulfide was low. The current output of a poisoned electrode declined by only 1% to 2% when, to continuously aerated 0.2 mol dm^{-3} HEPES buffer (pH = 7.0), sulfide was added up to a concentration of 10^{-4} mol dm^{-3}. When sulfide was added up to 10^{-3} mol dm^{-3}, the current output was lowered by 15% to 20%. In addition to the immediate decrease in the sensitivity to oxygen caused by the addition of sulfide, an increased drift in the electrode current was observed. Since sulfide rarely reaches a concentration of 10^{-4} mol dm^{-3} in normal O_2-containing marine sediments, corrections for interference by sulfide are unnecessary. Only microelectrodes with an active surface of platinum have been checked for the quantitative interference of H_2S. Other authors have, however, found gold cathodes to be less affected by sulfide than platinum cathodes ([10], Chap. I.6). The results I have obtained with gold-microelectrodes indicate that the interference by sulfide was negligible, and the gold microelectrodes also showed less current drift than the platinum microelectrodes. All the oxygen microelectrodes made at our laboratory are now gold-plated at the tip.

Changes in pH below pH = 7 may exert an influence on the O_2 readings (Chap. I.4). This is irrelevant in the surface layers of marine sediments where pH values range from 7.2 to 10.2 ([6], N.P. Revsbech, unpublished results).

Like all POS, oxygen microelectrodes are sensitive to changes in temperature with temperature coefficients of 2.1% to 2.7% per °C when calculated as oxygen tension (Chap. I.4), corresponding to about 4.8% per °C when calculated as oxygen concentration. Simultaneous temperature measurements may therefore be necessary if there is a temperature gradient in the profile.

6 Examples of Data Obtained Using Oxygen Microelectrodes

The oxygen microelectrodes described above were used to measure the oxygen profiles in undisturbed, dark incubated sediment [18, 21]. In addition, oxygen profiles under varying conditions of illumination and water flow above the sediment surface were recorded [13, 17]. It is possible to calculate from the oxygen profiles the rates of oxygen consumption and the diffusion coefficient of oxygen in the sediment [17]. Furthermore, a new method for the determination of benthic primary production by oxygenic photosynthesis was created [19]. The new method is based on microelectrode measurements of the change in oxygen concentration within the photic layer of the sediment when the sediment is darkened for a few seconds. In contrast to previous methods, this new method also shows the vertical distribution of photosynthetic activity.

The effect of stirring of the overlying water on the oxygen profile in dark-incubated sediment is shown in Fig. 3. During "vigorous stirring" some sediment material went into suspension, a new steady state oxygen profile was quickly attained, and oxygen penetration increased by only 2 mm. After 12 min of reduced stirring, the oxygen profile differed slightly from that at the beginning probably due, to some extent, to changes of the sediment surface during the experiment. The boundary between sediment and water is a continuum, rendering an exact definition of "sediment surface" difficult. The zero depths at Figs. 3 and 4 were defined as the depth at which the oxygen tension, in the dark and during gentle stirring, was about 20% less than that of the overlying water. These zero depths coincided reasonably well with the surface observed directly through a microscope. Since the zero depth is more or less influenced by the degree of stirring (Fig. 3), a standardized gentle stirring should be used to obtain reproducible zero depths.

The dip technique was also used to record the oxygen profiles during a light-dark cycle as shown in Fig. 4. The light intensity used in the experiment (700 μEinst m^{-2} s^{-1}) was much higher than the in situ intensity at a water depth of 4 m. During the ex-

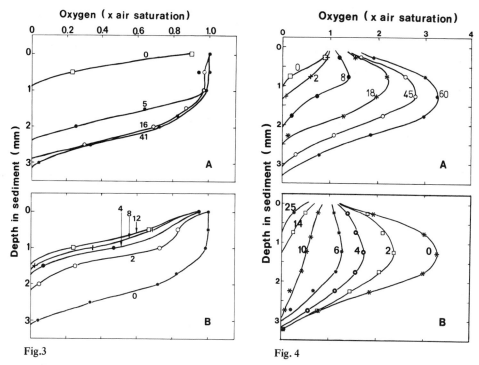

Fig.3   **Fig. 4**

Fig. 3A,B. The effect on the oxygen profiles of a dark incubated, sandy sediment produced by stirring of the overlying water. The time indicated for each curve refers to the start of the dip. The deepest part of the profile was recorded ca. 1 min later. Temperature, 10°C. **A** Oxygen profiles during vigorous stirring. The *numbers* indicate the duration in minutes of vigorous stirring. **B** Oxygen profiles during gentle stirring after vigorous stirring for 41 min. *Numbers* indicate the time in minutes after cessation of vigorous stirring

Fig. 4A,B. Oxygen profiles in a sandy sediment during a light-dark cycle. The light intensity was 700 μEinst m^{-2} s^{-1} and the temperature was 15°C. The time [min] indicated for each curve refers to the start of the dip. The deepest part of the profile was recorded ca. 1 min later. **A** Oxygen profiles in the light. Numbers indicate the time in minutes after the light was turned on. **B** Oxygen profiles in the dark following illumination for 60 min. *Numbers* indicate the time in minutes after the light was turned off

periment a gentle water current was maintained above the sediment. Illumination resulted in increased oxygen tensions in the surface layers of the sediment and the oxygen penetration increased from 1 to 3 mm. The steady-state dark profile was reestablished within 25 min of turning off the light. The profiles shown in Fig. 4 were recorded in the laboratory, but similar results have been obtained in the field [13].

The oxygen tension as a function of time can be continuously recorded by the "fixed point" technique. The results from such an experiment are shown in Fig. 5. The water depth at the sampling locality was only a few cm, and the experimental light intensity (1000 μEinst m^{-2} s^{-1}) was comparable to the maximum in situ intensity. In the uppermost millimeter, the oxygen tension rose immediately after commencement of illumination, and very high values were obtained within a few minutes. The resulting steep oxygen gradients in the surface layer led to an increased loss of oxygen to the

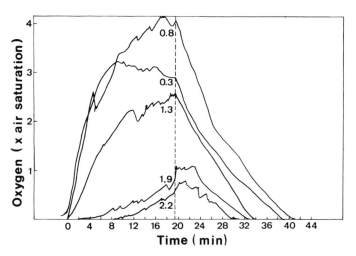

Fig. 5. Oxygen tension at various depths in the sediment during a light-dark cycle. *Numbers* indicate the depth in the sediment in mm. Light (1000 μEinst m^{-2} s^{-1}) was turned on at time 0 and turned off at 19.5 min. Temperature, 15°C

water and to the deeper sediment layers. The curves therefore only approximate straight lines during the first minute, after which a diminishing rate of increase was observed. At depths of 1.9 and 2.2 mm there was no initial rise in oxygen tension after illumination, and consequently no net oxygen production. After 2 and 8 min, respectively, the oxygen tension began to rise in these two layers, indicating that oxygen was supplied by diffusion from photosynthetically more active layers. The reverse sequence of events occurred during the initial dark period.

The application of oxygen microelectrodes in sediments reveals otherwise unobtainable data on the distribution and dynamics of oxygen in sediments. It would be valuable to follow, with similar spatial resolution, the changes of other chemical parameters. It is now possible to construct pH and sulfide microelectrodes. Their application in sediments (N.P. Revsbech, B.B. Jørgensen, and Y. Cohen, in preparation) will give us further information about sediment metabolism.

Acknowledgments. I thank H. Adler-Fenchel, E. Gnaiger, and Y. Cohen for reviewing this manuscript, and J.P. Lomholt, H. Baumgärtl, and D.W. Lübbers for help and advice during my work with microelectrodes. Financial support was provided by the Danish Natural Science Research Council, grant no. 511-1504.

References

1. Baumgärtl H, Lübbers DW (1973) Platinum needle electrodes for polarographic measurement of oxygen and hydrogen. In: Kessler M et al (eds) Oxygen supply. Urban & Schwarzenberg, München, pp 130–136
2. Brafield AE (1964) The oxygen content of interstitial water in sandy shores. J Anim Ecol 33: 97–116

3. Brafield AE (1965) Quelques facteurs affectant la teneur en oxygène des eaux interstitielles littorales. Vie Milieu 16:880–897
4. Fenchel T (1969) The ecology of marine microbenthos. 4. Structure and function of the benthic ecosystem, its chemical and physical factors and the meiofauna communities with special reference to the ciliated protozoa. Ophelia 6:1–182
5. Giere O (1973) Oxygen in the marine hygropsammal and the vertical microdistribution of oligochaetes. Mar Biol 21:180–189
6. Gnaiger E, Gluth G, Wieser W (1978) pH fluctuation in an intertidal beach in Bermuda. Limnol Oceanogr 23:851–857
7. Grunewald W (1973) How "local" is p_{O_2} measurement? In: Kessler M et al (eds) Oxygen supply. Urban & Schwarzenberg, München, pp 160–163
8. Hallberg RO (1968) 4. Some factors of significance in the formation of sedimentary metal sulfides. Stockholm Contrib Geol 15:39–66
9. Hargrave BT (1972) Aerobic decomposition of sediment and detritus as a function of particle surface area and organic content. Limnol Oceanogr 17:583–596
10. Hitchman ML (1978) Measurement of dissolved oxygen. Wiley, New York, p 115
11. Jansson BO (1967) The availability of oxygen for the interstitial fauna of sandy beaches. J Exp Mar Biol Ecol 1:123–143
12. Jørgensen BB (1977) The sulfur cycle of a coastal marine sediment (Limfjorden, Denmark). Limnol Oceanogr 22:814–832
13. Jørgensen BB, Revsbech NP, Blackburn TH, Cohen Y (1979) Diurnal cycle of oxygen and sulfide microgradients and microbial photosynthesis in a cyanobacterial mat sediment. Appl Environ Microbiol 38:46–58
14. Kessler M (1973) Problems with use of platinum cathodes for polarographic measurement of oxygen. In: Kessler M et al (eds) Oxygen supply. Urban & Schwarzenberg, München, pp 81–85
15. Lemon ER, Erickson AE (1952) The measurement of oxygen diffusion in soil with a platinum microelectrode. Proc Soil Sci Soc Am 16:160–163
16. McLachlan A (1978) A quantitative analysis of the meiofauna and the chemistry of the redox potential discontinuity zone in a sheltered sandy beach. Estuarine Coastal Mar Sci 7:275–290
17. Revsbech NP, Sørensen J, Blackburn TH, Lomholt JP (1980) Distribution of oxygen in marine sediments measured with microelectrodes. Limnol Oceanogr 25:403–411
18. Revsbech NP, Jørgensen BB, Blackburn TH (1980) Oxygen in the seabottom measured with a microelectrode. Science 207:1355–1356
19. Revsbech NP, Jørgensen BB, Brix O (1982) Primary production of microalgae in sediments measured by oxygen microprofile, $H^{14}CO_3^-$-fixation, and oxygen exchange methods. Limnol Oceanogr 26:717–730
20. Strickland JD, Parsons TR (1972) A practical handbook of seawater analysis. Bull Fish Res Board Can 167:310
21. Sørensen J, Jørgensen BB, Revsbech NP (1979) A comparison of oxygen, nitrate, and sulfate respiration in coastal marine sediments. Microb Ecol 5:105–115
22. Teal JM, Kanwisher J (1961) Gas exchange in a Georgia salt marsh. Limnol Oceanogr 6:388–399

Chapter III.3   Methods for Measuring Benthic Community Respiration Rates

P. Newrkla[1]

1 Introduction

Mortimer [29] pointed out the importance of sediments for nutrient recycling and for the oxygen balance of lakes, following which, oxygen uptake of benthic communities has been investigated by many authors. The overall aim was to use benthic community respiration as an indirect measure for trophic conditions of different lenitic and lotic water habitats. Hayes and MacAulay [19] related the oxygen consumption rates of sediments to the fish catch. Others [15, 35, 37] related the sediment oxygen uptake rates to the degree of eutrophication of the lakes investigated. Sediment respiration rates were also used to calculate the share of sediment in the mineralization of photosynthetically fixed carbon [17].

Generally, three methods are taken for obtaining oxygen uptake rates of benthic communities (see also Chaps. III.4, III.5).

1. The oxygen uptake of isolated portions of sediments is investigated under defined conditions, using the Warburg apparatus or its modifications [3, 4, 7, 8, 15, 16, 20, 41].

In the natural environment the sediment—water interface is characterized by steep gradients in physical and chemical parameters ([11, 22, 36, 37, 42]; Chap. III.2). Due to the release of reducing substances, such disturbed sediments show much higher oxygen uptake rates than natural stratified sediments. Dechev's [8] investigation of oxygen debts in sediment layers at different depths demonstrated the effect of the release of reducing substances on oxygen consumption (see also [31, 37]. Since this method results in an overestimation of actual oxygen uptake rates, it should be applied to non-stratified sediments only; e.g., waste water sludges.

2. Cores of more or less naturally stratified sediments were used for measuring benthic community respiration [9, 10, 13, 14, 18, 24, 26, 32, 40].

3. In situ measurements of benthic community respiration avoid the errors arising from pressure changes and the handling procedures connected with the former methods ([9, 21, 32, 34, 39]; Chap. III.5). However, the in situ method cannot be consider-

---

1 Institut für Zoologie (Limnologie) der Universität Wien, Althanstr. 14, A–1090 Wien, Austria

Polarographic Oxygen Sensors (ed. by Gnaiger/Forstner)
© Springer-Verlag Berlin Heidelberg 1983

ed as free of errors, since a disturbance of the uppermost sediment layer may occur, or the stirring speed within the jars may differ from natural water currents [18, 23].

The object of this investigation was to test the reliability of methods 2 and 3.

2 Methods

2.1 Sediments and Sampling Techniques

Methods were tested with sediments of the oligo- to mesotrophic lake Attersee (Upper Austria). In situ measurements and core samplings were undertaken just below the thermocline at a depth of 25 m, (water temperature max. 6°C, min. 4.2°C). Four different sediment strata could be distinguished vertically in the core. The uppermost layer, brownish in color and of light, fluffy consistence (1.5–2 cm thick, 75% water content), was followed by soft clay (4–5 cm thick, 65% water content). A black spotted clay layer of partially decomposed organic matter was followed by a clay stratum somewhat lighter in color (58% water content).

The most reliable way of obtaining sediment cores without disturbing sediment structures is to take them by hand [28, 40]. In deep lakes Jenkins or Kajak corers and their modifications are most often used [5, 13, 36]. Since disturbing the sediment along the walls of the core cannot be prevented, its effect should be minimized by using (a) tubes with large diameter; and (b) carefully sharpened edges which facilitate penetration into the sediment [1].

In this study samples were taken using a modified Kajak sampler (Fig. 1). Modifications were necessary, since sediment samples taken with the original model [25] were often lost. A heavy lead ring on the lower end of the frame and the four wings stabilize the corer in its vertical position and also provides more weight for penetration into the sediment. Penetration depth of the tube was always between 10 and 15 cm. The Plexiglas cores fitted into the frame have an outer diameter of 6 cm and an inner diameter of 5 cm. The upper end of the tube is reinforced by an aluminum ring, the lower end by a sharp-edged brass ring. A 2.4 mm wide slot along the side of the tube, sealed with silicon, permits convenient measurements of redox potential and oxygen concentration at different depths.

The cores were transported in styrofoam boxes to the laboratory within 1 h (4°–7°C), and were acclimated at 4°C in a water bath.

2.2 Eh Measurements

Redox potential measurements were carried out by inserting a platinum electrode (0.85 mm diameter) through the sealant 1.5 cm deep into the sediment. A Calomel electrode, placed on top of the tube, served as reference. The equilibration time was chosen so as to reduce drift of Eh to less than 10 mV/min (ca. 5 min). Calibration was carried out in a Fe^{2+}/Fe^{3+}-standard solution [43]: 3.3 mmol dm^{-3} $K_3[Fe(CN)_6]$ and 3.3 mmol dm^{-3} $K_4[Fe(CN)_6]$ in 100 mmol dm^{-3} KCl: 430 m V at 25°C.

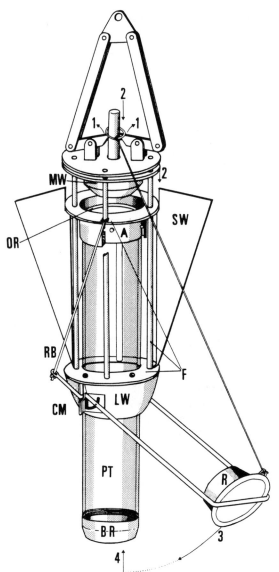

Fig. 1. Modified Kajak sampler: *A* aluminum ring for bayonet fitting, *BR* brass ring, *CM* closing mechanism, *F* frame, *MW* metal weight, *OR* O ring, *PT* Plexiglas tube, *R* rubber stopper, *RB* rubber band, *SW* stabilising wing, *numbers* indicate successive steps of closing mechanism

2.3 Oxygen Determinations Using Winkler Techniques

The micro Winkler method [12] with 5 cm^3 syringes was used for determination of the oxygen stratification in the water overlying the sediment. Replicate samples varied by ± 0.6 μmol O_2 dm^{-3}.

To determine the oxygen uptake of the benthic community, cores were sealed with a layer of paraffin oil (3 cm) and incubated in the dark for 3 h at constant temperature ($4° \pm 0.5°$C). The difference between initial and final oxygen concentrations was used for calculating repsiration rates. Prior to sampling, the water overlying the sediment was gently mixed to prevent errors due to a gradient in the unstirred cores.

In another set of experiments the water above the sediment was continuously stirred during incubation, using a slow-moving propeller inserted through the paraffin oil layer by means of a rod. In this way slight currents were produced without disturbing the sediment surface.

2.4 Laboratory Experiments Using Polarographic Oxygen Sensors (POS)

Mechanical disturbance and temperature changes due to handling and transport are associated with the core method [9, 16]. These problems may partly be avoided by incubating the cores in the lake at the sampling site [20, 30, 40]. If transport to the laboratory is inevitable samples should be kept at in situ temperatures, since cold-adapted, stenothermal biocoenoses of profundal sediments are damaged by high temperatures (see Fig. 7).

In order to investigate the influence of oxygen concentration on the oxygen uptake rate, respiration chambers made of the same tubes as the sampling cores were used (Fig. 2). To avoid disturbance of the sediment structure a rubber piston was used to push the sediment from the sampling core into the respiration tube. The closed respiration chamber was connected to a stirring chamber containing the POS (YSI 5750). Glass tubes were connected with natural rubber tubes.

At low oxygen concentrations the accuracy of the oxygen uptake measurements depends on the permeability of the material used for respiration and stirring chambers. Oxygen diffusion rates are high for 1.6 mm thick Tenite cores [33], but diffusion through 6 mm Plexiglas tubes is negligible [13].

For the respirometer described above, the rate of oxygen diffusion $-\dot{N}_{O_2}$ [μmol h^{-1}] (cf. Chap. II.3) into the respiration chamber was measured as

$$-\dot{N}_{O_2} = \exp\left(5.54536 - 0.03555 \times c_s\right), \tag{1}$$

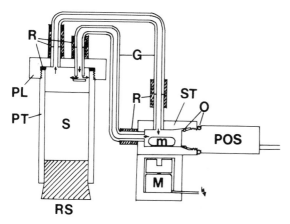

Fig. 2. Respiration chamber system for measuring oxygen uptake rates of undisturbed sediments at various oxygen concentrations, G glass tubes, M motor, m magnetic stirring bar, O O ring, Pl Plexiglas lid, PT Plexiglas tube, R rubber washers, RS rubber stopper, S sediment, *arrows* indicate flow direction

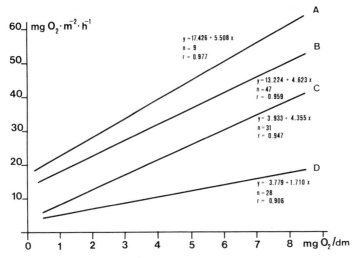

Fig. 3. Correlation between oxygen uptake rates and oxygen concentration for undisturbed sediments. *n* number of experiments, *r* confidence limits for 95%; temperature = 10°C. *A–D* indicate different sampling sites with different levels of oxygen uptake rates

where c_s is the oxygen concentration [μmol dm^{-3}]. Results obtained at oxygen concentrations below 80 μmol O_2 dm^{-3} were corrected accordingly. Decreasing oxygen concentrations during experiments exert considerable influence on oxygen uptake rates as shown in Fig. 3 (e.g., a reduction of oxygen tension from 250 to 220 μmol O_2 dm^{-3} causes a diminution of 9–15% in uptake rates. However, this effect, can be minimized by reducing the duration of the experiment. Otherwise, the error has to be corrected for [35].

Oxygen consumption of the water overlying the sediment was insignificant and was therefore neglected.

2.5 In Situ Measurements

The instrument used for the in situ measurements of benthic community respiration consists of a monitoring, a recording and an energy-supply unit (12 V battery) contained in waterproof aluminum boxes (Fig. 4). The fronts of the boxes are closed by 2-cm-thick Plexiglas plates, and a switch for the power supply can be operated from outside. The Plexiglas respiration chambers are fixed to a bar on each side of the instrument, and connected by rubber tubes to the stirring chambers into which the POS are inserted. The lids of the jars remain open during descent to the bottom, where they are closed by a messenger. The bell jars cover an area of 296 cm^2. They are painted black to avoid oxygen production by benthic algae. In situ measurements were carried out for periods of 3 to 5 h.

Obviously, in situ methods also involve errors due to a disturbance of the sediment surface along the walls of the jars, and to differences between the water currents in

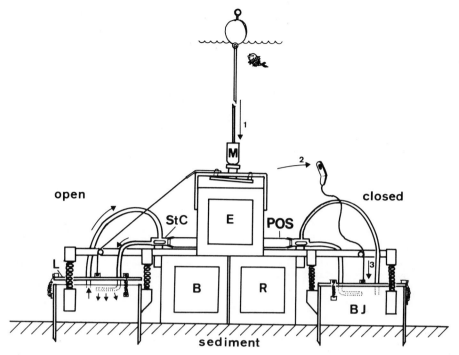

Fig. 4. Bell-jar for in situ measurements, *B* battery, *E* electronics, *BJ* bell jar, *M* messenger, *R* recording unit, *StC* stirring chamber, *arrows* indicate flow direction, *arrows with numbers* indicate sucessive steps in closing the bell jars

nature and within the bell jars. The first can be minimized by using bell jars of a large diameter, the latter was found to be of minor importance as long as the stirring speed does not disturb the sediment surface. Carey [6] found a definite effect of stirring speed on the oxygen uptake rate of the benthic community. This may be explained by the relatively low oxygen concentrations (down to 50% saturation) and the differences in sediment quality and structure of his samples as compared with Attersee sediments. Various investigations support the assumption that in situ measurements with the bell jar technique provide reasonable estimations of the natural oxygen exchange rates between sediments and water [2, 9, 13, 32, 38].

3 General Results

Oxygen uptake rates of sediments in unstirred cores change considerably with time of exposure (Fig. 5). During the first 3 h oxygen uptake rates were constant, between the 4th and 7th h a significant drop in oxygen uptake was observed, followed by a gradual decline during the subsequent hours. In an experiment which lasted for 72 h the development of an oxygen gradient and a shift of the discontinuity layer of redox potential was observed (Fig. 6, see also Chap. III.2).

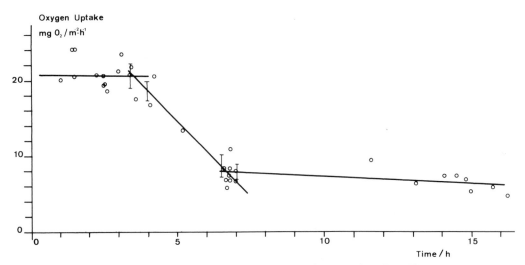

Fig. 5. Oxygen uptake rates of undisturbed sediment cores plotted against time. The water over-lying the sediment was not stirred. *Vertical lines* indicate 95% confidence limits

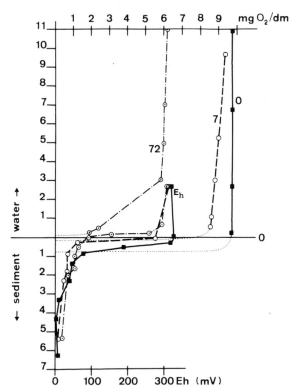

Fig. 6. Development of an oxygen gradient in the unstirred water overlying the sediment with time, and the expansion of anoxic conditions across the sediment water interface. *Black squares* indicate the initial situation, *circles* the situation after 7 h and *circles with dots* the situation after 72 h. The *dotted lines* indicate the course of the oxygen gradient, assuming the discontinuity as zero concentration (cf. Chap. III.2)

Table 1. Comparison between replicate measurements of respiration rates of undisturbed sediment cores. In one group the water overlying the sediment was stirred, the values for the other group were obtained in short-term experiments (3 h) without stirring. Three sites with differing oxygen uptake levels were investigated (1975-01-25). The means ± 95% confidence limits are given with the number of experiments in brackets

| | Respiration [μmol O$_2$ m^{-2} h^{-1}] | |
	Stirred	Unstirred
1.	128 ± 20 (8)	128 ± 13 (7)
2.	288 ± 38 (4)	306 ± 36 (4)
3.	709 ± 68 (4)	722 ± 43 (4)

Oxygen uptake rates measured in stirred cores were constant for at least 9 h. During the first three hours the oxygen uptake rates of stirred and unstirred cores were identical for three types of sediment differing in their intensity of oxygen uptake (Table 1).

The heigh of the sediment in the tube has no influence on the oxygen uptake rates if sediment columns are between 5 and 15 cm. This is in agreement with [10, 27].

A linear relationship between temperature and community respiration was confirmed for the temperature range of 4° to 19°C. Higher temperatures resulted in lower respiration rates and a higher variance (Fig. 7).

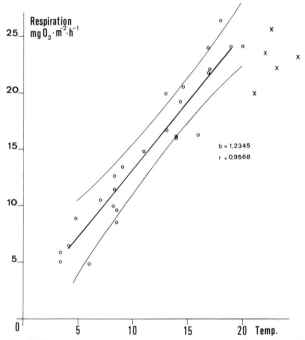

Fig. 7. Correlation between temperature and respiration rates of undisturbed sediments. The points indicated by *crosses* are not included in the regression since oxygen uptake rates decrease above 19°–20°C; 95% confidence limits are indicated by *thin lines*

Table 2. Comparison between replicate measurements of respiration rates [μmol O_2 m^{-2} h^{-1}] at 4°C of undisturbed sediments obtained by in situ and laboratory methods at different times of the year (see Table 1)

Date	In situ	Stirred	Unstirred
1975-10-05	416 ± 39 (5)	460 ± 18 (3)	–
1978-03-03	219 ± 24 (6)	232 ± 40 (3)	225 ± 19 (18)
1978-05-20	454 ± 56 (7)	459 ± 43 (4)	457 ± 22 (8)

For three types of sediments, differing in their level of oxygen uptake, the rates were linearly correlated with the oxygen concentration of the overlying water (Fig. 3).

Respiration measurements carried out with the in situ bell jar method, and on stirred and unstirred cores (short term) revealed similar results (Table 2).

4 Conclusions with Regard to Usage of POS

In order to give reliable values POS require stable water currents across their membrane which makes the use of a stirring chamber necessary. This increase in the total volume of the system is a disadvantage if, for any reason, only small volumes of water are used. Likewise, the inner surface is increased for bacterial growth. At present the application of chemicals inhibiting bacterial growth is considered to be inadvisable since high concentrations must be applied on account of the adsorption capacity of sediments and their effect on the other benthic organisms is largely unknown (Chap. II.9).

The great advantage of the continuous recordings obtained from POS measurements in situ is the possibility of evaluating each phase of the oxygen decline during an experiment.

Acknowledgments. My thanks are due to H. Forstner for placing at my disposal the preliminary construction of the bell jars, to A. Swoboda for abandoning his plans and data for further improvement of the instrument and to Dr. Machan for his help with the electronics. For help in statistics I thank H. Winkler. To F. Schiemer and M. Pamatmat I am grateful for constructive discussions, as well as to A. Gunatilaka for help in translation. This study is part of the international Man and Biosphere-Programme, Project Attersee.

References

1. Ambühl H, Bührer H (1975) Zur Technik der Entnahme ungestörter Großproben von Seesediment: ein verbessertes Bohrlot. Hydrologia 37 (1):175–186
2. Balzer W (1978) Untersuchungen über Abbau anorganischer Materie und Nährstoff-Freisetzung am Boden der Kieler-Bucht beim Übergang vom oxischen zum anoxischen Milieu. Rep Sonderforschungsbereich 95, Univ Kiel 36, pp 126
3. Bowman T, Delfino JJ (1980) Sediment oxygen demand techniiques: A review and comparison of laboratory and in situ systems. Water Res 14:491–499

4. Brewer WS, Abernathy AR, Paynter MIB (1977) Oxygen consumption by freshwater sediments. Water Res 11 (5):471
5. Brinkhurst RO, Chua RE, Batvosingh E (1969) Modifications in sampling procedures as applied to studies on the bacteria and tubificid oligochaetes inhabiting aquatic sediments. J Fish Res Can 26:2581–2593
6. Carey A (1962) An ecologic study of two benthic annual populations in Long Island Sound. Ph thesis, Yale Univ
7. Christensen JP, Packard TT (1977) Sediment metabolism from the northwest African upwelling system. Deep Sea Res 24 (4):331–345
8. Dechev G, Yordanov S, Matveeva E (1977) Oxygen consumption and oxygen debt in bottom sediments. Arch Hydrobiol Suppl 52 (1):63–71
9. Edberg N, Hofsten BV (1973) Oxygen uptake of bottom sediments studied in situ and in laboratory. Water Res 7:1285–1300
10. Edwards RW, Rolley HLJ (1965) Oxygen consumption of river muds. J Ecol 53:1–19
11. Fenchel T (1969) The ecology of marine microbenthos. IV Structure and function of the benthic ecosystem, its chemical and physical factors and the microfauna communities with special reference to the ciliated protozoca. Ophelia 6:1–182
12. Fox HM, Wingfiled CA (1938) A portable apparatus for the determination of oxygen dissolved in a small volume of water. J Exp Biol 15:437–445
13. Granéli W (1977) Measurements of sediment oxygen uptake in the laboratory using undisturbed sediment cores. Vatten 3:1–15
14. Granéli W (1978) Sediment oxygen uptake in South Swedish lakes. Oikos 30:7–16
15. Hanes BN, Irvine RL (1968) New technique for measuring oxygen uptake rates of benthal systems. J WPCF 40 (2/1):223–232
16. Hargrave BT (1972) Oxidation-reduction potentials oxygen concentration and oxygen uptake of profundal sediments in a eutrophic lake. Oikos 23:167–177
17. Hargrave BT (1973) Coupling carbon flow through some pelagic and benthic communities. J Fish Res Board Can 30:1317–1326
18. Hargrave BT (1976) Metabolism at the benthic boundary. In: McCave IN (ed) The benthic boundary layer. Plenum Press, New York London, p 323
19. Hayes FR, MacAulay MA (1959) Lake water and sediment. V Oxygen consumed in water over sediment cores. Limnol Oceanogr 4:291–298
20. Hunding C (1973) Diel variation in oxygen production and uptake in a microbenthic littoral community of a nutrient-poor lake. Oikos 24:352–360
21. James A (1974) The measurement of benthos respiration. Water Res 8:955–959
22. Jansson BO (1966) Microdistribution of factors and fauna in marine sandy beaches. Veröff Inst Meeresforsch Bremerhaven II:77–86
23. Jeppesen E (1979) An evaluation of methods used in measurement of the sediment oxygen uptake in lotic environments. In: Enell M, Gahnström G (eds) 7th Nordic symposium on sediments. Presentation of methods and analytical results. Aneboda, Sverige, Sweden, May 24–27
24. Johnson MG, Brinkhurst RO (1971) Benthic community metabolism in Bay of Quinte and Lake Ontario (Part III). J Fish Res Board Can 28:1715–1725
25. Kajak Z, Kacprzak R, Polkowski R (1965) Tubular bottom sampler. Ekol Pol SB 11:159–165
26. Knowles GR, Edwards RW, Briggs R (1962) Polarographic measurements of the rate of respiration of natural sediments. Limnol Oceanogr 7:481–484
27. Martin DC, Bella DA (1971) Effect of mixing an oxygen uptake rate of estuarine bottom deposits. J Water Pollut Contrib Fed 43:1865–1876
28. McDonnell AJ, Hall DS (1969) Effect of environmental factors on benthal oxygen uptake. J WPCF 41 (8/2):353–363
29. Mortimer CH (1941/42) The exchange of dissolved substances between mud and water in lakes. J Ecol 29:280–329; 30:147–201
30. Neame PA (1975) Oxygen uptake of sediments in Castle Lake, California. Verh Int Ver Limnol 19:792–799
31. Oertzen v JA, Sandberg H, Kahl G (1976) Sedimentaktivität als Eutrophierungsindikator für oligohaline Küstengewässer. Limnologica 10/2:427–435

32. Pamatmat MM (1979) Benthic community metabolism: A review and assessment of present status and outlook. In: Coull BC (ed) Ecology of marine benthos, vol VI. Belle W Baruch Libr Mar Sci, Columbia, SC, pp 89–111
33. Pamatmat MM, Banse K (1969) Oxygen consumption by the sea bed. II In situ measurements to a depth of 180 m. Limnol Oceanogr 14:250–259
34. Pamatmat MM, Fenton D (1968) An instrument for measuring subtidal benthic metabolism in situ. Limnol Oceanogr 13:537–540
35. Provini A (1975) Sediment respiration in six Italian lakes in different trophic conditions. Verh Int Ver Limnol 19:1313–1318
36. Rybak JI (1966) Method for analysing the micro stratifications in the near-bottom water layers. Bull Acad Pol Sci Cl II–XIV:321–325
37. Rybak JI (1969) Bottom sediments of the lakes of various trophic type. Ekol Pol A17:611–662
38. Smith KL, Burns KA, Teal JL (1972) In situ respiration of benthic communities in Castle Harbour, Bermuda. Mar Biol 12:196–199
39. Smith KL, Clifford CH, Eliason AM, Walden G, Rowe GT, Teal JM (1976) A free vehicle for measuring benthic community metabolism. Limnol Oceanogr 21:164–170
40. Teal JM (1957) Community metabolism in a temperate cold spring. Ecol Monogr 27:283–302
41. Viner AB (1975) The sediments of Lake George (Uganda). I. Redox potential, oxygen consumption and carbon dioxide output. Arch Hydrobiol 76:181–197
42. Wieser W (1975) The meiofauna as a tool in the study of habitat heterogeneity. Ecophysiological aspects. A review. Cah Biol Mar 16:647–670
43. Zobell CE (1946) Studies on redox potential or marine sediments. Bull Am Assoc Petrol Geol 30:477–513

Chapter III.4 In Situ Measurement of Community Metabolism in Littoral Marine Systems

A. Svoboda[1] and J. Ott[2]

1 Introduction

Following the pioneering work of the Odums [10, 11] ecologists concerned with energetics of multispecies systems became increasingly aware of the significance of "community respiration" or "community production". Spatial and temporal variations of
total community metabolism measured as oxygen consumption or production have
been related to the properties of forcing functions of the system (e.g., abiotic factors)
or to relative age and maturity of the system itself [4, 5, 8, 9, 12]. Only in very
simple systems can an additive estimation of community metabolism be attempted
from data gained by the study of its isolated constituents. In most cases the number of
components, their intricate connections and regulatory mechanisms render direct,
integrated measurement the only economic approach.

In situ measurements are performed on fully adapted organisms under ecological
conditions. To retain this advantage respirometer development has been directed
towards minimizing the influence of the experimental setup on the performance of
the system investigated.

In plankton studies measurements of community metabolism have been a standard
procedure for many years ("light-and-dark bottle" productivity measurements with
natural water samples belong to this category). Benthic systems, however, were studied
more intensively only after polarographic oxygen sensors (POS) became readily available.

We describe here a surface-independent underwater unit using POS for measuring
benthic community metabolism. Two examples representing a wide range of applications will be discussed: an essentially "two species" model system (symbiotic coelen-
terata) and a highly complex macrophyte system.

Only a few marine scientists have made in situ long-term continuous recordings of
oxygen production and consumption in corals and benthic macrophytes. Some systems are equipped with land-based measuring and recording units [1, 14, 19]. Others
used completely submersible surface-independent units of similar technical standard

1 Lehrstuhl für spezielle Zoologie, Ruhr-Universität Bochum, D–4630 Bochum, FRG
2 Institut für Zoologie der Universität Wien, A–1090 Wien, Austria

Polarographic Oxygen Sensors (ed. by Gnaiger/Forstner)
© Springer-Verlag Berlin Heidelberg 1983

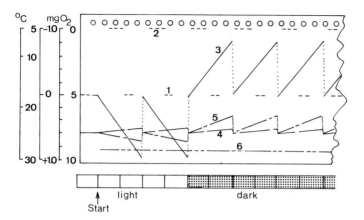

Fig. 1. Photosynthetic oxygen production (2 h) and respiration in the dark (3 h) in the sea anemone *Aiptasia diaphana* as an example for the recorder strip. *1* zero mark for the oxygen sensors,
2 zero mark for Δ O$_2$(O$_2$A $-$ O$_2$B), *3* Δ O$_2$, *4* O$_2$A, *5* O$_2$B, *6* temperature on sensor B. In this recording sensor B was connected with a respirometer containing an animal, sensor A with a control
chamber containing water only. The respirometers were flushed every hour. $\times$ 1/2

[2, 6, 20]. Safare-Crouzet (Nice, France) is now offering an automatic in situ oxygen
recorder commercially.

2 Design of Apparatus

2.1 Electronics, Recorder and Power Supply

The electronics unit consists of amplifiers and signal conditioning for the two tempera-
ture- and pressure-compensated POS (YSI-5739), as well as a power supply and control
for the two stirring chambers and the flushing pump. The following adjustments and
switches are provided for on the front panel: switches $-$ power on/off; stirrer on/off;
oxygen full scale range for both sensors 10 or 20 mg $\times$ dm^{-3}; output temperature from
sensor B or difference signal between POS A and B (Δp_{O_2}); range difference POS A
and B $\pm$ 5 mg (10 mg) or $\pm$ 1 mg (2 mg) $\times$ dm^{-3} depending on full-scale oxygen range;
start/end marker. Adjustments $-$ Stirring speed for both stirrers; calibration for both
sensors; flushing interval; flushing duration.

Five outputs can be selected: POS A and B; temperature sensor A and B, or Δp_{O_2}
(Fig. 1). 4 LED displays indicate the function being recorded at any one time. Output
of channels can be monitored on a small ammeter (100 μA) independently of the recorder.

A compact DC powered galvanometric strip chart recorder (Miniscript D, Goerz-
Electro) is used for plotting the oxygen and temperature values. The chart width is
59 mm and values are marked by imprinting on a pressure-sensitive wax paper, at 3 s
intervals. The paper is advanced by a low power, quartz controlled, synchronous
motor. Standard gearbox speeds of either 20 or 60 mm/h are available. At a speed of
20 mm/h one roll of paper lasts for more than one month of continuous recording.

Four 6 V lead gel batteries (Sonnenschein, Büdingen, Germany) are connected to units of 12 V/8 Ah or 12 V/13 Ah. The smaller unit contains the electronic charger, the larger unit has to be charged from an external low voltage supply. Both packs are equipped with electronic cut-off switches to guard against damage to the battery and electronics if discharged below 10.5 V. The power consumption of the electronics, recorder and both stirrer pumps is less than 120 mA, which allows continuous recording for 3 or 5 days without change of the battery pack. The flushing pump is powered by a separate NiCd battery pack (3.6 V/8 Ah) to avoid a voltage drop of the main battery during the high initial power consumption. On the average this pump consumes about 4 A which allows up to 120 flushings of 2 min duration (1 flush/h for 5 days) with one battery charge.

Each unit is contained in a PVC box (100 × 110 × 250 mm) to prevent corrosion at high humidity or exposure to saltspray. The weight of the electronic unit is 1.5 kg, the recorder unit 2.6 kg, and the battery pack 4.5 to 5 kg.

2.2 Underwater Housing

One type of underwater housing consists of three separate casings, one for each unit (electronics, recorder, and battery pack). It can be made even in a modestly equipped workshop because of the small diameter of the front window. If one of the chambers is flooded the other compartments are not affected. Water pressure scarcely represents a problem, thanks to the small front window. The second type contains all three units within one housing and has to be provided with reinforcing cast ribs, and an aluminum brace in the middle of the window to support the wall against water pressure. In addition, the thickness of the acryl glass has to be increased. The single unit housing is lighter (12.5 kg versus 19.5 kg) and smaller than the other type (430 × 160 × 375 mm versus 475 × 155 × 470 mm).

Both housings are cast in seawater-resistant aluminum alloy ($AlMg_3$). After milling the front surface and drilling the holes and threads for the bolts and electric plugs, the cast is first impregnated under pressure with liquid polyester resin to fill the fine pores and then anodized. With the exception of the front surface the housing was painted with primer and a water-resistant acryl paint. A cage of pure untreated aluminum, mounted above the electric plugs for mechanical shock protection, serves as an electrode protecting against corrosion.

The front is made of clear acryl glass so that the front panel of the electronic unit can be observed. This cover is sealed to the housing by a soft silicone rubber O-ring (50 shore) which is fixed to the acryl glass with silicone rubber cement which protects the O-ring from water and sand when the housing is opened. The window of the small units is secured by four bolts, and that of the large one by eight bolts and stainless steel nuts. Switches and potentiometers on the front panel can be operated from outside by O-ring sealed stainless steel axles in the front plate.

The waterproof electrical connectors [3] consist of a standard five-pin LF-connector fitted into a waterproof PVC housing. To avoid the suction effect of a piston type seal the main O-ring seal is on the face of the plug. The two halves of the connector are locked by a screw-cap. Wherever a cable enters a housing, additional precautions have

been taken to guard against flooding of electronics or batteries. The space between
the wires of the cable and the aperture in the housing is filled with epoxy-resin (CIBA,
Araldite D). The cables of YSI-POS are insulated with an elastic compound, probably
polyurethane, which forms a bond with epoxy adhesives. Other type of cable that do
not form such a bond (e.g., PVC) are best unsheathed down the bare metal. Care has
to be taken to keep the wires separated during sealing. With this type of seal no water
can enter the housing if the cable is damaged.

2.3 Stirring and Flushing

Mixing and flow in bell jars is usually maintained by magnetically driven stirring bars.
Sensors, stirrers, and bell jars may be assembled in one unit ([1, 15], McCloskey, pers.
comm.). We decided to separate the sensor and stirrer from the bell jar to obtain a
more flexible system.

Both stirrers are connected to one plug by a 3-m cable. Small DC tape recorder motors gave reliable service after an additional bearing was installed below the axle to
support the weight of the driving magnet (Fig. 2). Since the stirrer starts to slip at
more than 800 rpm, the motor is run at 6 V instead of 12 V. The maximum power
consumption is 500 mW at free flow conditions. The driving unit is mounted in a
pressure-proof acryl housing (a metal housing does not work well because of the brak-
ing effect of eddy currents). The Teflon stirrer bar is centered on a glass axle in a glass
centrifugal pump body (Fig. 3). The POS has a tapered shaft that fits into a ground

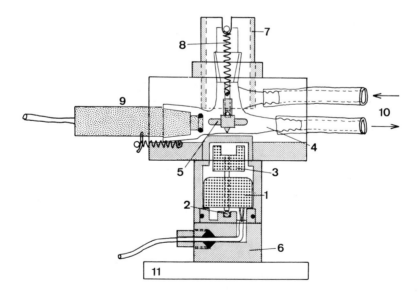

Fig. 2. Stirring unit. *1* DC tape recorder motor, *2* additional bearing for motor axle, *3* driving magnet, *4* glass centrifugal pump, *5* magnetic stirring bar, *6* acryl glass housing for the motor with raised side walls to hold the pump, *7* acryl glass tube to center the pump over the driving magnet,
8 springs to hold down the tube, *9* POS, *10* connections to respirometer, *11* lead plate. × 1/4

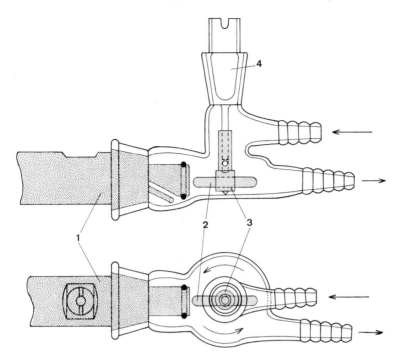

Fig. 3. Glass centrifugal pump. *1* YSI oxygen sensor with tapered ground glass joint, *2* Teflon-coated magnetic stirring bar, *3* Teflon bearing, *4* ground glass joint stopper with stirring axle. $\times$ 1/2

glass joint on the pump. Flow speed at the membrane of the POS is about 0.5 m/s, which slightly exceeds saturation conditions. The glass pump is protected by inserting it into an acryl cage. It is connected to the bell jar by rubber tubing (a nonpoisonous type is available for the beverage industry). If long connections are necessary, we use semitransparent polyamide tubes with rubber sleeves which can be checked for air bubbles. Silicone rubber tubing is highly permeable to oxygen and introduces errors at increased differences of internal and external p_{O_2}.

In experiments lasting for several hours, deteriorating water conditions within the respirometer may cause problems. For example, the production curves obtained for symbiotic coelenterates were strongly asymmetrical as compared with the light intensity curves, so that a pronounced inhibition of photosynthesis at higher light intensities had to be assumed. After ensuring sufficient water exchange, however, the production curves proved to be symmetrical with the light curves.

Increasing the size of the bell jars proportionally decreases the sensitivity of the measurement. Continuous-flow systems, on the other hand, need a second POS and expensive, complicated, and power-consuming constant flow pumps.

Rapid automatic intermittent flushing proved to be the best method of changing the water in the bell jar. A pressure-proof self-contained DC centrifugal pump was designed for this purpose (Fig. 4). A motor (3.6 V/4 A) drives an impeller through a magnetic coupling. Impeller and coupling were obtained from a commercially available centrifugal pump (EHEIM, Germany). Pumping capacity is 25 dm^3/min at a counterpressure

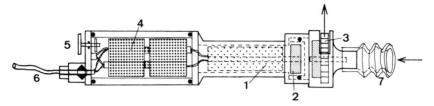

Fig. 4. Pressure proof flushing pump. *1* DC motor, *2* driving magnet, *3* turbine, *4* NiCd batteries, *5* hand switch, *6* cable to electronic switch, *7* connection to respirometer. × 1/8

of 0.1 bar. Water is sucked out by a steel-reinforced plastic tube through a sieve-covered outlet and is replaced in the bell jar through a spring-loaded large valve (30 mm diameter, 0.014 bar pressure) which is sealed by a soft V-ring. A small shield of acryl glass prevents direct sucking out of inflowing water and promotes rapid mixing (Fig. 5). No outlet valve is needed because water exchange stops as soon as the pump stops. A single large bell jar up to 16 dm^3 or several small ones can be sufficiently flushed by one pump. The pump is controlled from the electronic unit. Flushing intervals and duration are determined by experience. Since a portion of the recording is visible through the port, flushing rates can be adjusted by the diver 1−2 h after the start of the experiment.

2.4 Bell Jars

Acryl glass bell jars are preferred to those made of mineral glass because of their light weight and shock resistance, although they are not easy to sterilize. They are perfectly transparent to radiation in the visible part of the spectrum and there is no shading by the sealing.

For experiments with larger marine coelenterates containing symbiotic algae, hemispherical bell jars proved optimal. A transparent flat base of the chamber is useful where natural reflection from the sea bottom effects photosynthesis in symbiotic algae shaded from direct surface light. The chamber is sealed with a very soft O-ring of foam rubber, glued firmly to the rim of the hemisphere (Fig. 5). Several snap-on clamps keep the hemisphere on the acryl glass base. All fittings connect to the base to allow stacking of the hemispheres without scratching during transport or storage. Plastic fasteners with hooks securely anchor the bell jars even in strong currents or swell. Three sizes of hemisphere were used in our experiments: A bell jar of 46 cm diameter (16 dm^3) for large individuals of the mediterranean sea anemone *Anemonia sulcata,* two smaller sizes (27 and 20 cm diameter; 3 and 1.4 dm^3 respectively) for experiments with soft corals and hermatypic corals in the Red Sea (Fig. 5).

Water from the stirrer pump is injected tangentially and forces the water body to rotate with a maximum speed of 20 cm/s at the periphery. Flow may be reduced by clamping the rubber tube. Homogenous mixing takes place within a few minutes.

Small bell jars with no bottom (10 cm diameter) were used in experiments with large spherical coral heads (Fig. 6). The rim of the bell was sealed with soft nonvulcanized rubber. The device is held in place by elastic fasteners. The rubber cement squeezes

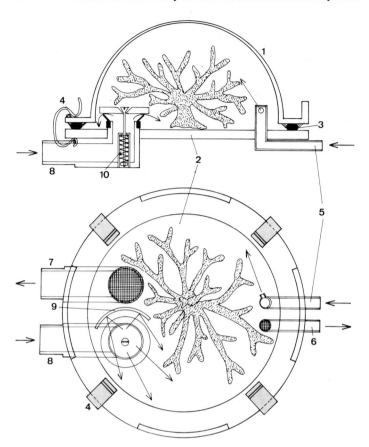

Fig. 5. Hemispherical bell jar for small corals, sea anemones etc. *1* acryl glass hemisphere, *2* transparent bottom, *3* soft O-ring seal, *4* fastening clip, *5* stirrer inflow, *6* stirrer backflow, *7* connection to flushing pump, *8* flush inflow, *9* reflection shield, *10* spring loaded valve. × 1/4

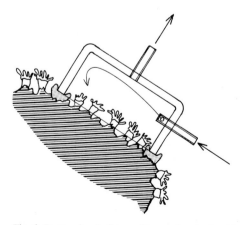

Fig. 6. Bottomless bell jar with unvulcanized rubber seal on a large coral head. × 1/4

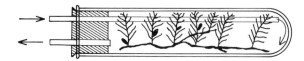

Fig. 7. Test tube respirometer for small hydroid colonies. × 1/2

the tissues deep into the skeleton and after the experiment the sample can be chiseled out of the coral head along the white circle remaining after removal of the bell jar.

Small centrifuge glass tubes (20 cm³) (Fig. 7) which could be acid-sterilized were used for respiration measurements with small hydroids (*Aglaophenia* ssp.). The total volume of the stirrer pump and tube was less than 50 cm³ [18].

Based on the experience gained from in situ work, a respirometer was designed for use in cultivation tanks containing the small symbiotic sea anemone *Aiptasia diaphana*. The respirometer volume is 70 cm³ including the stirring and respiration chamber (Fig. 8). A small dish serves as substrate for the animal and protects it from the stirrer. The animals can be removed and inserted through a large ground neck without disturbing them. With the exception of the POS all parts can be acid-sterilized. To flush the respirometer, water is sucked out through a capillary tube whose end is above the water level so that there is no need for a valve.

The seagrass respirometer consists of the chamber and the basal ring (Fig. 9). The bell jar is made of a clear acryl glass tube with an inner diameter of 14.2 cm covering a bottom area of 1/64 m². For measurements in *Posidonia* meadows we use tubes of 1 m length. Together with connecting tubes and stirring chamber the system has a volume of 15 dm³. At the lower end of the tube the outer wall slants slightly inward to facilitate the mounting of the tube into the basal ring. The upper end has a lid sealed with an O-ring. 10 cm from the lower end there are two fittings with an inner diameter of 4 mm for the connection of the bell jar to the stirring chamber. One fitting leads

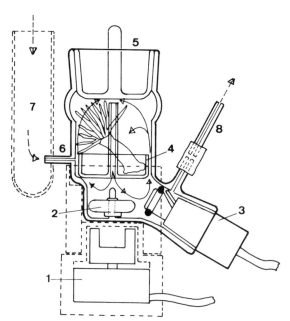

Fig. 8. Submersible respirometer for cultivation tanks. *1* motor and driving magnet in watertight housing, *2* stirring bar, *3* YSI POS, *4* specimen dish, *5* ground stopper with handle, *6* flush inflow capillary, *7* bubble trap, *8* flush outflow capillary. (After [17]). × 1/3

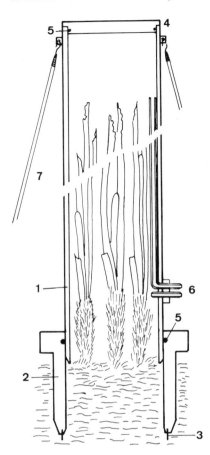

Fig. 9. Seagrass respirometer. *1* acryl glass chamber, *2* basal ring, *3* cutting edge, *4* lid, *5* O-ring seal, *6* connections to stirring pump, *7* rubber strings. × 1/8

straight into the bell jar, the other connects to a tube which runs on the inside up to a level of 10 cm below the lid to ensure water circulation through the whole length of the bell jar. Fluorescent dye introduced at the inlet appears after 5 min evenly distributed at the outlet. On the upper end of the bell jar are four rings to which elastic fasteners can be attached to anchor the jar.

Most seagrasses grow in sediments; roots and rhizomes form mats, which are especially tough in *Posidonia oceanica,* so that bell jars cannot simply be pushed into the ground. The basal ring is a polypropylene tube with a wall thickness of 20 mm. The lower rim is equipped with a serrated brass cutting edge. The upper rim is reinforced with an additional 20 mm thick polypropylene ring. The basal ring is pushed into the substrate using a combination of turning (= cutting) movements and heavy hammering. Polypropylene is used because it withstands such treatment, whereas PVC would break. The chamber is sealed to the basal ring by an O-ring on the inside of the latter. The effectiveness of the bottom seal was tested by introducing fluorescent dye. No obvious leakage could be observed around the basal ring and the concentration within the respirometer was unchanged after 24 h. A very similar type of respirometer was designed by Bay [1].

3 Experience in Use

Depending on the size of the bell jars, up to two units can be set up by an experienced diver in a single dive. The use of an inflatable buoyancy vest to compensate for the underwater weight of 5 kg for each unit is recommended. The possibility of taking large cumbersome bell jars to the bottom separately and assembling the unit there, even without surface support, makes the apparatus especially suited for expedition-type field work. For use below 20 m depth it is advisable to free the respirometers from air bubbles (which may take some time) in shallow depths.

The most time-consuming task is still the selection and preparation of specimens for measurement, and the subsequent collection of samples for determination of biomass, chlorophyll or leaf area index etc. Coral samples should be selected in a separate dive prior to the measurement, detached, cleaned from symbiotic shrimps, snails and fish and checked for boring bivalves and sponges. Also sea anemones should be detached well before measurements and be allowed to settle on a new substrate, e.g., a china dish, which may be inserted into the respirometer. Sea anemones can easily be handled with rubber gloves which have been smeared with mucus of neighboring individuals to avoid sticking or loss of tentacles during transport.

The time required to install one basal ring and chamber in a *Posidonia oceanica* mat varies from 15 to 30 min, increasing with density and height of the stand. Setting the basal ring is the most time-consuming part in the installation of the bell jar. The sea-grass leaves on the selected spot have to be carefully "combed" with the fingers to separate intermingled leaves and then the ring is slipped over. It is advisable to hold the leaves together with rubber bands to make space for the ring. After the ring has been driven in, the chamber is brought into position. The leaves are straightened by reaching in from the top and the rubber bands removed. After a final check for air bubbles the lid is placed on top and the bell jar connected to the stirring chamber. A second set of bell jars can be connected to the instrument package within 5 to 10 min. In this way up to five 24-h cycles may be run on different specimens before the unit has to be brought to the surface for exchange of batteries and recalibration of the POS.

The seagrass respirometers proved to be very sturdy but are somewhat susceptible to strong water movement. Prolonged exposure to oscillating water will shake the basal ring loose. Even when secured it is not advisable to use the setup when water speeds exceed 50 cm s^{-1}. It is advisable to camouflage the equipment as well as possible to prevent human interference.

4 Examples of Application

Both photosynthetic oxygen production and respiratory oxygen consumption may vary considerably during a daily cycle. Whereas this is well recognized for the production process with its dependence on light intensity, estimations of respiratory rates (e.g., for the calculation of gross production), are in most cases based on rates determined over short periods in experimentally darkened chambers. Continuous records,

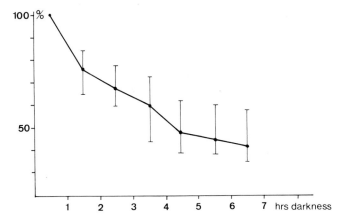

Fig. 10. Decrease in oxygen consumption in the dark (means and ranges, n = 9) in 1/64 m² of a *Posidonia oceanica* meadow. Oxygen consumption in the first hour of darkness was taken as 100%

both for symbiotic coelenterates and the seagrass community, however, showed that in the first hours of the dark period respiration rates are especially high. In the coelenterates this high rate drops rather rapidly over 1 to 2 h to a lower rate that is sustained throughout the rest of the night. This may be due to a rapid consumption of accumulated photosynthetic carbohydrates until the lower metabolic level of the respiration of storage fats and amino acids is reached [16]. In the seagrass community respiration rates decreased throughout the night. In the hours just before sunrise community respiration may only be 35% of the initial rate at the beginning of the night (Fig. 10). Measurements taken after experimental shading during the day also show similarly elevated respiration rates. In the sea anemona *Anemonia sulcata* these rates even exceed the natural rates at the beginning of the night [18]. This is a strong argument against an extrapolation of data gained in short term light-and-dark chamber measurements to daily gross production figures.

Both the coelenterate-zooxanthellae system and the seagrass community proved to be well acclimated to ambient light conditions. Seagrass shoots, transplanted together with their epigrowth from 30 m to 4 m depth, produce considerably more oxygen than control plants originally growing at 4 m depth, whereas in the reverse transplants production was much lower than in the controls (Ott, unpublished data). In the sea anemone *Aiptasia diaphana,* in addition to the effect on photosynthesis, a considerable increase in respiration rate was observed in the night following a day under light exceeding acclimation conditions. The reverse happened under lowered light intensities [17]. For ecologically valid estimations of metabolism from laboratory data care should be taken to simulate the conditions in the field as closely as possible. As this is hard to achieve in the case of organisms from highly structured environments, in situ measurements are to be preferred.

Figure 11 exemplifies how measurements of community net production may be used as an integrated parameter in energy budgets. Community net production is defined as the net primary production of the system minus consumer respiration. The net gain in oxygen measured under the jar should therefore indicate an equivalent in-

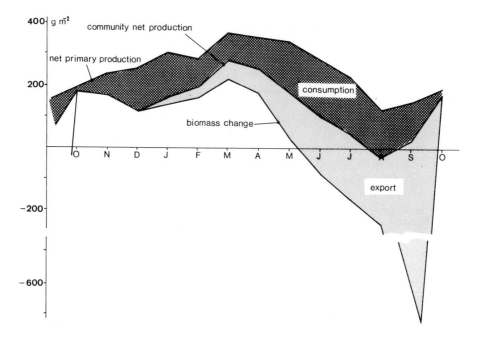

Fig. 11. Energy budget of a *Posidonia oxeanica* meadow. The community net production curve was determined by in situ community gas exchange measurement. (After [13])

crease of the energy content of the system, e.g., in the form of biomass increase. In systems with a considerable export of organic matter, as for instance in seagrass stands, the difference between community net production as estimated by oxygen production and the actual biomass increase of the system, as measured, e.g., by harvesting methods, thus allows an estimate of the exported biomass, which otherwise would be very difficult to determine.

References

1. Bay D (1978) Etude "in situ" de la production primaire d'un herbier de Posidonies [*Posidonia oceanica* (L.) Delile] de la Baie de Calvi-Corse. Thèse, Univ Liege
2. Jaubert J (1977) Un appareillage autonome et automatique permettant la mesure et l'enregistrement in situ du metabolisme d'organismes marins spécialment concu pour être utilisé en plongée. Rapp Comm Int Mer Medit 24 (6):149–151
3. Machan R (1973) Lichtenenergiemessung unter Wasser. Diplomarbeit, Inst elektrische Meßtechnik, TU Wien
4. Margalef R (1963) On certain unifying principles in ecology. Am Nat 97:357–374
5. Margalef R (1968) Perspectives in ecological theory. Univ Chicago Press, Chicago, 112 pp
6. McCloskey LR, Wethey DS, Porter JW (1978) Measurement and interpretation of photosynthesis and respiration in reef corals. In: Stoddard DR, Johannes RE (eds) UNESCO Monographs on oceanographic methodology, vol V. Coral reefs: research methods

7. Mergner H, Svoboda A (1977) Productivity and seasonal changes in selected reef areas in the Gulf of Aqaba (Red Sea). Helgol wiss Meeresunters 30:383–399

8. Odum EP (1969) The strategy of ecosystem development. Science 164:262–270

9. Odum HT (1956) Efficiencies, size of organisms, and community structure. Ecology 37:592–597

10. Odum HT (1957) Trophic structure and productivity of Silver Springs, Florida. Ecol Monogr 27:55–112

11. Odum HT, Odum EP (1955) Trophic structure and productivity of a windward coral reef community on Eniwetok Atoll. Ecol Monogr 25:291–320

12. Odum HT, Pinkerton RC (1955) Times speed regulator, the optimum efficiency for maximum output in physical and biological systems. Am Sci 43:331–343

13. Ott J (1980) Growth and production in *Posidonia oceanica* (L.) Delile. PSZN. Mar Ecol 1:47–64

14. Schramm W (1973) Langfristige in situ-Messungen zum Sauerstoffwechsel benthischer Meerespflanzen in einem kontinuierlich registrierenden Durchflußsystem. Mar Biol 22:335–339

15. Smith KL Jr, Burns KA, Teal JM (1972) In situ respiration of benthic communities in Castle Harbor, Bermuda. Mar Biol 12:196–199

16. Svoboda A (1978) In situ monitoring of oxygen production in cnidaria with and without zooxanthellae. In: McLusky DS, Berry AJ (eds) Physiology and behaviour of marine organisms. Pergamon Press, New York, pp 75–82

17. Svoboda A, Porrmann T (1980) Oxygen production and uptake by symbiotic *Aiptasia diaphana* (Rapp), (Anthozoa, Coelenterate) adapted to different light intensities. In: Smith DC, Tiffon Y (eds) Nutritition in lower metazoa. Pergamon Press, New York

18. Svoboda A (1981) 24 h in-situ Registrierung der O_2-Aufnahme und Produktion der symbiontischen mediterranen Coelenteraten *Anemonia sulcata* und *Aglaophenia tubiformis*. Verh Dtsch Zool Ges 1981:275

19. Weiner P, Kirkman H (1979) Continuous recording technique to measure oxygen release from a seagrass community within an acrylic insulation chamber. CSIRO Div Fish Oceanogr Rep 96:1–8

20. Wells JM (1974) The metabolism of tropical benthic communities: in situ determinations and their implications. Mar Technol Soc J (MTS) 8 (1):9–12

Chapter III.5 Deep-Sea Respirometry: In Situ Techniques

K.L. Smith, Jr. and R.J. Baldwin[1]

1 Introduction

The deep-sea environment occupies over half the earth's surface. However, inaccessibility has greatly restricted the study of this unique environment characterized by high pressure and low temperature. Technological advances over the past 15 years permit us to "enter" the deep-sea environment either directly with submersibles or indirectly with sophisticated instrumentation to make observations, collections and measurements and to conduct experiments. With these advances, the biological dynamics of the deep-sea ecosystem are slowly being resolved. Our contributions to this resolution are related to energy flow and, more specifically, the metabolism of deep-sea organisms. The integral component of our instrumentation used in these studies is the polarographic oxygen sensor (POS).

There are two basic approaches to studying the metabolism of deep-sea organisms, (1) a laboratory approach and (2) an in situ approach. The first method involves capturing the organisms in their environment and returning them to the surface to conduct measurements and experiments in the laboratory [2, 11, 19, 24]. Many limitations of this approach must be resolved or acknowledged. (A) The capture process, no matter how gentle, stresses the organism. This is evident from initial oxygen consumption rate increases noted immediately following entrapment [15, 17, 18]. (B) Organisms captured at depth and brought to the surface unprotected will undergo decompression and temperature change. Decompression problems are being addressed with pressure-retaining mechanisms that have been employed in traps [25] and net cod ends [9]; temperature control problems are being addressed with insulated traps [25] and net cod ends [3]. (C) Solar or artificial light sources can adversely affect visual pigments of deep-sea organisms accustomed to low ambient light levels [4, 10]. (D) Organisms are generally returned to a surface ship where they experience varying degrees of abnormal motion (rolling, yawing, pitching). (E) Confinement in containers for both laboratory maintenance and metabolic measurements places physical and biological constraints on organisms. (F) Organisms are generally held in surface seawater, ignoring the water quality differences between surface and deep sea.

---

1 Scripps Institution of Oceanography, A-002, University of California, San Diego, La Jolla, CA 92093, USA

Polarographic Oxygen Sensors (ed. by Gnaiger/Forstner)
© Springer-Verlag Berlin Heidelberg 1983

The in situ methodology is based on capturing and measuring the metabolism of organisms in their environment. There are limitations to this approach which must be considered. (A) The capture process, with its associated stress to the organism, is a problem pertaining to in situ work as well as to the laboratory approach mentioned above. (B) When submersibles are used for in situ operations, the artificial lighting will have a temporary blinding effect on animals with visual pigments which normally experience attenuated light levels [4, 10]. (C) There is a severe limitation on the number of measurements and the complexity of the experiments which can be performed. (D) Containment of organisms in either flowthrough or closed chambers is an abnormal situation. These effects are believed to be minimal from comparisons of the observed behavior of entrapped versus free animals [15, 17, 18].

After analyzing the limitations of the two methodologies, we chose to use the in situ approach because it allows measurements of metabolism to be made under conditions more closely approximating the natural environment from which the organisms have been removed.

2 In Situ Methodology

We have developed equipment for measuring the metabolism (oxygen consumption and excretion) of both individual animals and communities of organisms in situ. These instruments were developed for deployment with manned submersibles and with free vehicles (systems autonomous from the surface ship). The emphasis of our research has been directed toward the deep-sea benthos and the water column immediately overlying the sediment-water interface which together constitute the benthic boundary layer. The biota of this layer can be divided into two basic components, the sediment community and the pelagic community. Specific equipment has been developed to examine the metabolism of various components of these two communities.

2.1 General Equipment

The two parameters which we measure as indicators of the metabolic activity of deep sea organisms are rates of oxygen consumption and nutrient exchange within sealed enclosures. Changes in dissolved oxygen tension are measured with a POS [8]. Nutrient flux measurements are made by the chemical comparison of an initial and final water sample taken by a syringe withdrawal system from an enclosure incubated in situ. Organic enrichments, biological poisons, or radioactive tracers, useful in compartmentalizing O_2 uptake, may also be introduced into the respiration chambers by syringe injection systems. The electronic and mechanical systems designed for oxygen measurement and the syringe injection/withdrawal system are, with a few variations, the same for all experimental enclosures used. The dimensions and configuration of the enclosure vary depending on the organism (fish, amphipod, crustacean) or community (benthic or pelagic) being studied and the type of vehicle used for equipment deployment and recovery (deep-sea submersible or surface ship).

2.1.1 Polarographic Oxygen Sensor/Amplifier System

Each respirometer unit is capable of monitoring changes in dissolved oxygen tension by means of a POS of the type described by Kanwisher [8]. This sensor can easily be made in the laboratory at minimal expense. The basic unit consists of a $Ag\text{-}Ag_2O$ cylinder (1 cm diam., 1 cm length), which is concentric around a platinum disk (5 mm diam., 0.02 mm thick) and a thermistor (provides automatic temperature compensation). The whole assembly with attached electrical cables is cast in epoxy (Hysol resin R-2039, hardener HD 3561). Prior to use, the sensors are filled with electrolyte (1mol dm^{-3} KOH), covered with a Teflon membrane (0.03 mm), and calibrated using air saturated and nitrogen purged chilled seawater. The effects of hydrostatic pressure on the sensors are corrected for, using laboratory generated pressure/output curves for each sensor (see also Chap. I.8).

The amplifier and recorder, as well as the other associated electronic circuitry, are contained in an aluminum pressure-resistant housing. A POS is inserted into each respirometer enclosure to record the relative oxygen tension continuously. A power regulator provides a 1 V excitation voltage to each of two sensor anodes (Fig. 1). The resulting sensor current, which is proportional to dissolved oxygen tension, passes through a preamplifier before being recorded. These preamplified signals from both sensors are multiplexed so that the two outputs can be recorded on one chart recorder (Rustrak model 288/F137). This multiplexing is done by selecting one of two transmission gates with a switching signal.

For greater resolution, the preamplified signal is processed by an autoscaling circuit which splits the full scale of the recorder (zero to saturated dissolved oxygen solution) into five divisions, so that each chart width represents a 20% change in oxygen concentration. This circuit consists of a comparator, oscillator, counter, resistor ladder, step amplifier, and a reference voltage. Its function is to generate a step voltage which increases or decreases (by counting up or down in steps of 20% of full scale) until the sum of the step voltage and the preamplified sensor voltage is within the range of the meter movement. This sum is then amplified by a current follower to drive the recorder meter movement.

2.1.2 Stirring Motor

A stirring motor is inserted into each respiration chamber to circulate the water over the POS and prevent stratification within the enclosure. This stirring motor consists of three electromagnetic coils, symmetrically spaced about the axis of a ceramic magnetic rotor with attached Teflon impeller (Fig. 2). These coils, embedded in epoxy (Hysol resin R-2039, hardener HD 3561), are consecutively pulsed to attract the rotor.

The electronic circuitry which drives the coils is a simple waveform generator (Fig. 2c). An oscillator coupled with a monostable multivibrator creates the drive pulse and an oscillator (clock) signal. A sequencing circuit, built around a three-stage counter and driven by the oscillator, generates a holding pulse and sequentially routes it to each one of three outputs (A, B, and C, Fig. 2d). These holding pulses (with individually adjusted amplitudes) are summed with the power drive pulse at the bases of three

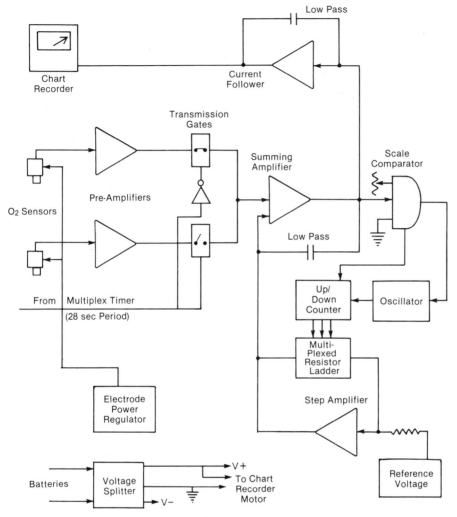

Fig. 1. Block diagram of autoscaling POS amplifier

transistor coil drivers, to activate the three coils. As each driver is sequentially selected, the corresponding coil is activated to attract and turn the rotor. A variable resistor adjustment permits water flow rates to be set between 0.28 to 5.65 cm^3/s.

2.1.3 Syringe Injection/Withdrawal System

Nutrient flux determinations (e.g., excretion and nitrification) are another useful measure of metabolic activity. For these measurements, we have developed a system capable of withdrawing 50 cm^3 of seawater from closed chambers for nutrient determination. An optional and interchangeable injection unit may be added for

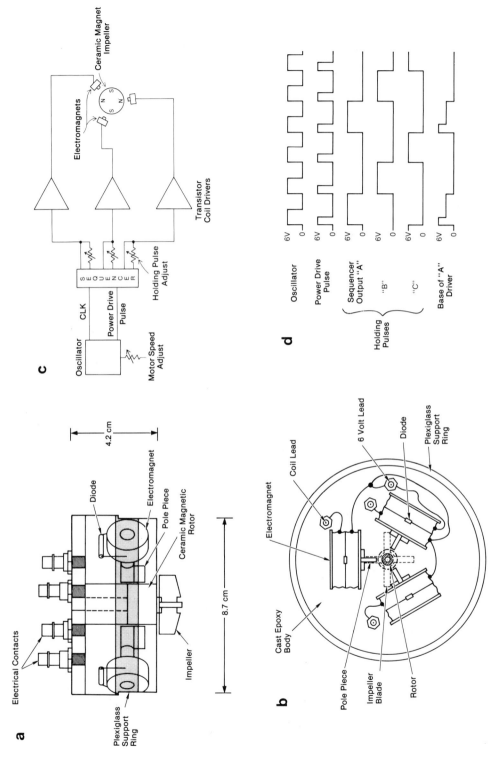

Fig. 2a–d. Stirring motor: **a** side view, **b** top view, **c** block diagram of circuitry, **d** wave form generated by stirring motor

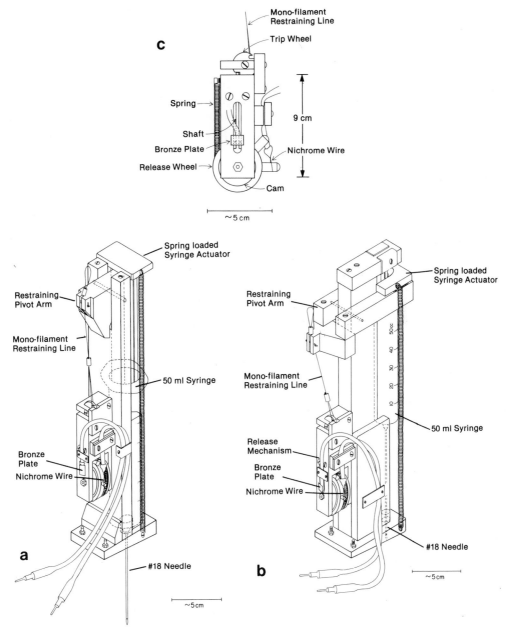

Fig. 3a–c. Syringe injection/withdrawal system: **a** injection syringe, **b** withdrawal syringe, **c** nichrome release

inoculating the chamber with organic enrichments or poisons. The nichrome releases associated with these syringes are controlled by a 12-event electronic timer with crystal oscillator. Selected activation times of the syringes may be preset in increments of 7.5 min for up to a maximum of 1250 h.

The injection unit consists of a 50 cm^3 plastic syringe (B-D Plastipak) supported by a Plexiglas stand (Fig. 3a). In a filled position, the syringe plunger is held by a restraining pivot arm. Atop the syringe plunger is a polyvinyl chloride plate (spring-loaded syringe actuator) with extension springs fastened on both sides and attached to the base of the stand. When the pivot arm is released, these springs drive the plunger down thus forcing the contents of the syringe into the grab chamber below through a self-sealing neoprene septum. The release mechanism for the pivot arm is mounted on the back of the stand and represents a modification of one designed by T.R. Folsom (Scripps Institution of Oceanography, personal communication) (Fig. 3c). It is constructed of Delrin and consists of a spring-loaded release wheel with an offset cam and fused shaft which engages a stainless steel trip wheel. A notch in the trip wheel holds a monofilament restraining line which prevents the pivot arm from releasing the syringe plunger while in the cocked position. The release wheel is held in the loaded position by a loop of nichrome wire (0.127 mm diam.) fastened around a pin on the periphery of the wheel. Located on the side of the release is a bronze plate ($7 \times 10 \times 0.3$ mm). When current is applied to the nichrome wire (cathode) an electrolytic cell is formed with the bronze (anode). A current of 300–400 mA applied for 10 to 20 s causes the oxidation of the nichrome wire. Upon breaking, the release wheel and cam rotate, disengaging the shaft and trip wheel and freeing the monofilament restraining line. The pivot arm than rotates and the syringe plunger is driven down by the resulting force on the spring-loaded syringe actuator.

The withdrawal system employs essentially the same principle as that of the injection system, however the plunger of the 50 cm^3 syringe is maintained in a fixed position by a bracket on the top of the stand (Fig. 3b). A pivot arm, much like that on the injection system, prevents the barrel of the syringe from being forced down by a spring-loaded actuator while in the cocked position. When released, the barrel of the syringe is driven down, but the plunger is held stationary, thus withdrawing a water sample from the respiration chamber. All samples are filtered in situ through a 25-mm filter holder containing a Whatman GF/C filter.

2.2 Specific Equipment

2.2.1 Sediment Biota Respirometers

Sediment Community Respirometers. The development of instruments to measure in situ oxygen consumption by the deep-sea sediment community has undergone a gradual progression toward more complex and auto-controlled systems. Our first instruments, bell jar respirometers, were placed with submersibles [20]. These units were subsequently designed into a free vehicle bell jar respirometer which could be deployed untethered from a surface ship in depths of water up to 7000 m [16, 21]. A similar respirometer was developed to measure the oxygen consumption of deep sea sediments using O_2 analysis of initial and final water samples [6].

One critical limitation of the bell jar respirometer was the inability to recover the sediments upon which the respiration measurements were made. Hence we developed a grab respirometer for use with a submersible [22]. The prohibitive cost, depth restrictions, and limited availability of the submersible prompted us to develop this system

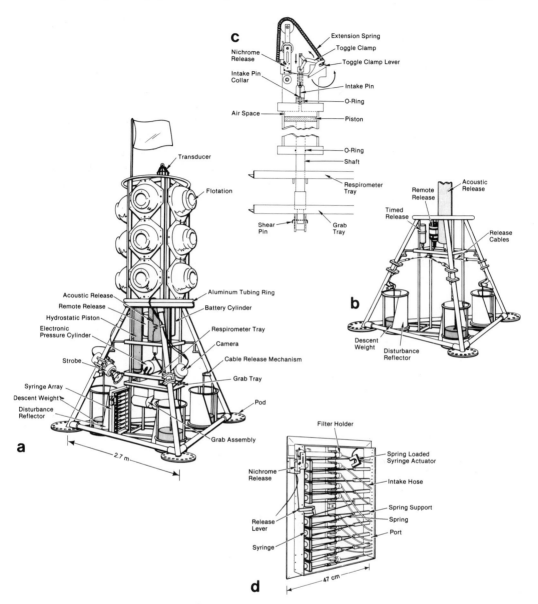

Fig. 4a–d. Free vehicle grab respirometer: **a** entire assembly, **b** descent weight release system, **c** hydrostatic piston assembly, **d** syringe array

further into a free vehicle [23]. This most recent stage of development in sediment community respirometers is described below.

The free vehicle grab respirometer (FVGR) consists of a tripodal frame which supports the acoustic control and grab respirometer instrumentation, and a flotation package (Fig. 4a). The tripod frame (2.4 m high × 2.7 m wide) is constructed of alu-

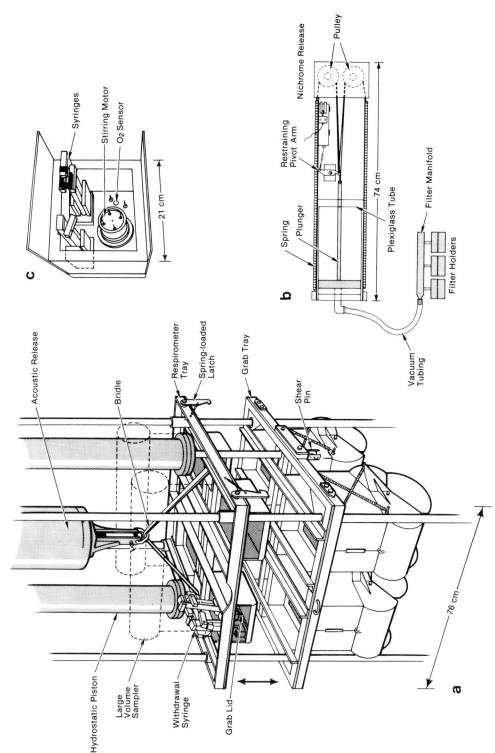

Fig. 5a–c. Detail of respirometer and grab tray of free vehicle grab respirometer: **a** respirometer and grab tray in deployment position, **b** large volume sampler, **c** respirometer tray with detail of grab lid

minum tubing (6.5 cm OD, 5 cm ID) which has been sleeved and bolted at all joints for easy disassembly. Each leg has a perforated disk pod at the base and is bolted to an aluminum tubing ring at the top (diam. 1.1 m). The upper ring also provides a base for the flotation array.

Central to the frame are the square grab and respirometer trays which are supported at each of four corners by a vertical guide rod (aluminum tubing, 2.5 cm ID). The grab tray consists of four stainless steel grabs (20.3 × 20.3 × 30 cm × 2.7 mm gauge) opened at the top and mounted on a rigid square stainless steel frame (Fig. 5a). Each grab has two spring-loaded jaws which can penetrate the sediment up to 20 cm and enclose a surface area of 413 cm^2. Two nichrome releases located on opposite sides of the frame control the closure of two grabs each through a mechanical linkage. Two hydrostatic pistons pass freely through the above positioned respirometer tray and attach to the perimeter of the grab tray on opposing sides. Each unit consists of an aluminum anodized cylinder (76.2 cm length, 10.2 diam.) with an internal stainless steel piston (76.2 cm throw).

The respirometer tray consists of a square stainless steel frame with cross bars supporting lids for each grab. Within each lid is a POS, a stirring motor and a pair of interchangeable injection/withdrawal syringes (Fig. 5c). This tray also supports the pressure cylinder with associated electronics for the respirometry system. Power for the respitometry electronics and releases is provided from a separate battery cylinder (Fig. 4a). The respirometer tray is held in place by a centrally positioned acoustic release (AMF, Model 324).

The cylindrical flotation rack is fastened to the top ring of the tripodal frame and consists of twenty spherical glass floats (eighteen 43.2 cm OD, two 25.4 cm OD Benthos floats) (Fig. 4a). These floats are enclosed in protective plastic covers and arranged in six vertical racks. On the top of the flotation rack is an acoustic transducer which relays coded messages between the shipboard transducer and the centrally located acoustic release.

For deployment to the deep-sea floor, the free vehicle grab respirometer is in a loaded configuration (Figs. 4a and 5a) i.e., both trays are in the raised position with the respirometer tray held by the acoustic release and the grab tray, with grab jaws open, held by the hydrostatic pistons.

Once on the bottom, a period of up to 1 h is allowed for the associated disturbance of the initial landing to dissipate. The disturbance reflectors on each tripod leg help prevent resuspended sediment from entering the central measurements area. After this initial period, the two hydrostatic pistons are actuated simultaneously by a pre-set nichrome release attached to the top of the piston (Fig. 4c). When the release fires, the toggle clamp lever swivels, thus freeing the toggle clamp which is pulled to a vertical position by the extension spring. The rotation of this clamp pushes the collar of the intake pin down, thereby allowing a fine stream of ambient seawater to enter the "air" space which can be backpressured between the top of the cylinder and the piston. The ambient pressure at depth is sufficient to slowly drive (1–2 cm/min) the pistons and the attached grab assembly into the sediment to a depth of approximately 15 cm. A high pressure line between the two pistons assures equal force on both plungers as the grabs are being driven down. In the event of a malfunction, shear pins at the point of attachment on the grab assembly prevent excessive force being directed to a single sides of the frame.

Once the hydrostatic pistons have had ample time to position the grab tray assembly in the sediment, a command from the surface ship activates the release of the upper respirometer tray which free falls down the four guide rods to engage the grab assembly. Four spring-loaded latches secure the two assemblies together and silicone gaskets assure a watertight seal between chamber units (Fig. 5a). A 35-mm camera and strobe system mounted on the tripod legs (Fig. 4a) are activated by the release of the respirometer tray through a magnetic reed switch. The self-winding camera with timer and solenoid activated shutter takes pictures at 0.6-s intervals before, during, and after release of the respirometer tray [1]. The strobe is synchronously triggered by the camera.

Once the grab respirometer chambers are sealed, the oxygen tension in the water overlying the enclosed sediment is monitored continuously by the POS system. At this time initial water samples may be withdrawn from the chambers for nutrient and dissolved oxygen analyses or injection syringes may innoculate the chambers with enrichments or poisons.

In addition to the syringe systems associated with the grabs, there are two other water samplers mounted on the FVGR and controlled by electronic timers: two large volume syringe samplers (LVS) and a vertical syringe array. The LVS consists of a 1.7 dm^3 syringe constructed of 10 cm ID Plexiglas cylinder with a spring-loaded PVC plunger (Fig. 5b). When released, the plunger inside the barrel is withdrawn by two sets of springs and the water is pulled into the LVS through three 4.7 cm filter holders, containing pre-combusted Whatman GF/C filters. The filter intake is located on the outside of the respirometer tray, 5 cm above the level of the intake for the grab syringes.

The syringe array consists of ten 50 cm^3 syringes (B-D Plastipak) mounted horizontally on a vertical frame, the bottom of which rests on the sediment surface (Fig. 4d). Each syringe is equipped with a filter (Whatman GF/C) and intake hose which can be adjusted from heights of 0 to 72 cm above the sediment surface at 2-cm intervals. The operating principle is similar to that of the individual withdrawal systems, except that all ten syringes are released at once by a single restraining pivot arm which is held by a nichrome release. The array is designed to collect water samples for the detection of chemical gradients (i.e., nutrients) immediately above the sediment-water interface. This information, coupled with the analysis of pore water extracted from the grab sediment provides a vertical profile from 13 cm below to 72 cm above the sediment-water interface for comparison with nutrient fluxes determined from the withdrawal syringe samples taken inside each grab.

At the termination of the grab respirometer incubation, which usually lasts from 2 to 6 days, final syringe withdrawal samples are taken from each grab and the grab jaws are closed by a pre-set electronic timer and nichrome wire-release system. Final water samples are also taken with one LVS and the vertical syringe array systems.

The final event is the acoustic command given from the surface ship to release the descent weights via the remote release. Suspended from each tripod leg is a cable release mechanism which holds a descent weight (99 kg). This cable release is controlled by the remote release through a series of three wire cables (Fig. 4b). The remote release, part of the main acoustic release system, is controlled by one of the two command modes via the shipboard transducer. This descent weight release system has a backup unit consisting of a pre-set timed release [13] connected in tandem with the remote release. On command, the descent weights are dropped and the free vehicle grab respirometer ascends to the surface at a rate of 50 to 60 m/min. The FVGR weighs 259 kg in water; the descent weights add another 297 kg (3 × 99 kg), yielding a net negative buoyancy (negative buoyancy minus positive flotation buoyancy) of 94 kg on deployment. Upon release of the descent weight, the FVGR achieves a positive buoyancy of 203 kg minus the additional weight of the sediment in the grabs. Surface recovery time has been minimized with the addition of a transmitter (OAR ST206-100), flasher (OAR SF500) and a meter square fluorescent flag mounted on a two meter mast above the float array. The time from surfacing to shipboard recovery is usually within 30 min, thus preventing significant surface warming of the water samples and sediment.

Megafauna Respirometer. The animals which occupy the sediment surface and are considered an integral part of the sediment community are the epibenthic megafauna. These animals include such taxa as ophiuroids, asteroids, holothurians, coelenterates, crustaceans, and gastropods [5] which are generally not enclosed in the grabs of the free vehicle grab respirometer because of their large size and spatial distribution.

The problems associated with locating and measuring the in situ respiration of these large animals have been solved by developing a respirometer for use with a submersible. The visual and manipulative capabilities of a submersible are essential for selecting and sampling specimens for placement in individual chambers of the megafauna respirometer.

The megafauna respirometer consists of dual Plexiglas cylinders (29 cm ID × 9.5 cm high) mounted on a 35 cm × 72 cm plate which has been reinforced with aluminum angle (3 × 3 cm) (Fig. 6). This plate is hinged at the back side and mounted on a 29 cm × 35 cm × 72 cm frame of fiberglass angle (5 cm × 5 cm). The top plate pivots so that the respiration chambers seal on two neoprene gaskets which are glued to the respirometer base plate. An aluminum pressure resistant cylinder containing the POS electronics, stirring motor circuitry, timers, and battery power is mounted horizontally beneath the respirometer base. Each chamber has a stirring motor, POS, and two injection/withdrawal syringe systems mounted on the top plate.

The manipulator arm of the submersible uses a 0.5 cm mesh basket (15.5 cm ID × 5.6 cm high) to scoop up animals from the sediment surface, thus sieving out the associated sediment. The basket with the captured animal is then placed on the respirometer base and the top portion of the animal chamber is manipulated to the closed position. A T-handle locking mechanism assures the integrity of the chamber seals. Usually the megafauna respirometer is rigged so that the syringes take initial and final samples from each chamber for excretion product analysis. Measurements are made for periods up to 3 days, depending on the size of the animal and suspected respiration rates.

2.2.2 Pelagic Biota Respirometers

The large numbers of animals that inhabit the water column of the deep-sea benthic boundary layer can be behaviorally divided into two groups: the mobile scavengers and

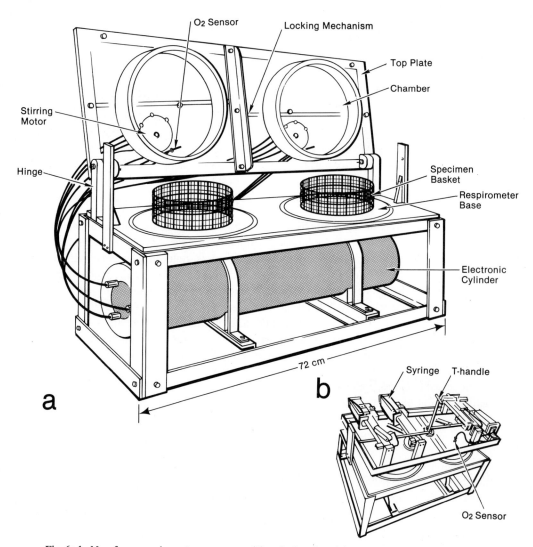

Fig. 6a,b. Megafauna respirometer: **a** open position, **b** closed position

the mobile nonscavengers. The mobile scavengers are those animals which are readily attracted to and captured in baited trap respirometers and include such organisms as rattail fishes, amphipods, and decapods. Mobile nonscavengers are those pelagic animals which are not attracted to such baited traps and include organisms such as myctophid and gonostomatid fishes, siponophores, chaetognaths, and copepods.

We have developed both free-vehicle and submersible-operated equipment to measure the respiration and excretion rates of both groups. The scavenging fishes and amphipods are lured into free-vehicle fish and amphipod trap respirometers. Nonscavenging organisms are actively captured using a submersible-operated slurp gun respirometer. These three respirometers and associated instrumentation are described below.

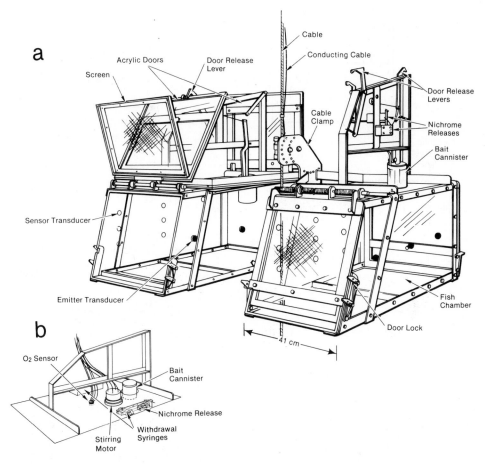

Fig. 7a,b. Fishtrap respirometer: **a** open and closed positions, **b** detail of top of trap

Fish Trap Respirometer. The fish trap respirometer (FTR) is a free-vehicle system designed for capturing pelagic scavengers attracted to bait and measuring their metabolic rates (Fig. 7a) [21]. Dual acrylic chambers of trapezoidal shape (91 × 74 × 41 × 41 cm) are joined together with a stainless steel frame. A shielded conducting cable connects the FTR with its electronic control center which is housed in a pressure cylinder 1 m above the trap. Each chamber has a total volume of 53.7 dm^3 and is equipped with independent POS, stirring motor and syringe withdrawal system (Fig. 7b). A bait canister is located in the top of each trap to attract a scavenger but is automatically withdrawn from the chamber once the animal is enclosed to prevent contamination from bait oxygen consumption and nutrient flux.

An acoustic sensing system is used to detect the presence of pelagic organisms in the trap and to record their movement during the metabolic measurement (Fig. 7a). A series of three emitter transducers is aligned on one wall of the trap and emits a pulse train of 200 kHz. Nine sensor transducers are positioned in an evenly spaced 3 × 3 array on the opposite wall and function as on/off switches. The sensors are sequentially

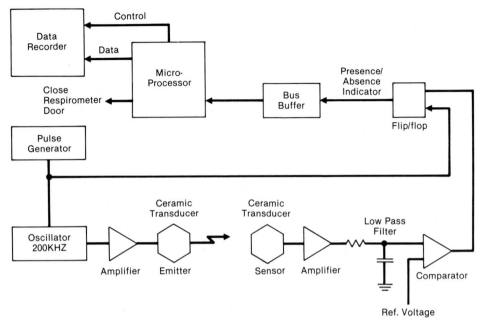

Fig. 8. Block diagram of acoustic sensor system for fish and amphipod trap respirometer

monitored by a microprocessor to check for a received signal (Fig. 8). The first continuous disruption of received acoustic pulses indicates an animal within the trap and a signal is given to the closing mechanism on the door. Once an animal is trapped the sensors are monitored sequentially and their on/off states are recorded to provide a record of the activity of the organism during the respiration measurements.

Dual chamber construction provides for a control to be run simultaneously with the measurement. Each chamber is equipped with a spring-loaded double-door system separate locking latches at the sloped end. During free vehicle deployment, the FTR descends with the control screen door closed and the experimental chamber screen door open; both bait canisters are filled and in the lowered position within the trap (Fig. 7a, left trap). When the acoustic sensor responds to the presence of a fish, the experimental chamber screen doors close via a battery-powered nichrome wire release system. After a selected period of time, all acrylic doors close and the spring-loaded bait canisters are withdrawn by electronic control. The delayed final closing of the acrylic doors allows the fish to become accustomed to the trap environment before initiation of respiration measurements and excretion sample withdrawal.

The combination acrylic/screen doors with separate timer-controlled release levers allow starvation studies to be performed in situ. Fish can be held for extended periods of time within the screened chamber, removed from their normal food supply. Free exchange of seawater through the screen eliminates metabolic waste accumulation and oxygen depletion within the chamber. The fine mesh covered bait canister prevents the animal from feeding on the bait.

A free vehicle mooring line system is used to deploy and recover the fish trap respirometer. The basic components of the mooring system are (1) a mast assembly,

(2) a main flotation package, (3) a nylon mooring line, and (4) a release system with disposable descent weights. The mast assembly is topmost, consisting of two 33 cm Benthos glass floats secured to a counter weighted 3.1 m fiberglass mast, a radio transmitter, submersible strobe light, and flag [7, 12, 14]. The mast system is connected to two 43 cm Benthos floats (25.5 kg total positive buoyancy) by a 7.6 m nylon line (Samson braid) of 0.95 cm diameter. The 43 cm spheres are connected to the mooring line (0.64 cm Samson nylon line) which supports the electronics pressure cylinder, the respirometer, the tandem timed-release system [2, 13] and the anchor weights. The mooring line is of variable length so that the FTR units can be positioned at any distance above the sediment-water interface.

The array is deployed by streaming the mast system off the stern of the ship followed by the 43 cm floats, electronics cylinder, respirometer and the mooring line while the ship is underway at 0.5 to 1 knot. The tandem timed-release system with anchor weight is deployed last.

On completion of the experiment, the preset releases are actuated, the anchor weight is dropped and the array ascends to the surface at approximately 50 m/min. Once on the surface, the free vehicle is located via the radio transmitter, strobe light and flag. The mast system and floats are recovered first, followed by the electronics cylinder, respirometer, mooring line and dual releases.

Amphipod Respirometer. A free vehicle amphipod respirometer (FVAR) was developed to capture amphipods or other small scavengers attracted to bait in four independent acrylic chambers. These chambers are mounted horizontally between two acrylic disks, perforated to decrease water resistance during descent and recovery operations (Fig. 9a). The disks are strengthened by two aluminum tubing rings (61 cm in diameter) and are held parallel to each other by four stainless steel support rods (22.5 cm long). The bottom disk supports four acrylic cylindrical chambers (16 × 7 cm ID – total volume of 615 cm^3) each attached to the center acrylic bait container and each having an independent acoustic sensing system, POS, stirring motor and closure device (Fig. 9b). The top disk supports five withdrawal syringes (Fig. 9c). The extra withdrawal provides the initial water sample for the four chambers, while the other four are taken as final samples in each of the chambers. The POS system, electronic control of nichrome releases and acoustic detection system have been described above. The aluminum pressure cylinder housing the control and sensing electronics and power supply is clamped onto the free vehicle mooring line 1 m above the respirometer and communicates via shielded conducting cables.

At the time of deployment, bait is placed in the center cage with nylon mesh (300 μm mesh opening) permitting water exchange between each chamber and the bait but preventing feeding. As an amphipod or other scavenger enters a chamber, the acoustic sensor system within that cylinder triggers the nichrome release system; a spring-loaded clamp constricts the flexible polyethylene tube opening to the chamber, thus sealing the animal in the respiration cylinder. This acoustic sensing system is identical in design to that described for the fish trap respirometer. When all four traps have been individually triggered and closed by the presence of an animal, the center bait container rotates 45° within the central housing which then excludes the bait from all four chambers.

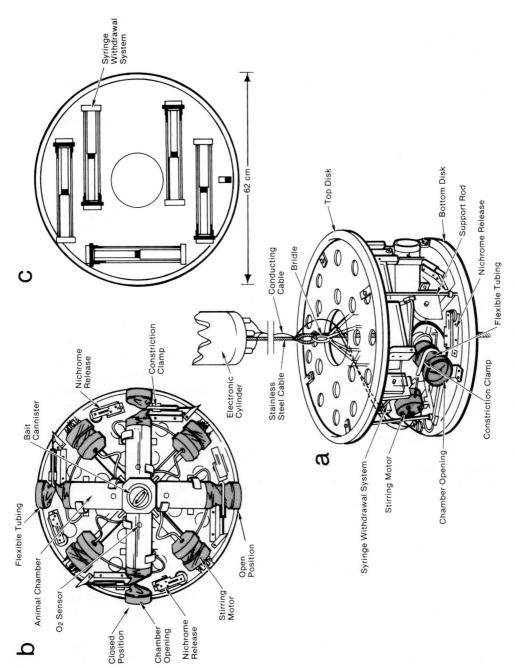

Fig. 9a–c. Amphipod respirometer: **a** complete assembly, **b** bottom disk, **c** top disk

The amphipod respirometers are deployed and recovered on free vehicle mooring lines as described above. These units with the accompanying electronics cylinder can be positioned at any desirable height above the bottom on this mooring line.

Plankton/Nekton Respirometer. The slurp gun respirometer was developed to capture pelagic animals (plankton/nekton) not attracted to the baited trap respirometers [21]. This instrument requires the visual and manipulative qualities of a manned submersible to collect such organisms in the water column of the deep-sea benthic boundary layer. The primary criteria considered in developing the slurp gun respirometer were: (1) collection of multiple discrete samples with continuous oxygen consumption measurement and withdrawal of water samples for excretion rate analyses, (2) gentle capture of organisms ranging from fast-swimming fishes to gelatinous zooplankton with minimum stress to the organism, (3) ability to change the volume of the animal chamber as warranted by the size or suspected respiration rate of the organism, and (4) ability to conduct these in situ metabolic measurements over periods in excess of a normal manned submersible dive (10 h). These four criteria were all considered in the final design.

The slurp gun respirometer consists of three to four modules, each of which includes a horizontal acrylic animal chamber (40.5 cm $\times$ 8.8 cm ID) with a right-angle valve at either end (Fig. 10). One valve is connected through a sliding manifold to a flexible reinforced intake hose (10.5 cm ID $\times$ 2 m) and collecting funnel 0.5 m diam. opening). This valve is closed by an acrylic spring-loaded plunger which has a volume-adjusting piston. A worm-gear driven piston can be turned into the animal chamber to adjust the volume to between 2460 and 200 cm^3. The other valve is connected through the sliding manifold to a variable speed pump (30 V dc; 0.3–3 dm^3/s) via a flexible reinforced vinyl hose (4 cm ID). The power for the pump is supplied through quick-disconnect cables from the submersible. This pump valve is closed by a spring-loaded acrylic plunger which has the stirring motor inserted into the end facing the animal chamber. An interchangeable nylon screen (50–300 μm mesh) is positioned between the animal chamber and the pump hose valve aperture to prevent animals from being sucked beyond the chamber or coming in direct contact with the stirring motor impeller.

Each animal chamber has a POS and two withdrawal syringes inserted through the bottom wall near the pump valve and stirring motor (Fig. 10b). The associated amplifying, recording, and timing electronics required by these systems are housed in a control electronics cylinder mounted on the respirometer frame. Battery power for these electronics is supplied from a contiguous pressure cylinder with all electronic communication via shielded cables.

A series of three to four of these modules are assembled in parallel on an acrylic platform reinforced with structural aluminum angle (the number of modules is dependent on the weight restrictions of the submersible) (Fig. 10a). Each module communicates with the intake hose and pump hose via a common sliding manifold which is driven by spring motors that sequentially engage and disengage each module.

The slurp gun respirometer with electronic cylinder, battery unit, and pump is mounted on a rigid aluminum tubing support frame which is then fastened to the front of the submersible with a solenoid release and slip-sleeves.

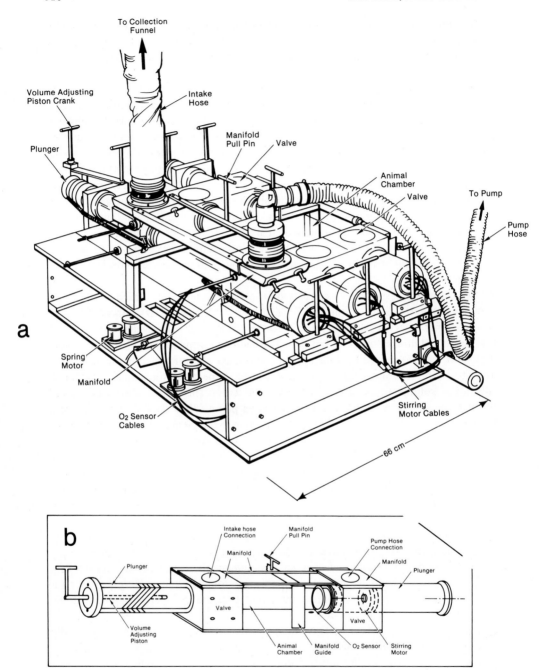

Fig. 10a,b. Slurp gun respirometer: **a** complete assembly, **b** single module

On deployment, the valve of each module is held open by pull pins, the manifold is positioned over the first module, and the volume-adjusting pistons are fully retracted into the valve plunger. Planktonic and nektonic organisms are visually located from the submersible and the intake hose and funnel are positioned for collection. The pump is activated and the speed adjusted so that suction is just sufficient to move the animal into the transparent animal chamber. At this point, the pump is turned off and the two right-angle valves on either end of the animal chamber are closed by releasing pull pins with the submersible manipulator. The volume of the animal chamber is then adjusted ($200-2460$ cm^3) for the size and the suspected metabolic activity of the enclosed organism by cranking the volume-adjusting piston. The common manifold is then released by another pull pin and it advances to angage the next respirometer module. The same procedure is followed in filling each animal chamber. In some cases, instead of collecting single organisms, the funnel is placed in a forward-looking position, the submersible driven forward at a known speed, and the pump activated for a specific time to supply a mixed plankton sample at a particular depth above the bottom. One module or an attached Niskin water sampler (5 dm^3) serves as a control.

When all the modules are filled, the slurp gun respirometer is detached from the submersible by the quick-release solenoid and tethered to a free-vehicle mooring line (similar to that described above) at the depth where the animals were collected. This allows the slurp gun respirometer to incubate for a period in excess of the normal dive time of the submersible (> 10 h). These incubations are generally for periods up to 3 days during which time the dissolved oxygen tension within each animal chamber is monitored continuously. The withdrawal syringe system is pre-set to take a chamber water sample (50 cm^3) at the initiation and termination of the incubation. A pre-set timed release or acoustic release on the mooring line is then activated, thus permitting the free vehicle with attached slurp gun respirometer to ascend to the surface for recovery.

3 Conclusions

We have presented our approach, with the associated instrumentation, to measuring the metabolism of organisms and communities in the deep sea. Inaccessibility of the deep-sea environment coupled with submersible and surface ship expense and availability serve to limit the number of measurements and the complexity of experiments which can be performed in situ. We feel the best direction for further research in deep-sea energetics is a combination of both in situ and laboratory studies, establishing meaningful comparisons between the two methodologies. However, in conducting such deep-sea metabolic studies, full cognizance of the artificial constraints being imposed on the organisms must be maintained.

Acknowledgments. We thank M. Unwin and E. Duffrin for design work and M. Unwin for meticulously building all of the in situ equipment. Other members of our laboratory including P. Klement, M. Laver, P. Rishel and G. White contributed much to equipment development. S. Hamilton critically reviewed the manuscript. This is a contribution of Scripps Institution of Oceanography and was supported by grants and contracts from the National Science Foundation (OCE-78-08640, OCE-76-10520, OCE-76-10535, OCE-76-10728), the Office of Naval Research (0152), and The Sandia Laboratories (13-2555, 13-9951).

References

1. Brown DM (1975) Four biological samplers: opening-closing midwater trawl, closing vertical tow net, pressure fish trap, free vehicle drop camera. Deep-Sea Res 22:565–567
2. Childress JJ (1977) Physiological approaches to the biology of midwater organisms. In: Andersen NR, Zahuranec BJ (eds) Oceanic sound scattering prediction. Plenum Press, New York, pp 301–324
3. Childress JJ, Barnes AT, Quetin LB, Robison BH (1978) Thermally protecting cod ends for the recovery of living deep-sea animals. Deep-Sea Res 25:419–422
4. Fernandez HRC (1978) Visual pigments of bioluminescent and nonbioluminescent deep-sea fishes. Vision Res 19:589–592
5. Grassle JF, Sanders JL, Hessler RR, Rowe GT, McLellan T (1975) Pattern and zonation: a study of bathyal megafauna using the research submersible Alvin. Deep-Sea Res 22:457–481
6. Hinga KR, Sieburth JMcN, Heath GR (1979) The supply and use of organic material at the deep-sea floor. J Mar Res 37:557–579
7. Isaacs JD, Schick GB (1960) Deep sea free instrument vehicle. Deep-Sea Res 7:61–67
8. Kanwisher J (1959) Polarographic oxygen electrode. Limnol Oceanogr 4:210–217
9. Macdonald AG, Gilchrist I (1972) An apparatus for the recovery and study of deep sea plankton at constant temperature and pressure. In: Brauer RW (ed) Barobiology and the experimental biology of the deep sea. Univ North Carolina, Chapel Hill, pp 394–412
10. O'Day WT, Fernandez HR (1976) Vision in the lantern fish *Stenobrachius leucopsarus* (Myctophidae). Mar Biol 37:187–195
11. Pamatmat MM (1973) Benthic community metabolism on the continental terrace and in the deep sea in the North Pacific. Int Rev Gesamte Hydrobiol 58:345–368
12. Schick GB, Isaacs JD, Sessions MH (1968) Autonomous instruments in oceanographic research. 4th Nat Mar Sci Instrum Symp, 28 pp
13. Sessions MH, Marshall PM (1971) A precision deep-sea time release. Scripps Inst Oceanogr Ref Ser 71-5, 18 pp
14. Shutts RL (1975) Unmanned deep sea free vehicle system. Scripps Inst Oceanogr Mar Tech Handb, Series TR-61, 50 p
15. Smith KL Jr (1978) Benthic community respiration in the N.W. Atlantic Ocean: in situ measurements from 40-5200 m. Mar Biol 47:337–347
16. Smith KL Jr, Clifford CH, Eliason AH, Walden B, Rowe GT, Teal JM (1976) A free vehicle for measuring benthic community metabolism. Limnol Oceanogr 21:164–170
17. Smith KL Jr, Hessler RR (1974) Respiration of benthopelagic fishes: in situ measurements at 1230 meters. Science 184:72–73
18. Smith KL Jr, Laver MB (1981) Respiration of the bathypelagic fish, *Cyclothone acclinidens*: Mar Biol 61:261–266
19. Smith KL Jr, Teal JM (1973) Temperature and pressure effects on respiration of thecostomatous pteropods. Deep-Sea Res 20:853–858
20. Smith KL Jr, Teal JM (1973) Deep-sea benthic community respiration: an in situ study at 1850 m. Science 179:282–283
21. Smith KL Jr, White GA (1982) Ecological energetics studies in the deep-sea benthic boundary layer: in situ respiration studies. In: Ernst WG, Morin J (eds) The environment of the deep sea. Prentice-Hall, Englewood Cliffs, NJ, pp 279–300
22. Smith KL Jr, White GA, Laver MB, Haugsness JA (1978) Nutrient exchange and O_2 consumption by deep-sea benthic communities: Preliminary in situ measurements. Limnol Oceanogr 23 (5): 997–1005

23. Smith KL Jr, White GA, Laver MB (1979) Oxygen uptake and nutrient exchange of sediments measured in situ using a free vehicle grab respirometer. Deep-Sea Res 26 (A):337–346

24. Teal JM (1971) Pressure effects on the respiration of vertically migrating decapod crustacea. Am Zool 11:571–576

25. Yayanos AA (1978) Recovery and maintenance of live amphipods at a pressure of 580 bars from an ocean depth of 5700 meters. Science 200:1056–1059

Appendix A Calculation of Equilibrium Oxygen Concentration

H. Forstner and E. Gnaiger[1]

1 Introduction

In the chemical literature solubility of a gas is generally given as the Bunsen absorption coefficient, α, which is a function of temperature and of the electrolyte content of the solvent. Hydrostatic pressure has little effect on solubility, but with increasing head the escaping tendency of the gas increases. Therefore, at constant concentration, a POS gives a higher reading at greater depth. This effect is negligible ($< 1\%$) at depths less than 100 m.

If α or an equivalent expression of solubility is known, the saturation concentration for any particular set of conditions can readily be calculated. Unfortunately, gas solubility is not a simple function of the controlling variables, and empirical data obtained by physical or chemical methods has to be used. Since the pioneering work by Winkler [21, 22], differing sets of such data have been published, and it has been a matter of much debate which particular set of values is more accurate. Mortimer [14] has reviewed the existing data and has convincingly recommended the solubility data recently published by Benson and Krause [2, 3] as a standard for limnological application. Extensive aids for determining oxygen solubility in the temperature range $0°-40°C$ and for atmospheric pressure from $61-111.5$ kPa ($0.6-1.1$ atm) are available in form of nomograms and equations [14]. Essential equations from [3, 14] are also presented here and their use is recommended for POS applications.

2 Solubility Measures

1. The Bunsen Absorption Coefficient, α, is the volume of gas (STP) dissolved in a unit volume of solvent at standard partial pressure of the gas $p_0 = 1$ atm (101.325 kPa or 760 Torr) at a specific temperature. The dimension is volume per volume per pressure.

2. The Solubility Coefficient, S_s, has the dimension of amount of substance per volume per pressure. If moles are taken as units of amount of substance, the relation to α is

---

1 Institut für Zoologie, Abteilung Zoophysiologie, Universität Innsbruck, Peter-Mayr-Str. 1A, A−6020 Innsbruck, Austria

Polarographic Oxygen Sensors (ed. by Gnaiger/Forstner)
© Springer-Verlag Berlin Heidelberg 1983

$$S_s = \frac{\alpha}{\text{molar volume} \times K} \, , \tag{1}$$

where the molar volume is 22.393 dm^3 mol^{-1} and K is a conversion factor depending on the unit of pressure chosen (1 for atm, 101.325 for kPa, and 760 for Torr).

Example: For pure water at 10°C, $\alpha = 0.0381726$ dm^3 dm^{-3} atm^{-1}, expressing pressure in kPa, the solubility of oxygen in the sample is then

$$S_s = \frac{0.0381726}{22.393 \times 101.325} = 16.824 \times 10^{-6} \text{ mol dm}^{-3} \text{ kPa}^{-1} \tag{2}$$
$$= 16.824 \ \mu\text{mol dm}^{-3} \text{ kPa}^{-1} \, .$$

3. Avoiding the term "solubility", Benson and Krause [2] define *unit standard concentration, C**, of atmospheric oxygen in pure water as the concentration of dissolved oxygen per unit volume of solution (measured at equilibrium temperature) "when it is in equilibrium with an atmosphere of standard composition and saturated with water vapor at a total pressure (including that of water vapor) of 101.325 kPa (1 atm)". When referred to unit mass of solution, the unit standard concentration is designated $C^\dagger$. C^* is converted to $C^\dagger$ by dividing by the density of the solution at the temperature concerned, or (to a very close approximation) by the density of pure water at that temperature.

Unit standard concentration C^* and the solubility coefficient S_s are related by

$$C^* = S_s \times (p_0 - p_{H_2O}) \times \phi_{O_2}^* \, , \tag{3}$$

where p_{H_2O} is the partial pressure of water vapor and $\phi_{O_2}^*$ is the volume fraction of oxygen in dry air which is 0.20946.

Example: At 10°C $S_s = 16.824 \ \mu\text{mol dm}^{-3}$ kPa^{-1} [Eq. (2)]
$$p_{H_2O} = 1.22763 \text{ kPa, [see Eq. (19)]}$$
then

$$C^* = 16.824 \times (101.325 - 1.22763) \times 0.20946 \tag{4}$$
$$= 352.739 \ \mu\text{mol dm}^{-3} \, .$$

In the biological and limnological literature concentration is generally the preferred measure of solubility, and since very accurate approximation formulas are available we will adhere to this usage. If required, C^* can easily be converted to S_s or α, using the relation of Eqs. (1) and (3). When C^* is given as μmol dm^{-3}, S_s in μmol dm^{-3} kPa^{-1} is

$$S_s = \frac{C^*}{(101.325 - p_{H_2O}) \times 0.20946} \tag{5}$$

and α in dm^3 dm^{-3} atm^{-1} is

$$\alpha = \frac{C^*}{(101.325 - p_{H_2O}) \times 92.315} \, . \tag{6}$$

3 Calculation of Unit Standard Concentration

In applications involving a POS the accuracy provided by the Benson and Krause solubility tables [3] is hardly necessary and the empirical equation [Eq. (7)] fitted by Mortimer [14] can be used. Over the temperature range $0°–37.5°C$ the interpolation error is not more than 0.05%. To be consistent with the S.I. $\mu mol\ dm^{-3}$ should be used as units, but the familiar $mg\ dm^{-3}$ (equivalent to $mg\ l^{-1}$) has also been included. To obtain the desired unit the constant A in Eq. (7) is selected from Table 1.

The unit standard concentration (at temperature θ, in $°C$) is

$$C* = \exp[A - 1.31403\ \ln(\theta + 45.93)]. \tag{7}$$

Table 1

Units O_2	A
$\mu mol\ dm^{-3}$	11.1538
$mg\ dm^{-3}$	7.7117
$cm^3\ dm^{-3}$ (ideal gas STP)	7.3557
$cm^3\ dm^{-3}$ (real gas STP)	7.3547

The calculation can be done quickly on any scientific pocket calculator and the concentration values are more accurate than many solubility tables hitherto in use.

The five term polynomial approximation from Eq. (3) will also be given, because it has been recommended as a standard [14]. The equation is presented in nested form, which is more efficient for programming on computers and calculators. The values for the B-coefficients and the constant A are taken from Table 2A or 2B, depending on whether $C*$ or $C^\dagger$ is to be calculated.

$$C^x = \exp\langle\{[(B_1 \times X + B_2) \times X + B_3] \times X + B_4\} \times X + A\rangle, \tag{8}$$

$$X = 1/T; \quad 273.15 < T < 313.15\ K \quad (0°–40°C).$$

4 Dependence on Salt Concentration

Due to interactions between the molecular species in a solution, solubility of a non-electrolyte in a salt solution is different from that in pure water. This dependence can be described by a logarithmic relationship proposed by Setschenow [17].

$$\ln\left(\frac{S_s}{S_e}\right) = \kappa_e \times I. \tag{9}$$

S_s and S_e are the solubilities in pure water and in the electrolyte solution respectively, κ_e is the salting coefficient and I the ionic strength. For oxygen dissolved in water, as for most aqueous solutions, κ_e is positive which means that with increasing salt con-

Table 2

(A) C^* = amount of oxygen per unit volume

B coefficients		Units O_2	A
B_1	-8.621949×10^{11}	μmol dm^{-3}	-135.90202
B_2	1.243800×10^{10}	mg dm^{-3}	-139.34410
B_3	-6.642308×10^{7}	cm^3 dm^{-3} (ideal gas STP)	-139.70011
B_4	1.575701×10^{5}	cm^3 dm^{-3} (real gas STP)	-139.70113

(B) $C^\dagger$ = amount of oxygen per unit mass

B coefficients		Units O_2	A
B_1	-8.621061×10^{11}	μmol kg^{-1}	-135.30002
B_2	1.243678×10^{10}	mg kg^{-1}	-138.74210
B_3	-6.637149×10^{7}	cm^3 kg^{-1} (ideal gas STP)	-139.09811
B_4	1.572288×10^{5}	cm^3 kg^{-1} (real gas STP)	-139.09909

centration solubility decreases. This phenomenon is known as the salting-out effect. The coefficient κ_e varies with temperature, but depends also on the particular ions and the nature of the nonelectrolyte.

For seawater, values of oxygen solubility for a customary range of temperatures and salinities have been tabulated [4, 10, 14, 19]. Equations for calculating a correction factor have also been proposed [8, 9]. In the equations salt concentration is defined as chlorinity [CL$^-$]‰ , which is g chlorine per kg of seawater. Since ionic composition of seawater remains highly constant throughout the oceans (± 0.002%), the relationship between salinity (g total salt content per kg of seawater) and chlorinity is defined by the Knudsen relationship [13]:

$$\text{Salinity} = 1.805 \, [\text{Cl}^-]‰ + 0.03. \tag{10}$$

Of the two equations most commonly cited in the literature, that proposed by Green and Carritt [9]

$$S_e = S_s \times \exp\left[-[\text{Cl}^-]‰ \times \left(-0.1288 + \frac{53.44}{T} - 0.04442 \ln T + 7.145 \times 10^{-4} \, T\right)\right] \tag{11}$$

is usually recommended. It is based on measurements for $273.1 < T < 308.16$ K and $0 < [\text{Cl}^-] < 30$‰ and is more extensive than the equation from Fox [8]. Both equations agree to better than ± 1% over the full range of Eq. (11) and with tabulated values [10]. It should be noted that the correction factor in Eq. (11) is for a solubility coefficient (or an absorption coeff.) and the effect of salinity on water vapor pressure is not accounted for. Applying a salinity correction to C^* or C_c, it can be shown, using Eqs. (3) and (5) that an additional correction factor

$$\frac{p_0 - p_{H_2O,e}}{p_0 - p_{H_2O}} \tag{12}$$

would be required, where $p_{H_2O,e}$ is the water vapor pressure in water with dissolved salts. In practice, over the range of salinity and temperature normally encountered, this correction will be near unity and it can be omitted.

5 Correction for Non-Standard Atmospheres

The standard concentrations C^* and $C^\dagger$ obtained from Eqs. (7) and (8) apply for water, in equilibrium with air saturated with water vapor and exerting a total pressure $p_0 = 101.325$ kPa. Since total pressure includes water vapor pressure

$$p_{0, \text{dry air}} = p_0 - p_{H_2O}. \tag{13}$$

The partial molar volume of oxygen is 0.20946, therefore at standard atmospheric pressure the partial pressure of oxygen is

$$p_{O_2}^* = (p_0 - p_{H_2O}) \times 0.20946. \tag{14}$$

If air and water are at equilibrium, water vapor is a function of temperature only. At constant temperature and any barometric pressure, $_bp$, the partial pressure of oxygen is therefore

$$p_{O_2} = (_bp - p_{H_2O}) \times 0.20946. \tag{15}$$

Combining Eqs. (14) and (15) we obtain

$$p_{O_2} = p_{O_2}^* \times \frac{_bp - p_{H_2O}}{p_0 - p_{H_2O}} \tag{16}$$

or separating the water vapor effect

$$p_{O_2} = p_{O_2}^* \times \frac{_bp}{p_0} \times \frac{1 - p_{H_2O}/_bp}{1 - p_{H_2O}/p_0}. \tag{17}$$

Since the dissolved oxygen concentration in equilibrium with the gaseous phase is proportional to p_{O_2}, the saturation concentration in solution, C_c (see Table 7), is

$$C_c = C^* \times \frac{_bp}{p_0} \times \frac{1 - p_{H_2O}/_bp}{1 - p_{H_2O}/p_0} \tag{18}$$

Although expressed somewhat differently, Eq. (18) is almost identical with Eq. (4) from [14], only the extremely small correction term for molecular interactions in the atmospheric gas has been omitted.

The water vapor pressure over the range 0°–40°C can be calculated to within 11 ppm of standard tabulated values by an approximation [14]. Again taking the appropriate constant from Table 3, p_{H_2O} can readily be determined in the three most commonly used pressure units,

$$p_{H_2O} = \exp\left[(-216961 \times X - 3840.7) \times X + A\right] \tag{19}$$

$X = 1/T;\ T$ in Kelvin; $0°C = 273.15$ K.

Table 3

Units p	A
kPa	16.4754
atm	11.8571
mm Hg	18.4904

With regard to the nonstandard atmospheric pressure, it is recommended that the value of $_bp$ to be used in Eq. (18) be obtained from a contemporary barometer reading. In a situation where this is not possible $_bp$ may be calculated from the pressure-altitude relation predicted by a standard atmospheric model, accepting an error of approximately $\pm 2.5\%$ due to varying local weather conditions. The power law

$$_bp = p_o \times (1 - h/44.3)^{5.25}, \tag{20}$$

where h is the altitude in km, reproduces the ICAO standard atmosphere to within $\pm 0.004\%$ from -0.5 to 3.0 km, and with a small underestimate at higher levels increasing to -0.03% at 5.0 km [14]. Other approximations widely in use, like $\ln\ _bp = h/7.986$ [16] are much less accurate and Eq. (20) should be used instead. However, considering the actual variations of the real atmosphere, as compared to a standard atmosphere, this is really a moot point.

6 Correction for Hydrostatic Pressure: Absolute Saturation

A water column of 1 m exerts a hydrostatic pressure, $_wp$, of 9.80 kPa (0.0967 atm). Since the actual pressure, p_h [kPa], at any depth h [m] is the sum of atmospheric (barometric) and hydrostatic pressure,

$$p_h =\ _bp +\ _wp \tag{21}$$

the amount of gas which can be dissolved in equilibrium with a gas phase is linearly dependent on depth. This equilibrium condition has been termed absolute concentration [12, 15] to distinguish it from the gas solubility corrected for barometric pressure, $_bp$, only. The calculation of absolute saturation is of ecological interest in three situations: (1) Air bubbles carried down to moderate depths cause supersaturation with respect to the pressure at the water surface. (2) If air-saturated water warms up at some depth, supersaturation occurs on account of lower solubility at higher temperature. (3) Photosynthetic oxygen production leads to supersaturation (Chap. III.1) which can, above a critical partial pressure of oxygen, $p_{O_2,c}$, at small depths, even surpass the absolute saturation with the possibility of gas bubble formation.

In cases 1 and 2 of abiogenic supersaturation, $p_{O_2,c}$ at depth h is simply given by

$$p_{O_2,c} = (p_h - p_{H_2O}) \times \phi^*_{O_2}. \tag{22}$$

According to Eqs. (16) and (18), the corresponding absolute saturation concentration, C_c, is

$$C_c = C^* \times \frac{p_h - p_{H_2O}}{p_o - p_{H_2O}}. \tag{23}$$

Evaluation of $p_{O_2,c}$ is more complex in case 3, where, at constant partial pressure of other constituent gases, the biogenic increase in p_{O_2} results in the possibility of bubble formation. At this critical point a bubble contains gases in proportion to $p_{O_2,c}$ and to the partial pressure of all other gases, p_X, since

$$p_h = p_{O_2,c} + p_{H_2O} + p_X. \tag{24}$$

If these bubbles rise they will remove the other gases and, while oxygen production continues, $p_{O_2,c}$ increases to the theoretical maximum, where only oxygen remains in solution

$$p_{O_2,c} = p_h - p_{H_2O}. \tag{25}$$

This demonstrates that, for calculating absolute oxygen saturation in general the partial pressures of all dissolved gases must be known. These are hardly ever measured in limnological and oceanographic routine investigations except in studies particularly devoted to the problem of gas bubble disease [20]. Therefore appropriate assumptions have to be made and, at high p_{O_2} in natural waters, only nitrogen (including argon) has to be considered quantitatively whence

$$p_{O_2,c} = {}_b p - p_{H_2O} - p_{N_2} + {}_w p. \tag{26}$$

At equilibrium with air at the water surface

$$p_{N_2} \approx {}_b p - p_{H_2O} - ({}_b p - p_{H_2O}) \times \phi^*_{O_2}. \tag{27}$$

If equilibration took place at the temperature still prevailing at depth h, we can insert Eq. (27) into (26)

$$p_{O_2,c} = ({}_b p - p_{H_2O}) \times \phi^*_{O_2} + {}_w p. \tag{28}$$

The first term incorporates the influence of gases other than oxygen (including water vapor) on $p_{O_2,c}$. This term becomes complicated, however, if temperature does not remain constant. Then the change in solubility of nitrogen and hence the change of p_{N_2} with temperature, and the dependence of p_{H_2O} on temperature have to be accounted for. Restricting ourselves to the simplifying assumption of constant temperature, the absolute saturation concentration is calculated as[2]

---

2 Hutchinson [12] discussed absolute saturation with reference to [15]. His equation for C_c contains a misunderstanding of the role played by water vapor, and hence C_c at 0 m differs erroneously from the surface saturation value

Table 4. Calculation of the critical p_{O_2} at depth h $(p_{O_2,c})$, of absolute saturation concentration (from surface concentration times the multiplication factor), and increase in absolute saturation concentration per metre of depth ($\Delta C_c \times m^{-1}$) as a percentage of air saturation in the range of temperatures between 4° and 25°C and altitudes between 0 and 2000 m

Source of supersatura-tion	$p_{O_2,c}$ [kPa]	Multiplication factor	$\Delta C_c \times m^{-1}$ [%]
Abiogenic	$(_bp - p_{H_2O}) \times \phi^*_{O_2} + 2.05 \times h$	$1 + \dfrac{9.80}{_bp - p_{H_2O}} \times h$	9.8 to 12.9
Biogenic	$(_bp - p_{H_2O}) \times \phi^*_{O_2} + 9.80 \times h$	$1 + \dfrac{46.79}{_bp - p_{H_2O}} \times h$	46.7 to 61.8
Pure O_2	$(_bp - p_{H_2O}) + 9.80 \times h$	$4.774 + \dfrac{46.79}{_bp - p_{H_2O}} \times h$	

$$C_c = C^* \times \frac{_bp - p_{H_2O} + _wp / \phi^*_{O_2}}{p_o - p_{H_2O}} \quad . \tag{29}$$

Table 4 compares expressions of the critical p_{O_2} for abiogenic [Eq. (22)] and biogenic sources of supersaturation [Eqs. (26–28)] and for the extreme case of saturation with pure oxygen [Eq. (25)]. It also lists the multiplication factors for obtaining absolute saturation at any depth from surface saturation concentrations [Eq. (18)] and shows the increase in absolute saturation concentration per metre of depth.

7 The Effects of Depth on p_{O_2} and Solubility

The effect of high hydrostatic pressure on p_{O_2} has been investigated by Enns et al. [3]. They found that with increasing hydrostatic head the partial pressure of a gas in solution, equilibrated at some lower reference pressure, increases exponentially. A number of other authors have discussed these results and an explanation for this effect, based on general thermodynamic principles, has been formulated [1, 5, 6]. A summary is given by Hitchman [11]. It has been shown that partial pressure varies according to

$$p_{O_2,h} = p_{O_2} \times \exp\left(\frac{M \times g_n \times h}{R \times T}\right) \tag{30}$$

where p_{O_2} and $p_{O_2,h}$ are the partial pressures at the surface and at depth h respectively, M is the molecular weight of the gas (32 for oxygen), g_n is the gravitational acceleration (9.81 N), h is the depth in metres, with R and T as defined in the gas law. Substituting p_o/p_{O_2} for $M/(RT)$ Eq. (30) becomes identical to the standard pressure/altitude relation. This means: for water saturated with a given gas at the surface the equilibrium partial pressure of that gas at any depth is equal to the partial pressure of that gas if it were contained in a gas column extending from the surface to that particular depth [5]. Table 5 gives values of $p_{O_2,h}/p_{O_2}$ for several depths, calculated with Eq. (30). The water is assumed to be equilibrated with air ($M = 28.8$) at the surface and the mean water temperature is $5°C$. It can be seen that at 500 m a correction of 6% will already be needed. Equation (30) shows that in this situation the increase in partial pressure is independent of the actual hydrostatic pressure and the density of the liquid and is entirely a function of the potential of the gas molecules dissolved in the water column. This effect is different from the case where liquid and gas phase coexist at some higher pressure, for instance, when air bubbles are carried downward. In the latter case equilibrium partial pressure is a linear function of hydrostatic pressure and is dependent on both the depth at which equilibration takes place and the density of the liquid (Table 4).

For gases in general, pressure also influences the equilibrium concentration (solubility). Following the same line of argument as before it has been shown [1, 6, 11] that

$$C_{c,h} = C_c \times \exp \frac{(M - v^\infty \times \rho) \times g_n \times h}{R \times T}. \tag{31}$$

C_c and $C_{c,h}$ are the equilibrium concentrations at the surface and at depth, h, v^∞ is the partial molar volume of the gas at infinite dilution and ρ the density of the solvent, with M, g_n, h, R and T as defined in Eq. (30). For oxygen both M and v^∞ have a value of 32. The density of water is always near 1 and $(M - v^\infty \rho)$ is practically zero in the case of freshwater, or very small for seawater ($\rho = 1.023$ g cm^{-3}). Therefore the ex-

Table 5. The increase of p_{O_2} with depth in water saturated with air at sea level pressure. Mean temperature of water column $5°C$

Depth h in m	$p_{O_2,h}/p_{O_2}$
100	1.01
500	1.06
1000	1.13
5000	1.84
10000	3.40

Table 6. Conversion factors for units of amount of oxygen

	= μmol	mg	cm^3 (ideal gas STP)	cm^3 (real gas STP)
1 μmol	= 1	0.031999	0.022414	0.022392
1 mg	= 31.251	1	0.70046	0.69978
1 cm^3 (ideal gas STP)	= 44.615	1.4276	1	0.99902
1 cm^3 (real gas STP)	= 44.659	1.4290	1.00098	1

Relative molecular mass of oxygen	$M_r(O_2) = 31.9988$
Standard molar volume of ideal gas, STPD	$V_o = 22.414$ dm^3 mol^{-1}
Standard molar volume of real gas, STPD	$V_m(O_2) = 22.392$ dm^3 mol^{-1}

Table 7. Solubility of oxygen, C_C, in air saturated pure water (μmol $O_2 \times$ dm^{-3})[1]

Temp °C	Atmospheric pressure in kilopascal									
	85	86	87	88	89	90	91	92	93	94
0.0	382.9	387.4	392.0	396.5	401.0	405.6	410.1	414.6	419.2	423.7
1.0	372.3	376.7	381.1	385.5	389.9	394.3	398.8	403.2	407.6	412.0
2.0	362.1	366.4	370.7	375.0	379.3	383.6	387.9	392.2	396.5	400.8
3.0	352.4	356.6	360.8	365.0	369.2	373.3	377.5	381.7	385.9	390.1
4.0	343.2	347.2	351.3	355.4	359.5	363.5	367.6	371.7	375.7	379.8
5.0	334.3	338.2	342.2	346.2	350.2	354.1	358.1	362.1	366.0	370.0
6.0	325.8	329.7	333.5	337.4	341.3	345.2	349.0	352.9	356.8	360.7
7.0	317.7	321.5	325.2	329.0	332.8	336.6	340.3	344.1	347.9	351.7
8.0	309.9	313.6	317.3	321.0	324.6	328.3	332.0	335.7	339.4	343.1
9.0	302.4	306.0	309.6	313.2	316.8	320.4	324.1	327.7	331.3	334.9
10.0	295.3	298.8	302.3	305.8	309.4	312.9	316.4	319.9	323.4	327.0
11.0	288.4	291.8	295.3	298.7	302.2	305.6	309.1	312.5	315.9	319.4
12.0	281.8	285.2	288.5	291.9	295.3	298.7	302.0	305.4	308.8	312.1
13.0	275.5	278.8	282.1	285.4	288.7	291.9	295.2	298.5	301.8	305.1
14.0	269.4	272.6	275.8	279.1	282.3	285.5	288.7	292.0	295.2	298.4
15.0	263.5	266.7	269.8	273.0	276.2	279.3	282.5	285.6	288.8	292.0
16.0	257.9	261.0	264.1	267.2	270.3	273.4	276.5	279.6	282.7	285.8
17.0	252.5	255.5	258.5	261.6	264.6	267.6	270.7	273.7	276.8	279.8
18.0	247.2	250.2	253.2	256.2	259.1	262.1	265.1	268.1	271.1	274.0
19.0	242.2	245.1	248.0	251.0	253.9	256.8	259.7	262.6	265.6	268.5
20.0	237.3	240.2	243.1	245.9	248.8	251.7	254.5	257.4	260.3	263.1
21.0	232.6	235.4	238.3	241.1	243.9	246.7	249.5	252.3	255.2	258.0
22.0	228.1	230.8	233.6	236.4	239.1	241.9	244.7	247.4	250.2	253.0
23.0	223.7	226.4	229.1	231.8	234.6	237.3	240.0	242.7	245.4	248.2
24.0	219.4	222.1	224.8	227.5	230.1	232.8	235.5	238.2	240.8	243.5
25.0	215.3	218.0	220.6	223.2	225.8	228.5	231.1	233.7	236.4	239.0
26.0	211.3	213.9	216.5	219.1	221.7	224.3	226.9	229.4	232.0	234.6
27.0	207.5	210.0	212.5	215.1	217.6	220.2	222.7	225.3	227.8	230.4
28.0	203.7	206.2	208.7	211.2	213.7	216.2	218.8	221.3	223.8	226.3
29.0	200.1	202.5	205.0	207.5	209.9	212.4	214.9	217.3	219.8	222.3
30.0	196.5	199.0	201.4	203.8	206.2	208.7	211.1	213.5	216.0	218.4
31.0	193.1	195.5	197.9	200.3	202.7	205.1	207.5	209.8	212.2	214.6
32.0	189.7	192.1	194.4	196.8	199.2	201.5	203.9	206.3	208.6	211.0
33.0	186.4	188.8	191.1	193.4	195.8	198.1	200.4	202.8	205.1	207.4
34.0	183.3	185.6	187.9	190.2	192.5	194.7	197.0	199.3	201.6	203.9
35.0	180.2	182.4	184.7	187.0	189.2	191.5	193.8	196.0	198.3	200.6
36.0	177.1	179.3	181.6	183.8	186.1	188.3	190.5	192.8	195.0	197.3
37.0	174.1	176.4	178.6	180.8	183.0	185.2	187.4	189.6	191.8	194.0
38.0	171.2	173.4	175.6	177.8	180.0	182.2	184.3	186.5	188.7	190.9
39.0	168.4	170.5	172.7	174.9	177.0	179.2	181.3	183.5	185.6	187.8
40.0	165.6	167.7	169.9	172.0	174.1	176.3	178.4	180.5	182.6	184.8

1 This table has been calculated from Eqs. (8) and (19) inserted into Eq. (18)
2 This column has been calculated for the standard pressure of 101.325 kPa to give unit standard concentration, C^*

95	96	97	98	99	100	101^2	102	103	104
428.2	432.8	437.3	441.8	446.4	450.9	456.9	460.0	464.5	469.0
416.4	420.8	425.2	429.6	434.0	438.4	444.3	447.3	451.7	456.1
405.1	409.3	413.6	417.9	422.2	426.5	432.2	435.1	439.4	443.7
394.2	398.4	402.6	406.8	411.0	415.1	420.7	423.5	427.7	431.9
383.9	388.0	392.0	396.1	400.2	404.2	409.6	412.4	416.5	420.5
374.0	378.0	381.9	385.9	389.9	393.8	399.1	401.8	405.7	409.7
364.5	368.4	372.3	376.1	380.0	383.9	389.0	391.6	395.5	399.4
355.5	359.2	363.0	366.8	370.6	374.4	379.4	381.9	385.7	389.5
346.8	350.5	354.2	357.9	361.5	365.2	370.1	372.6	376.3	380.0
338.5	342.1	345.7	349.3	352.9	356.5	361.3	363.7	367.3	370.9
330.5	334.0	337.5	341.1	344.6	348.1	352.8	355.1	358.7	362.2
322.8	326.3	329.7	333.2	336.6	340.0	344.6	346.9	350.4	353.8
315.5	318.9	322.2	325.6	329.0	332.3	336.8	339.1	342.4	345.8
308.4	311.7	315.0	318.3	321.6	324.9	329.3	331.5	334.8	338.1
301.7	304.9	308.1	311.3	314.6	317.8	322.1	324.2	327.5	330.7
295.1	298.3	301.5	304.6	307.8	310.9	315.1	317.3	320.4	323.6
288.9	292.0	295.1	298.2	301.3	304.4	308.5	310.6	313.7	316.7
282.8	285.9	288.9	291.9	295.0	298.0	302.0	304.1	307.1	310.2
277.0	280.0	283.0	286.0	288.9	291.9	295.9	297.9	300.8	303.8
271.4	274.3	277.3	280.2	283.1	286.0	289.9	291.9	294.8	297.7
266.0	268.9	271.7	274.6	277.5	280.4	284.2	286.1	289.0	291.8
260.8	263.6	266.4	269.2	272.1	274.9	278.6	280.5	283.3	286.1
255.7	258.5	261.3	264.1	266.8	269.6	273.3	275.1	277.9	280.7
250.9	253.6	256.3	259.0	261.8	264.5	268.1	269.9	272.6	275.4
246.2	248.8	251.5	254.2	256.9	259.5	263.1	264.9	267.6	270.2
241.6	244.3	246.9	249.5	252.1	254.8	258.3	260.0	262.7	265.3
237.2	239.8	242.4	245.0	247.6	250.1	253.6	255.3	257.9	260.5
232.9	235.5	238.0	240.6	243.1	245.6	249.0	250.7	253.3	255.8
228.8	231.3	233.8	236.3	238.8	241.3	244.6	246.3	248.8	251.3
224.7	227.2	229.7	232.2	234.6	237.1	240.4	242.0	244.5	247.0
220.8	223.3	225.7	228.1	230.6	233.0	236.2	237.9	240.3	242.7
217.0	219.4	221.8	224.2	226.6	229.0	232.2	233.8	236.2	238.6
213.3	215.7	218.1	220.4	222.8	225.2	228.3	229.9	232.2	234.6
209.7	212.1	214.4	216.7	219.1	221.4	224.5	226.1	228.4	230.7
206.2	208.5	210.8	213.1	215.4	217.7	220.8	222.3	224.6	226.9
202.8	205.1	207.4	209.6	211.9	214.2	217.2	218.7	221.0	223.2
199.5	201.7	204.0	206.2	208.5	210.7	213.7	215.2	217.4	219.6
196.3	198.5	200.7	202.9	205.1	207.3	210.2	211.7	213.9	216.2
193.1	195.3	197.4	199.6	201.8	204.0	206.9	208.4	210.5	212.7
190.0	192.1	194.3	196.4	198.6	200.7	203.6	205.1	207.2	209.4
186.9	189.0	191.2	193.3	195.4	197.6	200.4	201.8	204.0	206.1

ponential term will always be approximately 1. For seawater and a depth of 10,000 m, $C_{c,h}/C_c$ is 0.97.

This means that oxygen solubility at that depth is much the same as at the surface, while the partial pressure is several times higher [Eq. (30)]. A POS responds to partial pressure; if it is used for measuring oxygen concentration at greater depth, a correction according to Eq. (30) must be made.

References

1. Andrews FC (1972) Gravitational effects on concentrations and partial pressures in solutions: A thermodynamic analysis. Science 178:1199–1201
2. Benson BB, Krause D, Peterson MA (1979) The solubility and isotopic fractionation of gases in dilute aqueous solution: I Oxygen. J Solution Chem 8:655–690
3. Benson BB, Krause D (1980) The concentration and isotopic fractionation of gases dissolved in fresh water in equilibrium with the atmosphere: I Oxygen. Limnol Oceanogr 25 (4):662–671
4. Carritt DE, Carpenter JH (1966) Comparison and evaluation of currently employed modifications of the Winkler for determining dissolved oxygen in seawater; a NASCO Report. J Mar Res 24:286–318
5. Eckert CA (1973) The thermodynamics of gases dissolved at great depth. Science 180:426–427
6. Enns T, Scholander PF, Bradstreet ED (1965) Effect of hydrostatic pressure on gases dissolved in water. J Phys Chem 69:389–393
7. Fenn WO (1972) Partial pressure of gases dissolved at great depth. Science 176:1011–1012
8. Fox CJJ (1909) On the coefficients of absorption of nitrogen and oxygen in distilled water and sea water and of atmospheric carbon dioxide in sea water. Trans Faraday Soc 5:68
9. Green EJ, Carritt DE (1967) Oxygen solubility in sea water: thermodynamic influence of sea salt. Science 157:191
10. Gilbert W, Pawley W, Park K (1968) Carpenter's oxygen solubility tables and nomograph for seawater as function of temperature and salinity. Off Nav Res Data Rep no 29
11. Hitchman ML (1978) Measurement of dissolved oxygen. John Wiley & Sons and Orbisphere Corp, New York
12. Hutchinson GE (1957) A treatise on limnology. I. Geography, physics and chemistry. John Wiley, New York
13. Knudsen M, Forch C, Sörensen SPL (1902) Bericht über die chemische und physikalische Untersuchung des Meereswassers und die Aufstellung der neuen hydrographischen Tafeln. Wiss Meeresunters Abt Kiel 6:123–184
14. Mortimer CH (1982) The oxygen content of air-saturated fresh waters within ranges of temperature and atmospheric pressure of limnological interest. Mitt Int Ver Limnol 22:1–23
15. Ricker WE (1934) A critical discussion of various measures of oxygen saturation in lakes. Ecology 15:348–363
16. Schassmann H (1949) Die Sauerstoffsättigung natürlicher Wässer, ihre Ermittlung und ihre Bedeutung in der Hydrologie. Schweiz Z Hydrol 11:430–463
17. Setschenow J (1889) Concerning the concentration of salt solutions on the basis of their behaviour to carbonic acid. Z Phys Chem 4:117
18. Truesdale GA, Downing AL, Lowden GF (1955) The solubility of oxygen in pure water and sea water. J Appl Chem 5:53–63
19. Weiss RF (1970) The solubility of nitrogen, oxygen, and argon in water and sea water. Deep-Sea Res 17:721–735

20. Weitkamp DE, Katz M (1980) A review of dissolved gas supersaturation literature. Trans Am Fish Soc 109:659–702
21. Winkler LW (1888) Die Bestimmung des in Wasser gelösten Sauerstoffs. Ber Dtsch Chem Ges 21:2843–2854
22. Winkler LW (1889) Die Löslichkeit des Sauerstoffs in Wasser. Ber Dtsch Chem Ges 22:1764–1774

Appendix B Calculation of p_{O_2} in Water Equilibrated with a Mixture of Room Air and Nitrogen

E. Gnaiger[1]

When gas A is dry air and mixed with an oxygen-free dry gas or gas mixture B (e.g., N_2), then the p_{O_2} in water equilibrated with this gas mixture is

$$p_{O_2} = \left(_b p - p_{H_2O,T_s}\right) \times \phi_A \times \phi^*_{O_2}, \tag{1}$$

where $_b p$ is barometric pressure [kPa], p_{H_2O,T_s} is the saturation water vapor pressure at the temperature of the water sample, ϕ_A is the volume fraction of gas A in the gas mixture leaving the gas mixing pump,

$$\phi_A = \frac{p_A}{_b p} \quad \text{and} \quad 1 - \phi_A = \frac{p_B}{_b p}, \tag{2}$$

$\phi^*_{O_2}$ is the volume fraction of oxygen in dry air which can be taken as constant

$$\phi^*_{O_2} = 0.20946.$$

It may be more convenient to use room air instead of dry air. Then gas A contains water vapor proportional to the relative humidity, r_h, of room air,

$$p_{H_2O,A} = r_h \times p_{H_2O,T_A}, \tag{3}$$

where $p_{H_2O,A}$ is the water vapor pressure in room air, and p_{H_2O,T_A} is the saturation water vapor pressure at room temperature. The p_{O_2} in room air is therefore a function of r_h and T_A,

$$p_{O_2,A} = \left(_b p - p_{H_2O,A}\right) \times \phi^*_{O_2}. \tag{4}$$

According to Eq. (2), the water vapor pressure and oxygen pressure respectively in the gas mixture leaving the pump are

$$p_{H_2O,M} = \phi_A \times p_{H_2O,A} \tag{5}$$

and

1 Institut für Zoologie, Abteilung Zoophysiologie, Universität Innsbruck, Peter-Mayr-Str. 1A, A–6020 Innsbruck, Austria

Polarographic Oxygen Sensors (ed. by Gnaiger/Forstner)
© Springer-Verlag Berlin Heidelberg 1983

$$p_{O_2,M} = \phi_A \times p_{O_2,A}. \tag{6}$$

We cannot apply Eq. (1) if $p_{H_2O,M} > 0$.

To derive the appropriate algorithm we may regard room air as a mixture of three components,

$$_bp = p_{X,A} + p_{H_2O,A} + p_{O_2,A} \tag{7}$$

and we may express the pressure of the fraction of air not containing water vapor and oxygen, $p_{X,A}$, by combining Eqs. (4, 7),

$$p_{X,A} = (_bp - p_{H_2O,A}) \times (1 - \phi_{O_2}^*). \tag{8}$$

The pressure of gas not containing water vapor and oxygen in the gas mixture leaving the pump is

$$p_{X,M} = p_B + \phi_A \times p_{X,A}. \tag{9}$$

Recalling Eq. (2),

$$p_B = (1 - \phi_A) \times {_bp} \tag{10}$$

and inserting Eqs. (8, 10) into (9) yields after rearrangement

$$p_{X,M} = {_bp} \times (1 - \phi_A \times \phi_{O_2}^*) - p_{H_2O,A} \times \phi_A \times (1 - \phi_{O_2}^*). \tag{11}$$

The total pressure in the gas phase leaving the pump is then

$$_bp = p_{X,M} + p_{H_2O,M} + p_{O_2,M}. \tag{12}$$

What is the p_{O_2} after saturating this gas mixture with water vapor at T_s? When the water vapor pressure is brought from $p_{H_2O,M}$ to p_{H_2O,T_s} then the total pressure remains constant, hence

$$_bp = (p_{X,M} + p_{O_2,M}) \times x + p_{H_2O,T_s} \tag{13}$$

and

$$x = \frac{_bp - p_{H_2O,T_s}}{p_{X,M} + p_{O_2,M}} = \frac{_bp - p_{H_2O,T_s}}{_bp - p_{H_2O,A} \times \phi_A} \tag{14}$$

[see Eqs. (4, 6, 11) for obtaining the second expression]. The p_{O_2} in solution equals

$$p_{O_2} = p_{O_2,M} \times x. \tag{15}$$

Inserting Eqs. (4, 6, 14) in Eq. (15) we finally get

$$p_{O_2} = (_bp - p_{H_2O,T_s}) \times \phi_A \times \phi_{O_2}^* \times \frac{_bp - p_{H_2O,A}}{_bp - p_{H_2O,A} \times \phi_A}. \tag{16}$$

Thus we derived a correction factor, f_r for Eq. (1).

$$f_r = \frac{{}_bp - p_{H_2O,A}}{{}_bp - p_{H_2O,A} \times \phi_A} = 1 - p_{H_2O,A} \times \frac{1 - \phi_A}{{}_bp - p_{H_2O,A} \times \phi_A}. \tag{17}$$

f_r equals 1 when $\phi_A = 1$, i.e., for saturation with water saturated air [Eq. (1)]. Since ${}_bp \gg p_{H_2O,A} \times \phi_A$, we may neglect the term $p_{H_2O,A} \times \phi_A$ in Eq. (17) to obtain the simple approximation

$$f_r \approx 1 - p_{H_2O,A} \times \frac{1 - \phi_A}{{}_bp} = 1 - \phi_{H_2O,A} \times \phi_B. \tag{18}$$

Equation (18) approximates f_r to better than 1% for low volume fractions of air, $\phi_A < 0.4$, or high volume fractions of gas B, $\phi_B > 0.6$. This shows that the correction factor, f_r, is linearly dependent upon the volume fraction of water vapor in air, $\phi_{H_2O,A}$, and the relative volumes of the two gases mixed by the pump.

In practice this means that the error encountered by neglecting the correction factor for Eq. (1) increases with room temperature and relative humidity of air. It becomes relatively more important when working at low air saturations, and increases slightly with altitude (with decreasing barometric pressure). Under normal laboratory conditions and at air saturations set below 50%, errors from 1% to 3% result from disregarding f_r (Fig. 1). Hence the significance of the correction for $p_{H_2O,A}$ is similar to that of the correction for p_{H_2O,T_s} in Eq. (1) which is generally accounted for.

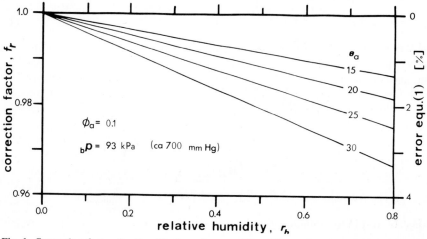

Fig. 1. Correction factor for Eq. (1), f_r, and error of Eq. (1) as a function of relative humidity of room air, r_h, and room temperature (numbers in °C) when 10% room air is mixed with 90% of an oxygen-free dry gas

Appendix C   Calculation of Energetic and Biochemical Equivalents of Respiratory Oxygen Consumption

E. Gnaiger[1]

1 Introduction

The evaluation of metabolic energy loss is the major aim of respiratory studies in ecological energetics and for the construction of energy budgets. The polarographic oxygen sensor is becoming the prominent tool for indirect calorimetry in laboratory as
well as in field investigations (Parts II and III). In this context a comprehensive outline
of the calculation procedures for conversion of oxygen consumption data to energy
equivalents is warranted. A more rigorous treatment of the subject will be presented
elsewhere [6].

2 Stoichiometry

2.1 Respiratory Quotient and Nitrogen Quotient

The respiratory exchange ratio or respiratory quotient, RQ or ν_{CO_2}/O_2 (moles of CO_2
liberated per mole of O_2 consumed), represents a well-known concept for the estimation of the oxycaloric equivalent in relation to the relative proportions of carbohydrate and lipid respired. The oxidation of carbohydrate involves equal amounts of carbon
dioxide and oxygen ($RQ = 1$), but, due to the more reduced state of lipids, less CO_2 is
liberated per mole of oxygen in the oxidation of fat ($RQ = 0.72$). The RQ for protein
varies as a function of the excretory product. Taking alanine as an example,

$$NH_2CHCH_3COOH + 3\,O_2 \rightarrow 2.5\,CO_2 + 1.5\,H_2O + 0.5\,CO(NH_2)_2 \qquad (1a)$$

$$NH_2CHCH_3COOH + 3\,O_2 \rightarrow 3\quad CO_2 + 2\quad H_2O + \quad NH_3, \qquad (1b)$$

the respiratory quotient for urea, RQ_{urea}, equals 0.83, whereas for ammonia RQ_{NH_3}
is 1.0.

1 Institut für Zoologie, Abteilung Zoophysiologie, Universität Innsbruck, Peter-Mayr-Str. 1A,
A–6020 Innsbruck, Austria
and Institute for Marine Environmental Research, Prospect Place, The Hoe, Plymouth PL1 3DH,
Great Britain

Polarographic Oxygen Sensors (ed. by Gnaiger/Forstner)
© Springer-Verlag Berlin Heidelberg 1983

The ammonia RQ ranged from 0.94 to 0.99 for bacterial, plant and animal proteins as calculated from their amino acid compositions [6].

In addition to the respiratory quotient, the nitrogen quotient, NQ [ν_{N/O_2}, moles of N excreted per mole of O_2 consumed, i.e., moles of NH_3 per O_2 or 2 × moles of $CO(NH_2)_2$ per O_2], must be known in order to attain accurate energetic and biochemical interpretations of oxygen consumption data. For proteins from various sources, NQ averages 0.27 ± 0.01 (Table 1).

RQ and NQ are often expressed in terms of volume ratios. These values are unclear if the use of ideal gas volumes (V_o = 22.414 dm^3 mol^{-1}), or of real gas volumes, V_m [dm^3 mol^{-1}], at STP [7] for oxygen (22.392), CO_2 (22.262), and NH_3 (22.117) is not specified. Real volume-based values of RQ should be multiplied by 1.006 and those of NQ by 1.012 to convert them to molar ratios.

2.2 Oxygen Consumption and Respiratory Substrates

The proportion of different respiratory substrates can be calculated on the basis of oxygen consumption measurements and accompanying RQ and NQ values. The principal approach for distinguishing between two substrate categories in maintenance metabolism has been outlined in classical textbooks [9]. Equation (2) summarises the stoichiometric analysis extended to separate the mass fractions of carbohydrate, lipid and protein, w_K, w_L and w_P respectively. With reference atomic compositions shown in Table 2 we obtain for ammonioteles (see Fig. 1),

$$w_K = \frac{-0.72 \quad + \quad RQ - 0.9259 \times NQ}{-0.3070 + 0.5870 \times RQ - 0.1298 \times NQ} \tag{2a}$$

$$w_L = \frac{1 \quad - \quad RQ - 0.1111 \times NQ}{-0.7432 + 1.4212 \times RQ - 0.3142 \times NQ} \tag{2b}$$

$$w_P = \frac{NQ}{-0.3646 + 0.6971 \times RQ - 0.1541 \times NQ} \ . \tag{2c}$$

Corrected RQ values have to be inserted into Eq. (2) if a mole fraction x_{urea} of nitrogen is excreted as urea.

$$RQ \text{ (corrected)} = RQ \text{ (measured)} + 0.5 \times y_{urea} \times NQ. \tag{2d}$$

Similarly, the loss of ash-free organic biomass catabolised per mole of oxygen respired, $\Delta_k W_{O_2}$ [g (mol O_2)$^{-1}$], can be assessed as a function of RQ and NQ (Fig. 2),

$$\Delta_k W_{O_2} = 29.63 - 56.66 \times RQ + 12.52 \times NQ. \tag{3}$$

If the fractions of the catabolised substrates, w_i, are known without information on CO_2 production and nitrogen excretion, then

$$\Delta_k W_{O_2} = 1/\sum_i \frac{w_i}{\Delta_k W_{O_2}(i)}, \tag{4}$$

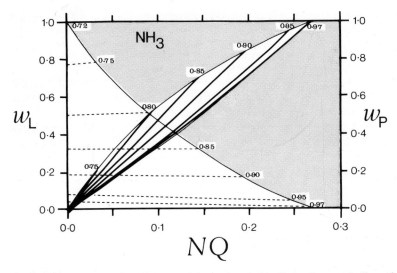

Fig. 1. Mass fractions of respiratory substrates in aerobic dissipative metabolism of ammoniotelic organisms as a function of the nitrogen quotient, NQ, and the respiratory quotient, RQ (*numbers*). *Full lines* mass fraction of protein, w_P, for different values of RQ; *broken lines* mass fraction of lipid, w_L, for different values of RQ. The mass fraction of carbohydrate, w_K, is obtained as

$$w_K = 1 - w_L - w_P$$

RQ/NQ combinations extending into the *hatched area* indicate the presence of anabolic or partial anoxic metabolism [see Eq. (2)]

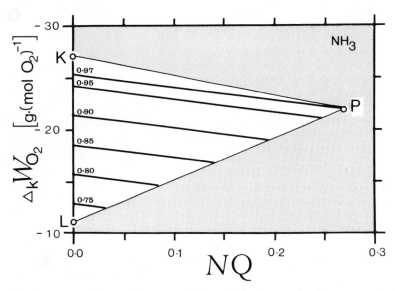

Fig. 2. Organic mass (ash-free dry weight) catabolised per mole of oxygen respired, $\Delta_k W_{O_2}$, as a function of the nitrogen quotient, NQ, and the respiratory quotient, RQ, with ammonia as the excretory product. The *full lines* display $\Delta_k W_{O_2}$ for different RQ values indicated by *numbers*. *Circles* (K, L, P) designate the pure reference substrates [see Eq. (3)]

Table 1. Substrate specific respiratory quotients, RQ_i, nitrogen quotients, NQ_i, catabolic weight equivalents of oxygen consumption, $\Delta_k W_{O_2}(i)$, and oxycaloric and oxyenthalpic equivalents, $\Delta_k H_{O_2}(i)$ and $\Delta_c H_{O_2}(i)$ respectively, in respiratory (dissipative) metabolism of aquatic organisms. For a protein substrate the values are given either for ammoniotelic organisms (protein $\rightarrow$ NH_3) or for ureotelic organisms (protein $\rightarrow$ urea) [6]

Substrate, i	RQ_i	NQ_i	$\Delta_k W_{O_2}(i)$ g mol^{-1}	$\Delta_k H^{\circ}_{O_2}(i)$ kJ mol^{-1}	$\Delta_k H'_{O_2}(i)$ kJ mol^{-1}	$\Delta_c H_{O_2}(i)$ kJ mol^{-1}
Carbohydrate	1.0	0.0	−27.02	−471	−478	−473
Lipid	0.72	0.0	−11.16	−440	−445	−441
Protein $\rightarrow$ NH_3	0.97	0.27	−21.95	−447	−451	−527
Protein $\rightarrow$ urea	0.84			−438	−443	

Table 2. Chemical and thermodynamic specification of standard carbohydrate (glycogen, starch), lipid (triacylglycerol) and protein (average amino acid composition). $\Delta_c h_i$ is the specific enthalpy of combustion obtained by bomb calorimetry [2]. $\Delta_c H_C(i)$ is the substrate specific molar enthalpy of combustion based on carbon

Substrate, i	Atomic composition	$\Delta_c h_i$ kJ g^{-1}	$\Delta_c H_C(i)$ kJ (mol C)$^{-1}$
Carbohydrate	$(C_6 H_{10} O_5)_n$	−17.5	−473
Lipid	$(C_{18.8} H_{33.0} O_2)_3$	−39.5	−611
Protein	$(C_{4.83} H_{7.58} O_{1.50} N_{1.35} S_{0.025})_n$	−24.0	−543

where i indicates the three substrates (K, L and P), and $\Delta_k W_{O_2}(i)$ are the substrate-specific weight equivalents of oxygen consumption (Table 1).

3 Indirect Calorimetry: Two Objectives

Insufficient attention has been paid to the conceptual disparity of two closely related functions of indirect calorimetry. The classical aim encountered a recent rennaissance with aquatic animals: Calorimetrically determined rates of heat dissipation are compared with heat changes of metabolic reactions (Chaps. II.3, II.4) [1, 5, 9]. Oxygen uptake measurements, $\dot{N}_{O_2}$, are converted to rates of *heat* dissipation, $\dot{Q}$, with appropriately derived oxy*caloric* equivalents, $\Delta_k H_{O_2}$,

$$_k\dot{Q} = \Delta_k H_{O_2} \times \dot{N}_{O_2}. \tag{5}$$

In the context of physiological and ecological energy balance studies, however, the oxy*enthalpic* equivalent, $\Delta_c H_{O_2}$, should be used to convert oxygen consumption to values of *enthalpies of combustion* of the aerobically catabolised biomass.

3.1 The Oxycaloric Equivalent, $\Delta_k H_{O_2}$

Chemical stoichiometries and the corresponding enthalpies of reaction form the basis of calculating the caloric equivalent of oxygen consumption in dissipative metabolism, $\Delta_k H_{O_2}$, in units [kJ $(mol\ O_2)^{-1}$] (for conversion factors between different units see Tables 3 and 4)[2]. Earlier calculations were based on enthalpies of combustion as determined in bomb calorimeters, e.g. [1, 4, 8, 9]. For calculating heat dissipation of aquatic organisms, consideration of the aqueous phase is more appropriate than reliance upon thermochemical values pertaining to dried solids or undissolved gases [5, 11]. Enthalpies of solution and dilution are very small in comparison to the large enthalpies of oxidation, but they were, together with enthalpies of dissociation, incorporated in the present calculations of the standard oxycaloric equivalent, $\Delta_k H^\circ_{O_2}$. An important side reaction is the conversion of ammonia to ammonium ion with an enthalpy of protonation of -52 kJ mol^{-1} [10] and neglect thereof [3] entails a significant error. While other neutralization reactions are very important in anoxic metabolism [5], they increase $\Delta_k H^\circ_{O_2}$ by $< 3\%$ under ecological conditions. Their effect is indicated by listing the oxycaloric equivalents, $\Delta_k H'_{O_2}$ for pH 7 and an effective enthalpy of neutralization, $\Delta_b H'_{H+}$, of -8 kJ $(mol\ H^+)^{-1}$ (Table 1),

$$\Delta_k H'_{O_2} = \Delta_k H^\circ_{O_2} + \nu'_{H^+/O_2} \times \Delta_b H'_{H+}. \tag{6}$$

An approximate calculation of the stoichiometric coefficient of protons released per mole of oxygen, ν'_{H+/O_2}, suffices for practical purposes,

$$\nu'_{H+/O_2} \approx RQ \times \frac{1}{1 + \exp(pK'-pH)} - x_{NH_3} \times NQ, \tag{7}$$

where pK' is the apparent dissociation constant of carbonic acid (6.37 in pure water [7]), and x_{NH_3} is the mole fraction of nitrogen excreted as ammonia ($= 1-x_{urea}$).

For aquatic ammonio-ureotelic organisms the oxycaloric equivalent, $\Delta_k H_{O_2}$ [kJ $(mol\ O_2)^{-1}$] can be accurately calculated on the basis of RQ, NQ and the nature of the excretory product (Fig. 3),

$$\Delta_k H^\circ_{O_2} = -360 - 111 \times RQ + (77 - 20 \times x_{urea}) \times NQ \tag{8a}$$

$$\Delta_k H'_{O_2} = -360 - 118 \times RQ + (87 - 27 \times x_{urea}) \times NQ. \tag{8b}$$

Equation (8b) is calculated for the conditions pH = 7 and $\Delta_b H'_{H+} = -8$ kJ $(mol\ H^+)^{-1}$. The oxycaloric equivalent can also be calculated from the mass fractions, w_i (carbohydrate, lipid and protein),

---

2 In $\Delta_k H_{O_2}$ the suffix "k" stands for "catabolic half cycle" [5] to make it clear that this expression does not apply to coupled catabolic reactions with a net gain of ATP from phosphorylation of ADP, but pertains to the dissipative conversion of substrates to carbon dioxide, water, and nitrogenous endproducts without storage of high energy intermediates (see App. E)

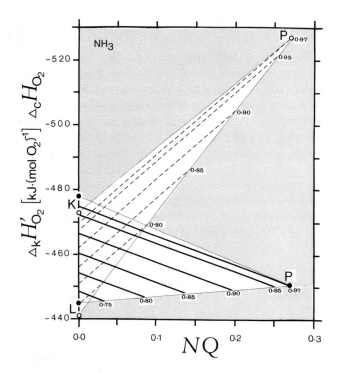

Fig. 3. The oxycaloric and oxyenthalpic equivalent, $\Delta_k H'_{O_2}$ and $\Delta_c H_{O_2}$ as a function of NQ and RQ with ammonia as the excretory product. *Full lines* $\Delta_k H'_{O_2}$ for different values of RQ indicated by numbers; *broken lines* $\Delta_c H_{O_2}$ for different values of RQ [see Eqs. (8, 14)]

$$\Delta_k H_{O_2} = \sum_i w_i \times \frac{\Delta_k W_{O_2}}{\Delta_k W_{O_2}(i)} \times \Delta_k H_{O_2}(i). \tag{9}$$

$\Delta_k W_{O_2}/\Delta_k W_{O_2}(i)$ is a correction factor for converting mass fractions, w_i, into molar fractions of oxygen consumption, $x_{O_2}(i)$ (= oxygen consumed for a particular substrate, i, per total oxygen consumption). $\Delta_k W_{O_2}$ is calculated from Eq. (4) and the substrate specific constants are listed in Table 1.

3.2 The Oxyenthalpic Equivalent, $\Delta_c H_{O_2}$

The fundamental importance of correct energy conversion factors for oxygen consumption data is reflected in the bioenergetic balance equation,

$$P = A + R + U. \tag{10}$$

Production, P, is the result of the net flow of energy into the open system, A (assimilation), and the flow of energy out of the system due to respiration, R, and excretion,

U. The flows of assimilated and excreted energy are bound to the exchange of organic matter. Dry biomass is converted into energy equivalents by bomb calorimetry, yielding specific enthalpies of combustion, $\Delta_c h$ [kJ g^{-1}]. Respiratory energy flow, R, relates to the flow of heat from an isothermal biological system. Actual heat effects occurring in the aqueous phase, however, are not at issue in this context. To satisfy the energy conservation law of thermodynamics in Eq. (10) the same reference state has to be chosen for metabolic loss and assimilatory gain. For this sake oxygen uptake, $\dot{N}_{O_2}$, is converted to the *enthalpy of combustion* equivalent of the catabolised organic mass, $_H R$, with consistently derived oxy*enthalpic* equivalents, $\Delta_c H_{O_2}$ [kJ (mol O_2)$^{-1}$],

$$_H R = R + U = \Delta_c H_{O_2} \times \dot{N}_{O_2}. \tag{11}$$

The oxyenthalpic equivalent is directly related to specific enthalpies of combustion, $\Delta_c h$,

$$\Delta_c H_{O_2} = \Delta_c h \times \Delta_k W_{O_2}. \tag{12}$$

The organic mass that is catabolised per mole of oxygen consumed, $\Delta_k W_{O_2}$ [g (mol O_2)$^{-1}$], is calculated from Eqs. (3) or (4), and $\Delta_c h$ [kJ g^{-1}] is obtained (Table 2) as

$$\Delta_c h = w_K \times \Delta_c h_K + w_L \times \Delta_c h_L + w_P \times \Delta_c h_P. \tag{13}$$

The nitrogen correction which is essential in the oxycaloric equivalent is omitted in deriving the oxyenthalpic equivalent [Eq. (12)]. This renders calculations of the total metabolic energy (enthalpy) flow, $_H R$, straightforward and correct. The fate of protein-nitrogen determines merely the partitioning of $_H R$ between R and U, without any influence on the conservative total value of $_H R$. Appreciation of the different concepts of indirect calorimetry as represented by the oxyenthalpic and oxycaloric coefficient obliterates a frequently encountered bias in energy budget calculations. Metabolic losses were usually calcualted with oxycaloric equivalents. If energy losses in excretion were then considered insignificant, the correction for ammonia or urea excretion, implicit in $\Delta_k H_{O_2}$, should have been avoided. In fact, due to the uncorroborated nitrogen correction, metabolic losses may have been underestimated by up to 15%, since protein is predominantly catabolised by many fish and aquatic invertebrates.

Table 3. Conversion factors for units of the oxycaloric and oxyenthalpic equivalent (energy per amount of oxygen). These factors are based on the ideal molar gas volume (see App. A, Table 6) and on the thermodynamic calorie (= 4.1868 J). Often the thermochemical calorie (= 4.184 J) is used

		kJ (mol O_2)$^{-1}$	kJ (g O_2)$^{-1}$	kJ (dm O_2)$^{-3}$
1 mJ μmol^{-1}	=	1	0.031251	0.044615
1 J mg^{-1}	=	31.9988	1	1.4276
1 J cm^{-3}	=	22.414	0.70046	1
1 kcal mol^{-1}	=	4.1868	0.13084	0.18679
1 cal mg^{-1}	=	133.97	4.1868	5.9772
1 cal cm^{-3}	=	93.843	2.9327	4.1868

Table 4. Conversion factors for units of oxygen consumption (amount of oxygen utilized per time) and heat dissipation (power, energy per time) on the basis of a generalized oxycaloric equivalent, $\Delta_k H_{O_2} = -450$ kJ (mol O_2)$^{-1}$. $_k\dot{W}$ is the rate of weight loss (dry organic matter) due to aerobic catabolism of substrates with a catabolic weight equivalent of oxygen, $\Delta_k W_{O_2} = -19$ g (mol O_2)$^{-1}$. No attention is paid to the sign of the conversion factors. Heat and weight loss, however, should always have a negative sign

		$\dot{N}_{O_2}$		$_k\dot{Q}$		$_k\dot{W}$
		μmol O_2 h^{-1}	nmol O_2 s^{-1}	mW	J h^{-1}	mg h^{-1}
1 μmol O_2 h^{-1}	=	1	0.27778	0.1250	0.450	0.019
1 nmol O_2 s^{-1}	=	3.600	1	0.450	1.620	0.068
1 mg O_2 h^{-1}	=	31.251	8.6809	3.906	14.06	0.594
1 cm^3 O_2 h^{-1}	=	44.615	12.393	5.577	20.08	0.848
1 mW = 1 mJ s^{-1}	=	8.000	2.222	1	3.600	0.152
1 J h^{-1}	=	2.222	0.6173	0.27778	1	0.042
1 cal h^{-1}	=	9.304	2.584	1.1630	4.1868	0.177

Analogous to Eq. (8) the oxyenthalpic equivalent can be derived from RQ, NQ and constants in Table 1 (Fig. 3),

$$\Delta_c H_{O_2} = -359 - 114 \times RQ - (213 + 56 \times x_{urea}) \times NQ. \tag{14}$$

The highest accuracy in respiratory energetics is accomplished only if the appropriate conversion coefficients are applied (Table 4). These complement the increased fidelity of continuous oxygen uptake measurements achieved by POS.

Acknowledgments. This study was in part supported by the *Fonds zur Förderung der wissenschaftlichen Forschung in Österreich,* project no. 3917.

References

1. Blaxter KL (1967) The energy metabolism of ruminants. Hutchinson Scientific and Technical, London, 332 pp
2. Domalski ES (1972) Selected values of heats of combustion and heats of formation of organic compounds containing the elements C, H, N, O, P and S. J Phys Chem Ref Data 1:221–277
3. Elliott JM, Davison W (1975) Energy equivalents of oxygen consumption in animal energetics. Oecologia 19:195–201
4. Gnaiger E (1977) Thermodynamic consideration of invertebrate anoxibiosis. In: Lamprecht I, Schaarschmidt B (eds) Application of calorimetry in life sciences. Walter de Gruyter, Berlin, pp 281–303
5. Gnaiger E (1980) Das kalorische Äquivalent des ATP-Umsatzes im aeroben und anoxischen Metabolismus. Thermochim Acta 40:195–223
6. Gnaiger E (in prep) Energy equivalents of oxygen consumption in relation to direct calorimetry and energy budgets in aquatic animals
7. Weast RC (1974–1975) Handbook of chemistry and physics. CRC Press
8. Ivlev VS (1934) Eine Mikromethode zur Bestimmung des Kaloriengehaltes von Nährstoffen. Biochem Z 275:49–55

9. Kleiber M (1961) The fire of life. An introduction to animal energetics. John Wiley, New York, 454 pp
10. Vanderzee CE, Mansson M, Wadsö I, Sunner S (1972) Enthalpies of formation of mono- and diammonium succinates and of aqueous ammonia and ammonium ion. J Chem Thermodynamics 4:541–550
11. Wilhoit I (1969) Selected values of thermodynamic properties. In: Brown HD (ed) Biochemical microcalorimetry. Academic Press, New York, pp 305–317

Appendix D   The Winkler Determination[1]

H.L. Golterman[2]

The most precise determination of dissolved oxygen can be carried out by a iodometric titration according to Winkler [15]. Although this method has been modified somewhat, the principle is unchanged and it is therefore probably the oldest method for water analysis still employed.

The dissolved oxygen in the water combines with $Mn(OH)_2$ forming higher hydroxides under alkaline conditions, which, on subsequent acidification in the presence of I^- libarate I_2 in an amount stoichiometrically equivalent to the original dissolved O_2 content of the sample. There is doubt on the oxidation state of the oxidized manganese hydroxide, but as eventually all electrons transported from the manganese to the oxygen will be donated from the iodide to the manganese, this argument is not essential for the practical employment of the method. The reactions are as follows:

$$2\,H_2O + 4\,Mn(OH)_2 + O_2 \rightarrow 4\,Mn(OH)_3$$

or

$$2\,H_2O + 2\,Mn(OH)_2 + O_2 \rightarrow 2\,Mn(OH)_4$$

followed by

$$4\,Mn(OH)_3 \text{ or } 2\,Mn(OH)_4 + (12 \text{ or } 8)\,H^+ + 4\,I^- \rightarrow (4 \text{ or } 2)\,Mn^{2+} +$$
$$+ (12 \text{ or } 8)\,H_2O + 2\,I_2 \,.$$

The I_2 is then determined by titration with (sodium)thiosulfate:

$$I_2 + 4\,Na^+ + 2\,S_2O_3^{2-} \rightarrow 2\,I^- + 4\,Na^+ + S_4O_6^{2-} + 2\,H^+.$$

However, any other method of titrating the liberated I_2 can be used equally well, especially when potentiometric apparatus is available. (As_2O_3, e.g., has the advantage of a stable stock solution and does not need to be standardized against another standard, as the thiosulfate; the pH has however to be between 4 and 9, which can be easily adjusted with borax buffer solutions).

Normally, however, thiosulfate is used. The endpoint is often determined with the blue color formed by I_2 with starch.

---

1 This article combines relevant parts of Golterman et al. (1978) and Golterman and Wisselo (1981)
2 Station Biologique de la Tour du Valat, Camargue, Le Sambuc, F−13200 Arles, France

Table 1. Concentrations of fixative and titrant solutions

Solutions	Concentrations		Add
Fixative	mol dm^{-3} ^{a}	g dm^{-3}	per 100–150 cm^3
MnSO$_4$ $\times$ 5 H$_2$O^b	2	500	1 cm^3
Alkaline iodide (with or without azide)	10	400 g NaOH 900 g NaI (10–25 g NaN$_3$)	1 cm^3
Titrants			
Na$_2$S$_2$O$_3$ $\times$ 6 H$_2$O (with 1 pellet NaOH)	0.0125 or 0.00250	6.2	Depending on O$_2$ conc. and volume of (sub)sample taken
1/6 KIO$_3$	0.100	3.567	

^{a} mol in the new S.I. sense is used
^{b} Most firms produce different hydrates; all can be used including MnCl$_2$. The concentration in mol dm^{-3} is decisive

In principle all compounds reacting with the I$_2$ will interfere. One of the most likely ones is nitrite. This interference can be eliminated by the use of NaN$_3$ (sodium azide), up to concentrations of 0.36 mmol dm^{-3} (= 5 mg dm^{-3} NO$_2$-N). It must be realized, however, that concentrations of 14 μg dm^{-3} of NO$_2$-N, which is a rather high value for unpolluted waters, is only equivalent to 0.1% of the normally occurring oxygen concentrations. This modification of the Winkler method is therefore only important for partially purified sewage waters.

An important modification of the Winkler method was given by Pomeroy and Kirschman [11], who advised a higher concentration of NaI than the reagent formerly used. The advantages are that errors due to volatilization of I$_2$ and to interference by organic matter are reduced (because of the reaction I$_2$ + I$^-$ $\rightleftharpoons$ I$_3^-$), that the hydroxides dissolve more readily, and that the starch end point is sharper. The possible interference of organic matter being oxidized by the liberated I$_2$ is often overlooked. Alsterberg [1] proposed a bromination of the organic matter before the oxygen determination, while the excess Br$_2$ is later combined with salicylate. This reagent must be prepared freshly and must be colorless; it is best prepared from reagent pure, crystalline sodium salicylate, which is commercially available. The Br$_2$ destroys NO$_2^-$ as well, so that NaN$_3$ is not required. The method is not often employed; its accuracy will be discussed below.

Greater precision than with the starch end point determination can be obtained with potentiometric end-point detections, of which the best is the socalled dead stop end-point technique (cf. Chap. III.1). In this method thiosulfate is added in an accurately measured excess and is back-titrated with KIO$_3$; an extra advantage is that the I$_2$ occurs only for a very short moment rendering oxidation of organic matter less likely. The reactions are as follows:

$$I_2 + 2 S_2O_3^{2-} \rightarrow S_4O_6^{2-} + 2 I^- \quad \text{and}$$
$$IO_3^- + 2 S_2O_3^{2-} \rightarrow I^- + S_4O_6^{2-} \quad \text{and}$$

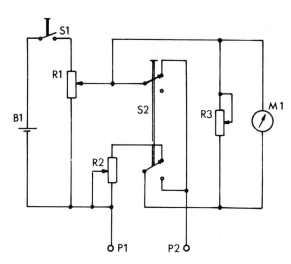

Fig. 1. Circuit diagram. (From Golterman et al. [5])

$$IO_3^- + 5\,I^- + 5\,H^+ \rightarrow 3\,I_2 + 3\,H_2O.$$

The dead-stop end-point technique is an amperometric or potentiometric titration with two identical Pt electrodes. One of these is made the indicator electrode by electrical polarization with a small constant voltage, normally 50–100 mV; the other one serves as a reference electrode. The electrodes are covered with very thin layers of H_2 and O_2 derived from the hydrolysis of water. Before the end point the current remains constant (practically zero). As the thiosulfate-tetrathionate couple is not reversibly oxidizable, the electrodes cannot be depolarised and the thin adsorbed gas layers serve as a barrier against the current. As soon as both IO_3^- and I^- are present, oxidation and reduction reactions become possible at the electrodes, the gas barrier (the electric resistance) disappears, the adsorbed oxygen being reduced and the adsorbed hydrogen being oxidized. Thus a sharp change in current and thus in voltage is found exactly at the end point.

Dead stop titrators may be bought, but they can be made relatively easily. A circuit diagram is given in Fig. 1. Relatively large Pt electrodes (gauze) are used here. Their large surface enables the usage of low voltages and small currents, so that the apparatus can easily be used in the field. The use of a precise micro burette is advisable. Best results will be obtained by graphing the current readings against volume of titrant as the trace is always slightly curved at the end point. A linear extrapolation is possible using points sufficiently distant from either side of the end point. With the apparatus of Fig. 1 we found that this, however, hardly improved the accuracy.

1 Sampling

It is important to realize that the precision of the method depends as much on the sampling procedure as on the chemical determination. For accurate work a displacement sampler is necessary. The lower end of the outlet tube of the sampler is placed above the bottom of the Winkler bottle. The contents of the sample bottle are displaced about three times, before the sample is collected. The bottle must always be completely filled and reagents or preservatives must be added immediately after sampling. If the samples cannot be titrated directly, they should be kept alkaline and be stored in a cool place, but avoiding temperature changes. The samples must never be kept after acidification; not even during the titration of other samples of a series.

For very shallow waters the sample may be taken in a syringe and analyzed by a small-scale technique [4]. See further discussion in [5] and [10].

Recently Golterman and Wisselo [6] presented a ceriometric method for dissolved oxygen and COD, following an older procedure using Ce^{3+} [7]. The principle is the same as for the Winkler method. The advantages are that no volatile I_2 is produced and that the Ce^{4+} may be measured either titrimetrically or colorimetrically. Combination with a COD estimation prevents all difficulties with organic matter. The chemical principles behind this combined COD-O_2 technique are not so straightforward as the Winkler method, the method is therefore not so easily understandable in the beginning.

2 Precision and Accuracy

The accuracy depends very much on the sampling precision and on the accuracy of the concentration of the titrant, i.e., the thiosulfate. Normally the concentration of the thiosulfate decreases slightly; this can be diminished by the addition of 0.5 g dm^{-3} of Na_2CO_3. The accuracy is high, as the method is based on stoichiometric reactions [3] and not on a calibration curve, while other possible factors are insignificantly small.

There are several papers about the precision of the Winkler determination; seldom, however, is the precision of this determination discussed for samples of natural waters (Chap. III.1).

Rebsdorf [12] found that the Alsterberg modification gave the same results as the Winkler method in unpolluted waters and higher results – but still too low – in polluted waters, which he attributed to oxidation of organic matter in the normal Winkler technique. He found the Alsterberg technique to have a lower precision than the Winkler.

A comprehensive review of modifications of the Winkler method [8] pointed out that several modifications lead to no improvement because they introduce new sources of error, decreasing the precision.

Bryan et al. [2] used a photometric titration by measuring the changing absorption of UV radiation. They mention a coefficient of variation of 0.1% for seawater; it seems likely that this will be less good in (eutrophic) freshwater.

Mackereth et al. [9] found a relative standard deviation of 0.6% ($n = 7$) for the normal Winkler titration for standard solutions. Talling [13] used an amperometric endpoint detection and found a reproducibility (mean minus extreme) for iodine solutions equivalent to 0.6 μmol O_2 dm^{-3}.

Carpenter [3] found an accuracy of 0.1% for the Winkler method, proving a stoichiometric reaction. He discussed precision of different iodometric end-point determinations, but not the precision of the Winkler method.

Tschumi et al. [14] mention that the precision of the Winkler method is often quoted as 0.6 μmol dm^{-3} and believed that this value refers to the standard deviation. The standard deviations in their study (using a photometric end-point detection) were 0.06 μmol dm^{-3} (0.02%) for H_2O, 0.5 μmol dm^{-3} (0.15%) for lake water containing 0.76 mm^3 phytoplankton per dm^3 and 1.0 μmol dm^{-3} for lake water containing 7.07 mm^3 dm^{-3}. The influence of the phytoplankton density on the precision is striking: 1.6 μmol dm^{-3} as largest difference for diluted lake water and 3.4 μmol dm^{-3} for their concentrated lake water. We have found the same phenomenon for the classical Winkler determination and think that this precision is controlled by the adsorbance of I_2 on the phytoplankton, which, although reversible, may not be sufficiently rapid for high precision. Therefore I feel that much more attention should be given in the discussions on the precision of this method to the phytoplankton density and that generalizations may be dangerous. The I_2 adsorption influence is overcome by an electrochemical end-point detection of a back titration of an accurately known excess of thiosulfate, which thus presents optimal conditions. Results that are possible with this method in lake water containing a high phytoplankton density are given in Table 2. From this table it can be seen that in 11 out of 15 cases the mean value differs by 0.6 to 0.9 μmol dm^{-3} from the extreme value. If a manual dead stop method is used (Fig. 1) this difference is about 0.6 μmol dm^{-3}. The pooled standard deviation of the series

Table 2. Results of Winkler determinations chosen at random from routine measurements. The Winkler determinations were carried out with an automated dead stop titration following procedure 8.1.4 of the I.B.P. manual Nr 8 (Golterman et al. [5])
All samples were from Tjeukemeer, and had chlorophyll a concentrations of about 100 mg m^{-3}

Results in mg dm^{-3} of O_2

9.66	11.25	11.68	9.70	11.38	11.60	11.71	10.78	
9.63	11.30	11.68	9.70	11.38	11.78	11.78	10.82	
9.73	11.27	11.63	9.68	11.44	11.77	11.68	10.81	
0.06	0.027	0.033	0.013	0.04	0.12	0.057	0.23	mean − extreme
10.88	11.59	11.26	11.60	12.90	13.34	10.12		
10.91	11.53	11.26	11.65	12.91	13.36	10.11		
10.91	11.60	11.32	11.63	12.93	13.35	10.13		
0.02	0.043	0.04	0.027	0.017	0.017	0.01		mean − extreme

From these values it can be seen that in 11 out of 15 cases the mean value differs by 0.02–0.03 mg dm^{-3} from the extreme value. If a manual dead stop method is used the difference is about 0.02 mg dm^{-3}. The pooled standard deviation of this series is 0.04 mg dm^{-3} or 0.025 after exclusion of columns with s > 0.05. The standard deviation was also measured directly on 12 samples. The following values have been found:

11.75   11.73   11.74   11.80   11.74   11.78

$\overline{m} = 11.77$

11.72   11.72   11.83   11.77   11.81   11.80
In this series s = 0.038 and the coefficient of variation 0.3%

of Table 2 is 1.3 μmol dm^{-3} or 0.8 after exclusion of columns with s $>$ 1.5. The standard deviation was also measured directly on 12 samples, in which the standard deviation s = 1.2 and the coefficient of variation = 0.3% (mean value was 367.8 μmol dm^{-3}).

Therefore I do not believe that the quoted 0.6 μmol dm^{-3} is the standard deviation, but that it is the likely difference between lowest (or highest) value and the mean value in a triplicate series under good, but not optimal conditions. It seems furthermore that Tschumi et al.'s precision is higher than the normal Winkler technique, but not higher than the most precise versions. For the normal technique a precision of 0.5%–0.6% seems to be likely, but may be influenced by phytoplankton density.

With a titrimetric method we found no difference in standard deviation of tapwater and of lake water with little or much phytoplankton.

The standard deviation for the proposed Cerium technique [6] was about 1%, but the method was designed to be rapid and to be used in organically rich waters.

References

1. Alsterberg (1926) Die Winklersche Bestimmungsmethode für in Wasser gelösten, elementaren Sauerstoff sowie ihre Anwendung bei Anwesenheit oxydierbarer Substanzen. Biochem Z 170: 30–75
2. Bryan JR, Riley JP, Williams PJLeB (1976) A Winkler procedure for making precise measurements of oxygen concentration for productivity and related studies. J Exp Mar Biol Ecol 21: 191–197
3. Carpenter JH (1966) New measurements of oxygen solubility in pure and natural water. Limnol Oceanogr 11:264–277
4. Fox HM, Wingfield CA (1938) A portable apparatus for the determination of oxygen dissolved in a small volume of water. J Exp Biol 15:437–443
5. Golterman HL, Clymo RS, Ohnstad MAM (1978) Methods for physical and chemical analysis of freshwaters, 2nd edn. Blackwell Scientific Publ, Oxford, 231 pp
6. Golterman HL, Wisselo AG (1981) Ceriometry, a combined method for chemical oxygen demand and dissolved oxygen (with a discussion on the precision of the Winkler technique). Hydrobilogia 77:37–42
7. Graaf Bierbrauwer IM de, Golterman HL (1967) The determination of oxygen in fresh water with trivalent cerium salts. In: Proceedings IBP-symposium; held at Amsterdam and Nieuwersluis, October 1966, pp 158–165
8. Legler Ch (1972) Methoden der Sauerstoffbestimmung und ihre Bewertung. Fortschr Wasserchem Ihrer Grenzgeb 14:27
9. Mackereth FJH, Heron J, Talling JF (1978) Water analysis. Freshwater Biol Assoc Sci Publ No 36, 120 pp
10. Montgomery HAC, Cockburn A (1964) Errors in sampling for dissolved oxygen. Analyst (London) 89:679
11. Pomeroy R, Kirschman HD (1945) Determination of dissolved oxygen; proposed modification of the Winkler method. Ind Eng Chem Anal Ed 17:715–716
12. Rebsdorf A (1966) Evaluation of some modifications of the Winkler method for the determination of oxygen in natural waters. Verh Int Ver Theor Angew Limnol 16:459–464
13. Talling JF (1973) The application of some electrochemical methods to the measurement of photosynthesis and respiration in freshwaters. Freshwater Biol 3:335–362
14. Tschumi PA, Zbären D, Zbären J (1977) An improved oxygen method for measuring primary production in lakes. Schweiz Arch Hydrol 39 (2):306–313
15. Winkler LW (1888) Die Bestimmung des im Wasser gelösten Sauerstoffes. Ber Dtsch Chem Ges Berlin 21:2843–2854

Appendix E Symbols and Units: Toward Standardization

E. Gnaiger[1]

In attempting to combine the expertise of specialists in various disciplines under the title of polarographic oxygen measurement, it became necessary to reach conformity as to the usage of different terms, units, and symbols. Despite the recommendations periodically published by international committees as to what constitutes a standardized scientific terminology [3, 5, 6], agreement is rather poor. Interdisciplinary communication and clarity in the documentation of scientific literature can be improved by a stricter adherance to the fundamental guidelines at present available. Some conflicting terms and some apparently problematic units are discussed briefly in the following.

"Polarographic oxygen sensor" (POS) has been proposed instead of the term "oxygen electrode" [1]. "Electrode" is reserved for "half cell" and should not be confused with the set of two electrodes making up the unit of a POS. The term "oxygen electrode" is retained only if the oxygen reducing cathode and the reference electrode are employed separately (Chap. III.2).

A basic quantity related to the signal of a POS is the partial pressure of oxygen (symbol: p_{O_2}, although subscripts to subscripts should be avoided where possible). The SI unit for pressure is the pascal (1 Pa = 1 N m^{-2}) (Table 1). Although the bar (1 bar = 10^5 Pa) is also retained for the time being, it does not belong to the International System of Units. The use of the torr (symbol: Torr) and the conventional millimeter of mercury (symbol: mmHg; the mmHg differs from the Torr by less than 2×20^{-7} Torr) "is to be progressively discouraged and eventually abandoned" [6].

Another basic quantity relevant to POS is the permeability coefficient, and two permeability coefficients of different dimensions, P and P, have been defined: if the standard states are carefully chosen, these are numerically identical (Chap. I.3). In the physiological literature the product $D \times S$ = P has been termed Krogh's constant of diffusion (where S includes physical solubility and chemical binding of a gas in a liquid), while D/δ is usually called "permeability" [8]. The latter term is particularly ambiguous if applied to a membrane separating compartments of different solubility (as is the case in POS and usually in cellular systems). In this case the effect that solubility and solubility changes along the diffusion path exerts on the diffusion impedance is ignored (Chaps. I.1, I.3).

---

1 Institut für Zoologie, Abteilung Zoophysiologie, Universität Innsbruck, Peter-Mayr-Str. 1A, A–6020 Innsbruck, Austria

Polarographic Oxygen Sensors (ed. by Gnaiger/Forstner)
© Springer-Verlag Berlin Heidelberg 1983

Table 1. Selected SI Base Units (a), SI Derived Units (b), and units in use with the International System (c) [6]

(a)			(b)			(c)		
Symbol	Name	Physical quantity	Symbol	Name	Physical quantity	Symbol	Name	Physical quantity
m	metre	l, h, z, δ	Pa	pascal	p, p_{O_2}, bp	h	hour	t
kg	kilogram	$W, {}_wW, {}_dW$	J	joule	Q	min	minute	t
s	second	t	W	watt	$P, \dot{Q}$	°C	degree	
A	ampere	I	C	coulomb	e		Celsius	θ
K	kelvin	T	V	volt	U	bar	bar	p
mol	mole	n, N	Ω	ohm	R			

The POS is most frequently applied for measurement of oxygen concentration, i.e., amount of oxygen divided by the volume of the solution or mixture. The SI Base Unit relating to amount of substance (oxygen), N_{O_2}, is the mole (symbol: mol) (Table 1). The SI Unit relating to volume, V, is the cubic metre (symbol: m^3). Instead of the proper unit for V, the litre is still found predominantly ($1\ l = 1\ dm^3$; $1\ ml = 1\ cm^3$; $1\ \mu l = 1\ mm^3$), although neither the word litre nor its symbol should be used [6]. The basic quantity for oxygen concentration is therefore the "amount-of-substance concentration" of oxygen, $c_{O_2} = N_{O_2} \times V^{-1}$, with units [mol m^{-3}] which may be used with any SI prefix, e.g., $1\ \mu$mol dm^{-3} = 10^3 mol m^{-3}. Frequently, oxygen concentration is given in cm^3 O$_2$ dm^{-3} (relating to the units of the Bunsen absorption coefficient, α). In limnology and oceanography the use of mass concentration, ρ_{O_2} [mg dm^{-3}] is still very common. However, c_{O_2} should always be given in units [mol m^{-3}]. For conversion of units see App. A, Table 6.

The conceptual advantage of expressing basic quantities in gas exchange physiology in terms of molar units was convincingly outlined in 1971 [8]. However, this was neglected by most respiratory physiologists and even by a physiological Committee on Nomenclature which, two years later [4], again recommended volume units for expressing oxygen consumption, $\dot{V}_{O_2}$ [cm^3 h^{-1}]. Many ecological physiologists express oxygen uptake rates in [mg h^{-1}], and a confusing variety of symbols exists. The proper symbol for "molar oxygen uptake rate" is $\dot{N}_{O_2} = dN_{O_2} \times dt^{-1}$ (or often more strictly $\Delta N_{O_2} \times \Delta t^{-1}$) [nmol s^{-1} or μmol h^{-1}]. In the steady state (of a POS or an organism) the reduction of the external amount of oxygen is equal to the internal (electrochemical or metabolic) conversion of oxygen. If the internal oxygen store changes (non-steady state, e.g., Chap. II.4) the oxygen consumption rate should be distinguished from the oxygen flux through the oxygen transducing area, $J_{O_2} \times A$ (with the same units as $\dot{N}_{O_2}$). For conversion of units see App. C, Table 2.

Oxygen consumption rates can be converted to rates of heat dissipation, $\dot{Q}$ [mW] (Chaps. II.3, II.4). The joule is progressively replacing the calorie. The symbol $\Delta_k H_{O_2}$ [kJ mol^{-1}] is used for the oxycaloric equivalent expressing the enthalpy change per mole of oxygen consumed in dissipative metabolism (App. C). The subscript k designates the "catabolic half cycle" [2]; dissipative metabolism corresponds to the metabolic state when net energy liberated in the sequence of reactions in the catabolic half cycle is neither conserved in catabolic energy coupling nor in anabolism. In a recent recommendation on biothermodynamic nomenclature [5] "catabolism" was confused with the "catabolic half cycle", thus neglecting that "catabolism and anabolism consist of two simultaneous and interdependent processes which may be analyzed separately" [7]. The enthalpy equivalent of catabolism may vary with the degree of energy-coupling and net accumulation of "high-energy intermediates" [2]. Any ambiguity is avoided in the term $\Delta_k H_{O_2}$ as defined above. Since thermochemical quantities are usually expressed on a mole basis, this practice should be adopted for the oxycaloric equivalent. For conversion factors see App. C, Table 3.

In addition to the obvious advantage of using the generally applicable unit "mole" in quantities involving "amount of oxygen", the adoption of the mole simplifies expressions and calculations of stoichiometric, i.e., molecular relationships (e.g., App. C). This may encourage functional interpretations of oxygen consumption measurements and contribute to an alignment of concepts in bioenergetic ecology, respiratory physiology, biochemistry and biothermodynamics. Several ad hoc symbols encountered in the physiological literature (e.g, $\sim P/O$ ratio, RQ, NQ) can be replaced by commonly accepted stoichiometric coefficients, v, which are unambiguously understood in a wider interdisciplinary context (Table 2, pp. 355–358).

References

1. Fatt I (1976) Polarographic oxygen sensors. CR C Press, Cleveland, p 2
2. Gnaiger E (1980) Das kalorische Äquivalent des ATP-Umsatzes im aeroben und anoxischen Metabolismus. Thermochim Acta 40:195–223
3. Weast RC (ed) (1974–75) Handbook of chemistry and physics, 55th edn. CR C Press, Cleveland
4. International Union of Physiological Sciences (1973) Glossary on respiration and gas exchange. Prepared for publication by Bartels H, Dejours P, Kellogg RH, Mead J. J Appl Physiol 34:549–558
5. Interunion Commission on Biothermodynamics (IUPAC, IUB, IUPAB) (1982) Calorimetric measurements on cellular systems. Recommendations for measurements and presentation of results. Prepared by Belaich JP, Beezer AE, Prosen E, Wadsö I. Pure Appl Chem (in press)
6. IUPAC (1979) Manual of symbols and terminology for physicochemical quantities and units. Prepared for publication by Whiffen DH. Pergamon Press, Oxford, 41 pp
7. Lehninger AL (1975) Biochemistry. Woerth Publ, New York
8. Piiper J, Dejours P, Haab P, Rahn H (1971) Concepts and basic quantities in gas exchange physiology. Respir Physiol 13:292–304

Table 2. List of symbols, description and corresponding units following the IUPAC recommendations [6]. Only symbols frequently used in the text or relevant to POS are listet. When the same symbol is used with different subscripts, then these are not explained separately. The subscripts are: a air; b barometric (in $_bp$); b blood; e electrolyte; h actual or total (pressure as a function of water depth h); m membrane (in V_m: molar); s sample medium; w hydrostatic (pressure of water column). Page numbers refer to the text where the respective symbol is used, pages in italics refer to tables or figures

Symbol	Description and units	Page
a_{O_2}	$= y_{O_2} \times c_{O_2}/c^{\circ}_{O_2} = f_{O_2}/f^{\circ}_{O_2}$; relative activity of oxygen, concentration based; a_m, a_s	4, 31ff., *33*
A	area [m^2]	5ff., 18f., 28, 34, *39*, 45, 77, *80*
c_{O_2}	$= N_{O_2} \times V^{-1} = p_{O_2} \times S_{O_2}$; amount-of-substance concentration of oxygen [μmol dm^{-3}]; c_a, c_b, c_e, c_m, c_s	15, 18, 23, 27f., 32ff., *104*, 112, 114, 123, 146ff., 153f., *154, 160*, 215, *224*, 227, *250f.*, *254ff.*, 277
$c^{\circ}_{O_2}$	standard concentration of oxygen, 1 mol dm^{-3}; c°_m, c°_s	4, 32ff.
C^*	unit standard concentration of oxygen (molar concentration) [μmol dm^{-3}]	322ff., *331*
$C^{\dagger}$	unit standard concentration of oxygen (molal concentration) [μmol kg^{-1}]	322f.
C_c	saturation concentration of oxygen as a function of temperature, salinity and barometric pressure [μmol dm^{-3}]	325, 327ff., *330f.*
D_m	diffusion coefficient of oxygen in the membrane [m^2 s^{-1}]; D_a (20°C) = 3 × 10^{-5} m^2 s^{-1}; D_e (25°C) = 2 × 10^{-9} m^2 s^{-1}; D_s	5ff., *6*, 11, 15, 19, 23f., 28, 32ff., 50f., 56, *57*, 77, 238f.
e	elementary charge, 1.60219 × 10^{-19} C	8
E°	electromotive force	22, 66, 68, 73, *74*
f	$= dV \times dt^{-1}$; flow rate [cm^3 s^{-1} or cm^3 h^{-1}]	112, 148, 150, 153
f_{O_2}	fugacity of oxygen [kPa]	4f., 7f., 19
F	$= L \times e$; Faraday constant, 96485 C mol^{-1} or As mol^{-1}	5ff., 18f., 28, 34, 56
F_c	calibration factor of a POS, concentration based [μmol dm^{-3} μA^{-1}]	147f.
F_p	calibration factor of a POS, partial pressure based [kPa μA^{-1}]	146
g_n	standard acceleration of free fall, 9.80665 m s^{-2}	328f.
ΔG°	difference or change in molar standard Gibbs energy [kJ mol^{-1}]	32
h	height or depth [m]	123, *124*, 326, *328f.*
$\Delta_c h$	specific enthalpy of combustion [kJ g^{-1}]	*340, 343*
$\Delta_k H_{O_2}$	molar caloric equivalent of respiratory oxygen consumption in dissipative metabolism [kJ mol^{-1}]	*160f., 163*, 167, *340ff., 343f.*
I_l	electric current, diffusion limited signal of a POS [μA], proportional to A (large cathode) or to r_o (microcathode)	5ff., *8*, 18ff., *20*, 34, *44f.*, *45f., 51, 54f.*, 56, 77, *78*, 93ff., *105*, 146ff., 215

Table 2 (continued)

Symbol	Description and units	Page
I_{O_2}	$= I_1 - I_r$; oxygen current of a POS, signal corrected for the residual current [μA]	146
I_r	residual (zero) current of a POS in an oxygen free medium [μA]	15, 46, 146ff.
I_r	radial diffusion current [μA]	27f.
J_{O_2}	flux of oxygen [mol m^{-2} s^{-1}]	5, 32ff., 236, 238, *281f.*
k	$= R \times L^{-1}$; Boltzmann constant, 1.3807×10^{-23} J K^{-1}	*357*
k_S	Henrys law constant [kPa]	36
K	equilibrium constant or distribution coefficient	32ff., 73
l	length [m]	23f., 123, *124*
L	Avogadro constant, 6.0220×10^{23} mol^{-1}	*355*
M_r	relative molecular mass of a substance; M_r (O$_2$) = 31.9988	328f.
n	charge number of a cell reaction, number of electrones added to each oxygen molecule at the cathode	5ff., 18f., 34, 56
N_{O_2}	molar amount of oxygen [mol]	148, 153f., 226f., *329*
$\dot{N}_{O_2}$	$= dN_{O_2} \times dt^{-1}$; molar oxygen uptake rate [nmol s^{-1} or μmol h^{-1}]	112, 123, 148, *213ff.*, 277, *344*
$\dot{N}^o_{O_2}$	reference (blank) rate of oxygen consumption [nmol s^{-1} or μmol h^{-1}]	*149*
$\dot{n}_{O_2}$	$= dN_{O_2} \times dt^{-1} \times W^{-1}$; specific rate of oxygen consumption [nmol s^{-1} g^{-1} or μmol h^{-1} g^{-1}]	114, *124*, 137, 156ff.
NQ	$= \nu_{N/O_2}$; nitrogen quotient [moles N in excretory products per mole O$_2$]	338ff., *340ff.*
p	pressure [kPa] (1 Pa = 1 N m^{-2}; 1 bar = 100 kPa; 1 mmHg = 0.133322 kPa; 1 atm = 101.325 kPa; 1 m H$_2$O = 9.80638 kPa; 1 psi = 6.89 kPa; 1 inchHg = 3.38 kPa)	85, 325
p_o	$= 1$ atm = 101.325 kPa; normal atmosphere, standard atmospheric (barometric) pressure	322, 324ff.
$_bp$	barometric (atmospheric) pressure [kPa]	*106, 238, 325ff., 328,* 334ff.
p_h	$= {_bp} + {_wp}$; actual (total) pressure at water depth h [kPa]	326f.
$_wp$	hydrostatic pressure under a column of water [kPa]; 1 m H$_2$O = 0.09678 atm = 9.80638 kPa at 4°C	12, *144*, 326ff.
p_A	$= x_A \times p$; partial pressure of gas A [kPa]; p_{N_2}; p_X	327, 334f.
p_{H_2O}	water vapor pressure at experimental temperature [kPa]	322, 324ff., *328*, 334ff.
p_{O_2}	partial pressure of oxygen; nearly identical to f_{O_2} for oxygen pressures $< p_o$ [kPa]; p_b, p_m, p_s	18, 34ff., 47, *54f.*, 56, *59ff.*, 77, *104f.*, 130, 146, *149*, 168, *172f.*, 222, 224, 226f., 236, *238ff.*, 325, 327f., *329*, 334f.

Table 2 (continued)

Symbol	Description and units	Page
$p_{O_2,c}$	critical (saturation) partial pressure of oxygen as a function of barometric and hydrostatic pressure [kPa]	327
P	$= dQ \times dt^{-1}$; power [μW]	
P_m	$= S_m \times D_m$; permeability coefficient of the membrane, based on unit partial pressure [mol m^{-1} s^{-1} kPa^{-1}]; P_e, P_s	31ff., 49f., *50, 57*, 77
P_m	$= K \times D_m$; permeability coefficient of the membrane, based on unit activity [m^2 s^{-1}]; P_e, P_s	18, 24, 31ff.
POS	polarographic oxygen sensor	
Q	heat [J], a positive sign indicates increase of heat of the system under discussion	*159*
$\dot{Q}$	$= dQ \times dt^{-1}$; power, rate of heat dissipation [1 J s^{-1} = 1 W] a negative sign indicates heat dissipation out of the system under discussion (inaccurately called "heat production")	*161ff.*, 340, *344*
$\dot{q}$	$= dQ \times dt^{-1} \times W^{-1}$; specific rate of heat dissipation [mW g^{-1}]	*159f.*
r_E	horizontal distance from the center of the electrode (cathode) [m]	47, *48*
r_o	radius of the cathode [m]	27f., *45*, 47, *48ff., 80, 266*
R	$= K \times L$; gas constant, 8.31441 ± 0.00026 J K^{-1} mol^{-1}	*8, 12, 31f.,* 36, 328f.
R	electrical resistance [Ω]	91, 93ff., 95, 98f., *101, 180, 251*
RQ	$= \nu_{CO_2}/O_2$; respiratory quotient or respiratory gas exchange ratio [moles CO_2 per mole O_2]	337ff., *340ff.*
S_s	solubility of oxygen in the sample medium [μmol dm^{-3} kPa^{-1}]; S_s (pure water, 25°C) = 12.5 μmol dm^{-3} kPa^{-1}; S_e (1.5 mol dm^{-3} KCl, 25°C) = 8.5 μmol dm^{-3} kPa^{-1}; S_m	5, *6ff.,* 19, 34ff., *50, 56, 57*, 77, *104,* 146, *204,* 226f., 236, 238f., 322ff.
STPD	standard temperature (T_o = 273.15 K = 0°C) and pressure (p_o = 101.325 kPa = 1 atm), dry	
t	time [s or min or h]	112, 114, 123, 153f.
T	thermodynamic (absolute) temperature [K]	8, 12, 31f., 36, 95ff., 105, *252,* 323f., 326, 328f., 334
T_o	zero of the Celsius scale, standard temperature, 273.15 K	
U	$= I \times R$; electric potential difference [V]	8, *46,* 93ff., *147, 161, 172f., 251*
V	volume [m^3]	112, 114, 123, 153f., 168, 226, *230, 251*
V_o	$= R \times T_o \times p_o^{-1}$; standard molar volume of ideal gas, 22.41383 ± 0.00070 dm^3 mol^{-1}	*329*
V_m	molar volume at STP; V_m (O_2) = 22.392 dm^3 mol^{-1}	11, *329*
$\dot{V}_{O_2}$	$= dV_{O_2} \times dt^{-1}$; volume based oxygen consumption [mm^3 h^{-1}]	168, *182*

Table 2 (continued)

Symbol	Description and units	Page
w_i	mass fraction of compound i	338f., 342f.
W	weight [g]; $_dW$ dry weight; $_wW$ wet weight	114, *137*, 150, 338, *339f., 344*
x_i	mole fraction of compound i	338, 341f., 344
y_{O_2}	activity coefficient of oxygen defined in terms of amount-of-substance concentration; y_m, y_s	4, 32ff.
z_m	thickness of the membrane [m]; z_e, z_s	5, *6ff.*, 18f., 24, 28, *33ff., 48ff., 50*, 77, 236, *238f.*
z_m/P_m	diffusion impedance or mass transfer impedance of the membrane [m² s kPa mol⁻¹]	6, 77
α	Bunsen absorption coefficient, solubility [cm³ cm⁻³ atm⁻¹]	*57, 322*
δ	thickness of diffusion layer [m]; $\delta_s = z_s$	24
θ	Celsius temperature [°C]	*96, 137, 168, 204, 250, 254ff., 323, 336*
μ	chemical potential [kJ mol⁻¹]; μ_m; μ_s	31f., 36
ν_{CO_2}	stoichiometric coefficient for carbon dioxide (positive for carbon dioxide as a product)	337
ν_{O_2}	stoichiometric coefficient of oxygen (negative for oxygen as a substrate)	337
$\nu_{CO_2/O_2} = -\nu_{CO_2}/\nu_{O_2}$; respiratory gas exchange ratio or respiratory quotient, RQ		337
$\nu_{H^+/O_2} = -\nu_{H^+}/\nu_{O_2}$; hydrogen ion-oxygen exchange ratio		341
$\nu_{N/O_2} = -\nu_{NH_3}/\nu_{O_2}$; or $= -2 \times \nu_{CO(NH_2)_2}/\nu_{O_2}$; nitrogen-oxygen exchange ratio or nitrogen quotient, NQ		338
ν_s	kinematic viscosity of the sample fluid [m² s⁻¹]	5, 7, 27
ρ_o	density at standard temperature [mg dm⁻³]	328
ρ_{O_2}	mass concentration of oxygen [mg dm⁻³], compare c_{O_2}	*254ff., 278, 280*
ϕ_A	volume fraction of gas A	*20, 149, 157, 271f., 334ff.*
$\phi_{O_2}^*$	volume fraction of oxygen in dry air, 0.20946	322, 327, *328*, 334ff.
τ	time constant, response time [s or min or h]	15, 52, 78, 80, 153, *154f., 192*

Subject Index

For symbols and units of physical quantities see also Appendix E, Table 2. *Italic expressions* refer to keywords where complete entries are found. *Italic numbers* refer to relevant figures and tables.

Fish Diseases

Third COPRAQ-Session
Editor: **W. Ahne**
1980. 120 figures. X, 252 pages
(Proceedings in Life Sciences)
ISBN 3-540-10406-2

The Third COPRAQ Fish Disease session drew
together over 100 scientists from all ove the world
to discuss the major problems affecting the health
of cultured fish. The papers presented at this
session pointed out that bacterial diseases such as
vibriosis, furunculosis and carp erythrodermatitis
are beginning to replace viral diseases as the
predominant threats to piscine health. They also
show the effectiveness of immunization in
preventing many fish diseases.
With its information on recent reseach findings
and on emerging problems in and approaches to
fish diseases, this volume will be of considerable
interest to all scientists engaged in ichthyopatho-
logy.

U. Förstner, G.T.W. Wittmann

Metal Pollution in the Aquatic Environment

With contributions by F. Prosi, J.H. van Lierde
Foreword by E.D. Goldberg
2nd revised edition. 1981. 102 figures, 94 tables.
XVIII, 486 pages. ISBN 3-540-10724-X

"This is an impressive and useful compilation of a
large amount of data on heavy metals in the
natural environment. The book contains nearly
500 pages and 70 of these are taken up by the
references, of which the majority were published
in the last ten years. I found the book lucidly
written and on the whole easily readable...
... In conclusion this is a very valuable book that
has been well produced. It will be, as I expect,
consulted regularly by many scientists interested
in the contamination of the natural environment
by metals..."   *Hydrobiologia*

S.A. Gerlach

Marine Pollution

Diagnosis and Therapy
Translated from the German by R. Youngblood,
S. Messele-Wieser
1981. 91 figures. VIII, 218 pages.
ISBN 3-540-10940-4

Wide-ranging in scope, *Marine Pollution* covers
household and industrial wastes, ocean dumping,
the effects of oil and radioactivity, the influence of
toxic materials on geochemical and biochemical
processes, and the dangers of wide-spread pollution by heavy metals and organic substances. This
English translation of the highly acclaimed
German original has been revised and expanded
to keep pace with the rapid process of research in
the field.

Intense Atmospheric Vortices

Proceedings of the Joint Symposium held at
Reading (United Kingdom) 14–17 July 1981
Editors: **L. Bengtsson, Sir J. Lighthill**
1982. 205 figures, approx. 11 tables. Approx.
360 pages. (Topics in Atmospheric and Oceanographic Sciences, Volume 1)
ISBN 3-540-11657-5

The concept of vorticity is of central importance
in fluid mechanics, and the change and variability
of atmospheric flow is dominated by transient
vortices of different time and space scales. Of
particular importance are the most intense
vortices, such as hurricanes, typhoons and torna-
does, which are associated with extreme and
hazardous weather events of great concern to
society.
This book examines the different mechanisms for
vorticity intensification that operate in two different kinds of meteorological phenomena of great
importance, namely the tropical cyclone and the
tornado. The understanding of these phenomena
has grown in recent years due to increased and
improved surveillance by satellites and aircraft, as
well as by numerical modelling and simulation,
theoretical studies and laboratory experiments.
The book summarizes these recent works with
contributions from observation studies (from
radio sonde data, aircraft, satellites and radar) and
from studies concerning the physical mechanism
of these vortices by means of theoretical, numerical or laboratory models. The book contains
articles by the leading world experts on the
meteorological processes and on the fundamental
fluid dynamics mechanism for vorticity intensification.

Springer-Verlag Berlin Heidelberg New York

Marine Biology

International Journal on Life in Oceans and Coastal Waters

ISSN 0025-3162   Title No. 227

Editor in Chief: O. Kinne, Biologische Anstalt Helgoland

Marine Biology publishes contributions in the following fields:

Plankton Research: Studies on the biology, physiology, biochemistry and genetics of plankton organisms both under laboratory and field conditions. Biological and biochemical oceanography. Environment-organism and organism-organism interrelationships.
Experimental Biology: Research on metabolic rates and routes in microorganism, plants and animals. Respiration. Nutrition. Life cycles.
Biochemistry, Physiology and Behavior: Biochemical research on marine organisms, photosynthesis; permeability; osmoregulation; ionregulation; active transport; adaption; analyses of environmental effects on functions (tolerances, rates and efficiencies of metabolism, growth and reproduction) and structures; migrations; orientation; general behaviour.
Biosystem Research: Experimental biosystems and microcosms. Energy budgets: flow routes and balances sheets of energy and matter in the marine environment. Interspecific interrelationships food webs. Dynamic and structures of microbial, plant and animal populations. Use, management and protection of living marine resources. Effects of man on marine life, including pollution.
Evolution: Investigations on speciation, population genetics, and biological history of the oceans.
Theoretical Biology Related to the Marine Environment: Concept and models of quantification and mathematical formulation; system analsysis; information theory.
Methods: Apparatus and techniques employed in marine biological research, underwater exploration and experimentation.

Subscription Information and/or **sample topies** are available from your bookseller or directly from Springer-Verlag, Journal Promotion Dept., P.O. Box 105280, D-6900 Heidelberg, FRG

Bulletin of Environmental Contamination and Toxicology

ISSN 0007-4861   Title No. 128

Editor in Chief: Y. Iwata

Editor of **Archives of Environmental Contamination and Toxicology:** A. Bevenue

The **Bulletin of Environmental Contamination and Toxicology** provides rapid publication of significant advances and discoveries in the fields of air, soil, water and food contamination and pollution as well as methodology, and other disciplines concerned with the introduction, presence, and effects of toxicants in the total environment.

It is the hope of the Editors that this Bulletin provides a meeting ground for research workers who daily encounter problems related to the contamination of our environment and who welcome opportunities to share in new discoveries as soon as they are made.

Oecologia

In Cooperation with the International Association for Ecology (Intecol)

ISSN 0029-8549   Title No. 442

Editor in Chief: H. Remmert

Oecologigia reflects the dynamically growing interest in ecology. Emphasis is placed on the functional interrealtionship of organisms and environmental rather than on morphological adaptation. The journal publishes original articles, short communications, and symposium reports on all aspects of modern ecology, with particular reference to physiological and experimental ecology, population ecology, organic production, and mathematical models.

Fields of Interest: Autecology, Physiological Ecology, Population Dynamics, Production Biology, Demography, Epidemiology, Behavioral Ecology, Food Cycles, Theoretical Ecology – including Population Genetics.

Springer-Verlag Berlin Heidelberg New York